Physical, Chemical and Biological Processes in Energy Geoscience

Physical, Chemical and Biological Processes in Energy Geoscience

Editors

Ping Gao
Yidong Cai
Yingfang Zhou
Quan Gan

MDPI

Basel • Beijing • Wuhan • Barcelona • Belgrade • Novi Sad • Cluj • Manchester

Editors
Ping Gao
China University of Geosciences
Beijing, China

Yidong Cai
China University of Geosciences
Beijing, China

Yingfang Zhou
University of Aberdeen
Aberdeen, UK

Quan Gan
Chongqing University
Chongqing, China

Editorial Office
MDPI
St. Alban-Anlage 66
4052 Basel, Switzerland

This is a reprint of articles from the Special Issue published online in the open access journal *Processes* (ISSN 2227-9717) (available at: https://www.mdpi.com/journal/processes/special_issues/Energy_Geoscience).

For citation purposes, cite each article independently as indicated on the article page online and as indicated below:

Lastname, A.A.; Lastname, B.B. Article Title. *Journal Name* **Year**, *Volume Number*, Page Range.

ISBN 978-3-0365-9407-1 (Hbk)
ISBN 978-3-0365-9406-4 (PDF)
doi.org/10.3390/books978-3-0365-9406-4

Contents

processes

MDPI

Article

Effect of Formation Pressure on Pore Structure Evolution and Hydrocarbon Expulsion in Organic-Rich Marine Shale

Xianglong Fang [1,2], Yidong Cai [1,2,*], Qinhong Hu [3], Ping Gao [1,2,*], Dameng Liu [1,2] and Yujing Qian [1,2]

1 School of Energy Resources, China University of Geosciences, Beijing 100083, China
2 Coal Reservoir Laboratory of National Engineering Research Center of CBM Development & Utilization, China University of Geosciences, Beijing 100083, China
3 Department of Earth and Environmental Sciences, The University of Texas, Arlington, TX 76019, USA
* Correspondence: yidong.cai@cugb.edu.cn (Y.C.); gaoping1212@cugb.edu.cn (P.G.)

Abstract: Exploring the relationship between formation pressure and shale pore evolution is helpful for the enrichment and development of marine shale gas accumulation theory. The thermal evolution experiment was carried out on the Xiamaling Formation (Pr3x) lowly matured marine shale, which has a similar sedimentary environment to the Longmaxi Formation (S1l) highly matured marine shale. Comparative experiments of open and semi-closed pyrolysis and multiple pore structure characterization techniques, including CO_2 and N_2 physisorption, mercury intrusion porosimetry, and field emission scanning electron microscopy, were conducted. The marine shale pore evolutionary model under formation pressure is proposed by characterizing pore evolution, and hydrocarbon expulsion and retention for shales under and without formation fluid pressures. The results show that the existence of formation pressure increases the percentage of quartz and reduces the content of clay minerals. The change in formation pressure has no obvious effect on the maturity evolution of shale samples. With the increase of formation pressure, the pore morphology of shale gradually changes from narrow slit pores to ink bottle-shaped pores. The retained hydrocarbons in shale mainly occupy the mesopore space, and the existence of formation pressure promotes hydrocarbon expulsion, especially the hydrocarbon expulsion in the mesopore. In addition, formation pressure improves pore connectivity, especially in the high-over mature stage of shale. With the increase of formation pressure, the micropore volume decreases slightly, the mesopore volume increases significantly, and the macropore volume changes have two stages.

Keywords: marine shale; open and semi-open system pyrolysis; formation pressure; pore structure evolution

Citation: Fang, X.; Cai, Y.; Hu, Q.; Gao, P.; Liu, D.; Qian, Y. Effect of Formation Pressure on Pore Structure Evolution and Hydrocarbon Expulsion in Organic-Rich Marine Shale. *Processes* **2023**, *11*, 1007. https://doi.org/10.3390/pr11041007

Academic Editor: Youguo Yan

Received: 26 February 2023
Revised: 19 March 2023
Accepted: 24 March 2023
Published: 27 March 2023

1. Introduction

China has huge shale gas reserves, and the marine shale of the Longmaxi Formation in South China has a large thickness and high organic content, which contributes to high gas production [1–3]. In recent years, increasing attention on deep fossil energy, and especially shale oil and gas exploration, have raised concerns regarding pore evolution as well as hydrocarbon retention in organic-rich shales [4–6]. Therefore, it is necessary to investigate pore evolution, gas occurrence, and the diagenetic evolution mechanism of organic-rich marine shale reservoirs [7–10]. In the process of shale thermal maturity evolution, the variation of pore structure directly influences the flow and transport capacity of shale oil and gas [11–14]. Thus, it is vital to research the pore structure evolution characteristics of shale after being influenced by high temperatures and high pressures. The storage system in shales consists of organic matter (OM) pores, mineral pores, and micro-fractures, which can be divided into three categories according to the aperture range, according to the IUPAC classification: micropore (<2 nm), mesopore (2–50 nm), and macropore (>50 nm) [15]. Pore characterization methods of shale can be divided into two categories: quantitative analysis methods and morphological visualization methods. Quantitative analysis methods can be used to obtain pore structure parameters, such as pore volume (PV), specific surface area (SSA) and pore size distribution (PSD), which includes CO_2/N_2 gas adsorption [16,17], mercury intrusion porosimetry (MIP) [1], nuclear magnetic resonance (NMR) [18], small-angle neutron scattering (SANS) [19], and small-angle X-ray scattering (SAXS) [20]. Morphological visualization methods are used for the intuitive description of the shape, size, number, and distribution of pores and fractures, and includes a series of analysis techniques, including scanning electron microscope (e.g., FE-SEM, FIB-SEM and HIM) [21], atomic force microscopy (AFM) [22], transmission electron microscope (TEM) [23] and X-ray computed tomography (CT) [11]. The applicability of characterization methods is different. Therefore, multiple testing methods should be integrated to obtain a full-scale of pore parameter.

The key factors influencing the pore evolution of shale include thermal evolution, organic matter type, TOC, mineral and chemical composition, compaction, and burial depth [5,24–27]. Thermal maturity is one of the most controlling factors for pore evolution and the production of gaseous and liquid hydrocarbon in shales [5,6,8,25]. Recently, various experimental research on shale evolution was carried out under different conditions, such as electromagnetic radiation [28], pyrolytic inert and oxic environments [29], high-temperature water vapor [6], pyrolysis under formation pressure [2,30], pyrolysis under effective stress conditions [12], convection and conduction heating [31], and microwave heating [32]. According to system openness, shale thermal evolution experiments can be divided into three categories: open system, semi-closed system, and closed system [33].

There have been few comparative studies on these experimental results. Zhang et al. [34] studied the influencing factors of lacustrine shale pore evolution by conducting a comparative experiment between closed and semi-closed system pyrolysis, and it was found that in the semi-closed system, a large number of pre-oil and hydrocarbons are discharged periodically, which is conducive to the development of organic pores. Zhao et al. [6] investigated the pore structure and seepage characteristics at different temperatures using oil shale pyrolysis by water vapor injection and found that the increase in pore volume due to pyrolysis temperatures mainly affected pores ranging from 10 nm to 100 nm and occurred in the specific temperature range (400 °C to 425 °C). Gao et al. [5] studied the pore structure evolution characteristics of continental shale in China through open system pyrolysis, and found that the filling of pores by oil generated at a lower temperature resulted in the reduction of the volume of macropores in shale samples, and these filled macropores were released through the thermal cracking of oil at a higher temperature. Song et al. [35] carried out a semi-closed thermal evolution experiment to study pore evolution characteristics at different temperatures by controlling lithostatic pressure and formation pressure, and found that evolutionary scenarios of nanopores can be divided into three stages with the increasing of pyrolysis temperature, including pore decreasing, increasing, and transforming stages. Zhao et al. [6] studied the pore evolution characteristics of shale under high temperature fluid without involving fluid pressure, and it was found that the seepage channel formed at 450 °C, and the best pore connectivity was at 550 °C;. Li et al. [12] studied the relationship between the permeability of marine shale and temperature and fluid pressure, and found that the apparent permeability decreased approximately linearly with the increase of pore pressure and temperature. Shao et al. [30] studied the effects of pressure on gas generation and pore evolution in shale using gold-tube pyrolysis, and indicated that high pressure inhibits residual oil cracking but simultaneously favors the generation of methane-rich gas. Geng et al. [11] conducted a pyrolysis experiment under the coupled effect of temperature and pressure, and reported that the pores and fractures in shale developed increasingly with both the temperature and pressure increasing. However, these studies do not systematically investigate the effect of formation pressure on pore structure evolution and hydrocarbon expulsion, nor clarified the differences in pore evolution with or without formation pressure.

Most of the marine shales in the Sichuan Basin are of high maturity. The low maturity Xiamaling Formation (Pr_3x) marine shales from northern China have a similar sedimentary environment to the high maturity Longmaxi Formation (S_1l) marine shale, and thus they were selected to conduct thermal evolution experiments. In this work, a comparative experiment of open and semi-closed pyrolysis experiments was carried out, and the pore structure of the solid product was characterized to study the thermal evolution, hydrocarbon expulsion and retention, and pore evolution characteristics of shale with and without formation fluid pressure. Meanwhile, the development models of shale pores under formation pressure are proposed. This research is helpful for the enrichment and development of marine shale gas accumulation theory.

2. Samples and Methodology

2.1. Sample Preparation and Basic Information

An organic-rich and immature shale sample with a total organic carbon (TOC) of 5.74% and an equivalent vitrinite reflectance (EqVR_o) of 0.67% was collected from the Neoproterozoic Xiamaling Formation (Pr_3x) in Xiahuayuan region of northern China. Before the tests, samples were crushed into small blocks with a size of 1–2 cm. Then the samples were evenly divided into fifteen parts after being mixed to minimize sample errors. One was used for basic analysis (Ro, XRD, and geochemical analysis), seven were used to carry out open-system pyrolysis, and the remaining parts were used for semi-closed thermo-compression pyrolysis experiments. The same experimental temperature points were selected for the pyrolysis experiments of the open and semi-open systems, and the only difference was whether there was formation pressure or not. After the pyrolysis experiments, all of the solid products from the two contrast experiments were provided for Ro, XRD, TOC, geochemical analysis, and pore characterization experiments (including CO_2 and N_2 adsorption), MIP, and FE-SEM.

2.2. Thermal Evolution Contrast Experiments

2.2.1. Open System Thermal Evolution Experiments

According to the principle of time-temperature compensation [36], a series of high temperatures with short time can simulate the long-time and low-temperature geological process [37]. In this research, the open system pyrolysis experiments were conducted using an OTF-1200X high-temperature pyrolysis simulator, as shown in Figure 1. The open system pyrolysis instrument contains a control system, reaction system with a furnace tube, heating system, gas supplement system that provided N_2, and a tail gas treating unit. Seven paired low-mature Longmaxi shale samples were selected for thermal evolution experiments at 300 °C, 350 °C, 400 °C, 450 °C, 500 °C, 550 °C, and 600 °C (Table 1). Each sample was heated to a preset experimental temperature with a constant rising rate of 5 °C/min, after which they remained at a constant temperature for 72 h.

Table 1. Experimental conditions for open and semi-closed pyrolysis experiments.

Sample No.	**XHY-1**	**XHY-2**	**XHY-3**	**XHY-4**	**XHY-5**	**XHY-6**	**XHY-7**
T, (°C)	300	350	400	450	500	550	600
Sample no.	**XHY-01**	**XHY-02**	**XHY-03**	**XHY-04**	**XHY-05**	**XHY-06**	**XHY-07**
T, (°C)	300	350	400	450	500	550	600
H, (m)	2500	3000	3500	4000	4500	5000	5500
P_f, (MPa)	33.15	39.78	46.41	53.04	59.67	66.3	72.93
P_l, (MPa)	56.75	68.1	79.45	90.8	102.15	113.5	124.85

Note: T = Simulated temperature, °C; H = Simulated burial depth, m; Pf = Formation pressure, MPa; Pl = Lithostatic pressure, MPa.

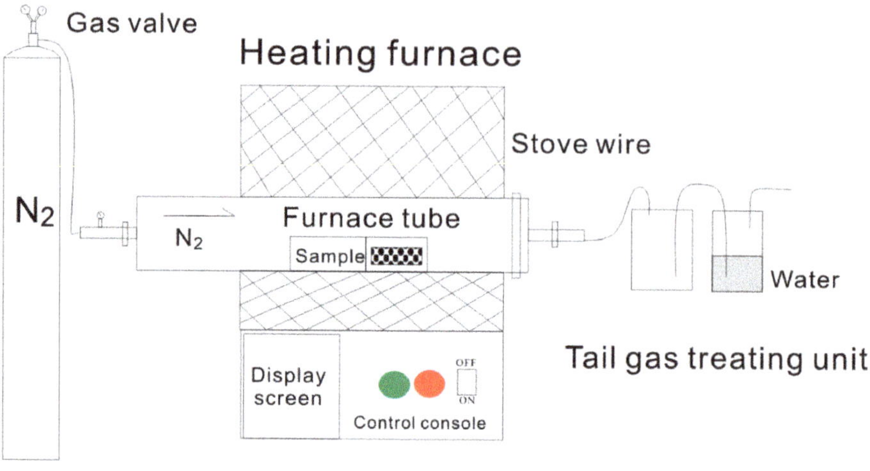

Figure 1. Schematic diagram of open system pyrolysis experiment.

2.2.2. Thermocompression Pyrolysis Experiments

The semi-closed thermocompression pyrolysis experiments were carried out using HXHTST-II and HXSRDE-II high-pressure pyrolysis and diagenetic evolution simulators. As shown in Figure 2, the semi-closed pyrolysis system includes six parts: heating system, fluid supplement system, reaction system, hydraulic control system, software control system, and collecting system [24,38]. The same low-mature marine shale samples that were used for open system pyrolysis were used for semi-closed thermocompression pyrolysis under formation pressure and lithostatic pressure corresponding to seven temperature points (Table 1). The lithostatic pressure and formation pressure were set as the burial and evolution conditions of organic-rich Longmaxi Formation marine shales from the eastern Sichuan Basin of southern China. Each sample was heated to a preset experimental temperature with a constant rising rate of 5 °C/min, after which they were kept at a constant temperature for 72 h.

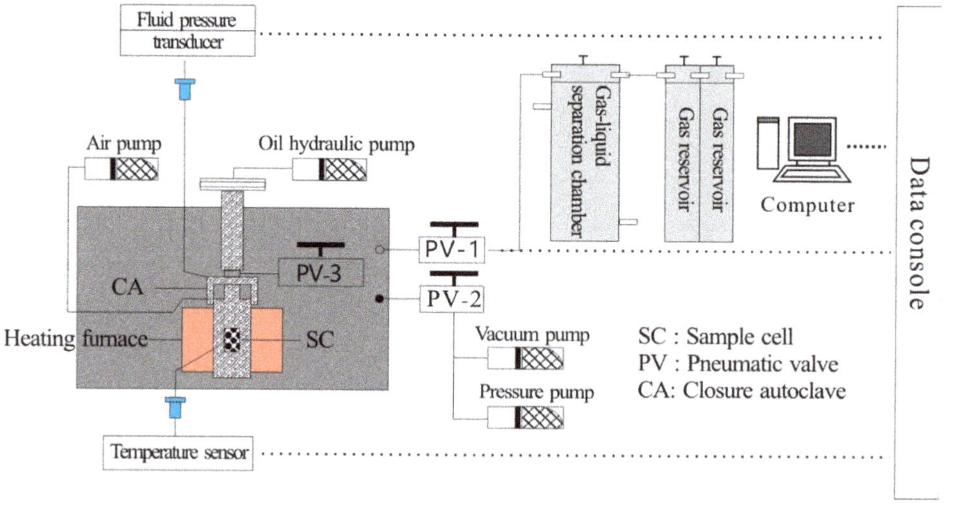

Figure 2. Schematic diagram of thermocompression pyrolysis experiment system [21].

2.3. Methodologies for Pore Structure Characteristic

Morphological characteristics of micro-nano pores in shales were performed by FESEM. The tests were carried out using the MERLIN FESEM system. Integrated MIP, and N_2 and CO_2 adsorption methods were employed to quantify the pore volumes (PV), surface area (SA), and pore size distribution (PSD) for full-scale pore sizes. Tests for CO_2 physisorption were used to determine pore data of micropores (<2 nm), N_2 physisorption was adopted to detect the pore structure parameter of mesopores (2–50 nm), and MIP measurements were carried out to obtain the pore date of macropores (>50 nm) [16,17,39]. The experimental methods were carefully described in our previous works [16,21,40].

3. Results

3.1. Open and Semi-Closed Pyrolysis Experiments

Vitrinite reflectance (R_o), representing a measure of the percentage of incident light reflected from the surface of vitrinite particles in sedimentary rock, is usually used to acquire shale maturity [41]. However, since there were no higher plants in the Neoproterozoic Era and only algae and fungi developed in the ocean, biogenic vitrinite could not be found in these marine shales. In this work, the equivalent vitrinite reflectance (EqVR_o) was used to characterize shale maturity. Wang et al. [42] have studied the relationship between equivalent vitrinite and bitumen reflectance (BR_o) of the marine shale of Longmaxi Formation: EqVR_o =1.125 × BR_o−0.2062. After the thermal evolution experiment, the bitumen reflectance (BR_o) was measured by high solution laser Raman spectroscopy analysis, and the EqVR_o was calculated. The results show that the EqVR_o increases from 0.68% to 3.09% in the open system pyrolysis while the temperature rises from 300 °C to 600 °C, and EqVR_o varied from 0.69% to 3.02% in the semi-closed system pyrolysis from 300 °C to 600 °C. The comparison of these two systems' pyrolysis is shown in Figure 3. The EqVR_o increases with increasing temperature, however, formation pressure has no obvious effect on the maturity evolution of shales. To obtain the quantitative relationship between formation pressure and shale maturity, a comparative experiment of different formation pressures at the same temperature point should be conducted.

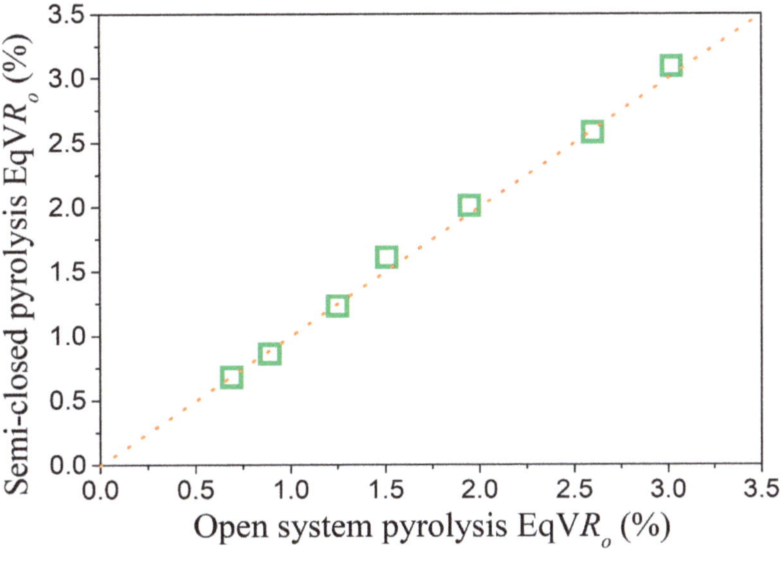

Figure 3. Comparison of shale thermal maturity based on open and semi-closed system pyrolysis experiment. Data of semi-closed pyrolysis are cited from Fang et al. [21].

The total organic carbon (TOC) decreased from 5.71% to 2.93% in the open system pyrolysis and varied from 5.03% to 1.19% in the semi-closed system when the experimental temperature increased from 300 °C to 600 °C. The quantities of gaseous and liquid hydrocarbon products during open and semi-closed system pyrolysis experiments are listed in Table 2. Due to no pressure being applied in the open system pyrolysis, uncollected liquid hydrocarbons were sporadically stored in the pore, fracture, and sample surface. Gaseous hydrocarbon products, which vary from 0.1 mg/g to 2.68 mg/g, are estimated by the thermal weight loss method. In a semi-closed system, kerogen generates oil and gas as thermal evolution proceeds. Due to the existence of formation pressure, part of the liquid hydrocarbon product is collected after discharge, namely expelled oil. The other part of the liquid hydrocarbon product retained in the pore-fracture system is called residual oil. With the increase of thermal maturity, the liquid hydrocarbon yield increases from 1.64 mg/g to 6.60 mg/g when EqVR_o < 1.51%, and then decreases from 6.60 mg/g to 8.81 mg/g when 1.51% < EqVR_o < 3.02%. Gaseous hydrocarbons were collected by drainage method, and the gaseous product yield of shale increased gradually from 0.01 mg/g to 3.23 mg/g with the increase of shale thermal maturity [21].

Table 2. Results of open and semi-closed pyrolysis experiments and Rock-Eval pyrolysis parameters.

Sample Type	Sample No.	*T* (°C)	*EqVRo* (%)	*L* (mg/g)	*G* (mg/g)	*TOC* (%)	*Tmax* (°C)	*HI* (mg/gTOC)	*S1* (mg/g)	*S2* (mg/g)
Original	XHY	-	0.67	-	-	5.74	432	308.06	0.89	10.55
Open system	XHY-1	300	0.68	-	0.03	5.71	436	295.3	2.45	6.31
	XHY-2	350	0.86	-	0.21	5.68	446	285.19	3.21	4.29
	XHY-3	400	1.23	-	0.58	5.32	466	232.1	4.32	3.19
	XHY-4	450	1.61	-	1.49	4.89	487	201.59	6.13	3.21
	XHY-5	500	2.01	-	2.33	4.01	509	178.58	5.17	2.14
	XHY-6	550	2.58	-	3.68	3.44	541	141.46	5.24	1.19
	XHY-7	600	3.09	-	4.41	2.93	569	114.51	4.32	0.32
Semi-closed system	XHY-01	300	0.69	1.64	0.00	5.03	440	268.6	0.32	4.62
	XHY-02	350	0.89	2.57	0.01	4.32	447	181.21	0.39	2.99
	XHY-03	400	1.25	6.12	0.22	2.15	535	175.91	0.07	2.41
	XHY-04	450	1.51	6.60	1.69	1.83	560	121.8	0.06	1.83
	XHY-05	500	1.95	5.42	1.81	1.53	559	84.96	0.05	1.1
	XHY-06	550	2.6	5.56	2.61	1.24	434	27.42	0.04	0.34
	XHY-07	600	3.02	3.81	3.23	1.19	449	3.48	0.03	0.04

Notes: *T* = experimental temperature, °C; *EqVRo* = equivalent vitrinite reflectance, %; *L* = liquid hydrocarbon yield, mg/g; *G* = gaseous hydrocarbon yield, mg/g; *TOC* = total organic carbon, %; *Tmax* = the temperature at the maximum release of hydrocarbons occurs during Rock-Eval pyrolysis, °C; *HI* = hydrogen index, mg/gTOC; *S1* = hydrocarbon released from source rock under 300 °C, mg/g; *S2* = hydrocarbon released from source rock under 300–600 °C, mg/g. Data of semi-closed pyrolysis are cited from Fang et al. [21].

3.2. Mineral Evolution Characteristics

According to XRD analysis, the shale samples are mainly composed of quartz, clay mineral orthoclase, apatite, and pyrite. For both experiments, quartz is the most abundant mineral, followed by clay, and pyrite is the least abundant (Figure 4a). Overall comparison of mineral content between open and semi-closed system pyrolysis experiments are shown in Figure 4b. The mineral contents at each temperature point are distributed near the straight line of 1:1. To make the comparison of mineral content by the two pyrolysis experiments more obvious, a separate comparison was provided for each mineral (Figure 5). For both open and semi-closed system pyrolysis experiments, the quartz content increases with the thermal maturity, while the content of clay minerals decreases with the thermal evolution. Previous studies showed that organic matter produces a large amount of organic acid during thermal evolution. Clay and feldspar can be dissolved by organic acids [43], and the catalytic activity of clay minerals can be strengthened by treatment with weak organic or inorganic acids [27,44]. During thermal evolution, organic acid accelerates the transformation from smectite to illite [27]. After being dissolved by organic acids, feldspar releases K ions and forms quartz; this results in the quartz-filling phenomenon [27,45], which explains the increase in quartz content with the shale thermal evolution. For each of the experimental temperature points, the existence of formation pressure increases the percentage of quartz and decreases the content of clay minerals (Figure 5a,b). This indicates that the formation pressure inhibits the formation of quartz. At the same time, high pressure is conducive to the conversion of montmorillonite and kaolinite to illite. This is because the higher formation pressure promotes the formation of quartz. At the same time, high liquid pressure is conducive to the transformation of montmorillonite and kaolinite to illite [46].

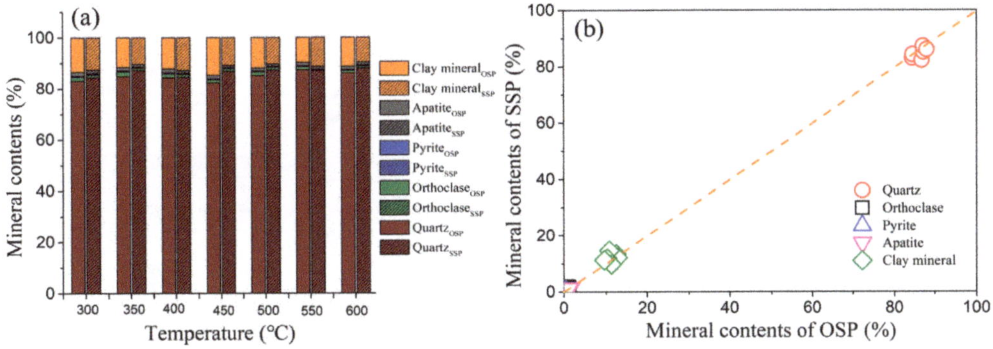

Figure 4. (**a**) Columnar diagram of mineral contents between two pyrolysis experiments, and (**b**) Mineral contents comparison between two pyrolysis experiments. Notes: OSP-Open system pyrolysis; SSP- Semi-closed system pyrolysis. Data of semi-closed pyrolysis are cited by Fang et al. [21].

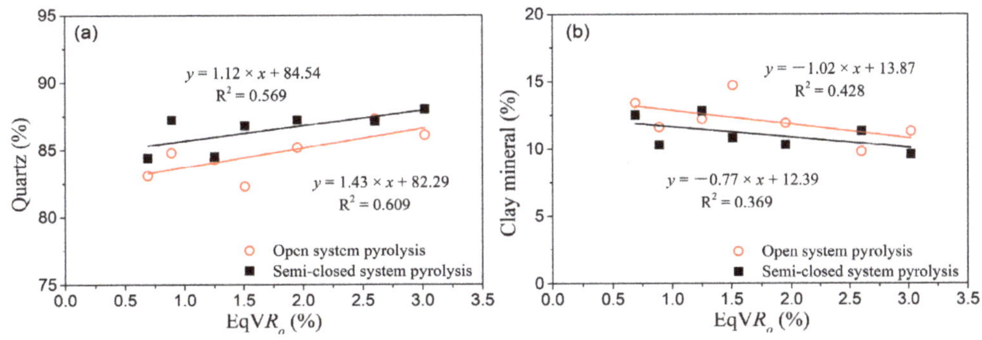

Figure 5. Changes of (**a**) quartz and (**b**) clay mineral content during thermal evolution of shale under two pyrolysis experiments Data of semi-closed pyrolysis are cited from Fang et al. [21].

3.3. Pore Structure Evolution Characteristics

To intuitively observe the development of organic pores, FESEM was conducted using shale samples after two pyrolysis experiments (Figure 6). For shales based on open system pyrolysis experiments, there are a large number of inorganic pores and micro-cracks at the lowly matured stage (300 °C–350 °C), and no primary pores can be seen in solid kerogen. During the oil generation stage (350 °C–400 °C), mineral shrinkage fractures occur and smaller pores are difficult to identify due to oil filling because there is no pressure exerted to drive hydrocarbon expulsion in the open system pyrolysis. At the wet gas stage (400 °C–500 °C), solid bitumen and petroleum were generated, pores and cracks developed, gas was released, and the smaller pores appeared. The proportion of smaller pores in the identifiable pore volume in the visual field increased, and secondary minerals were generated due to recrystallization. At the dry gas stage (550 °C–600 °C), organic matter and liquid hydrocarbon were further cracked to produce gas, the pore size of identifiable organic pores became smaller, and the connectivity noticeably increased.

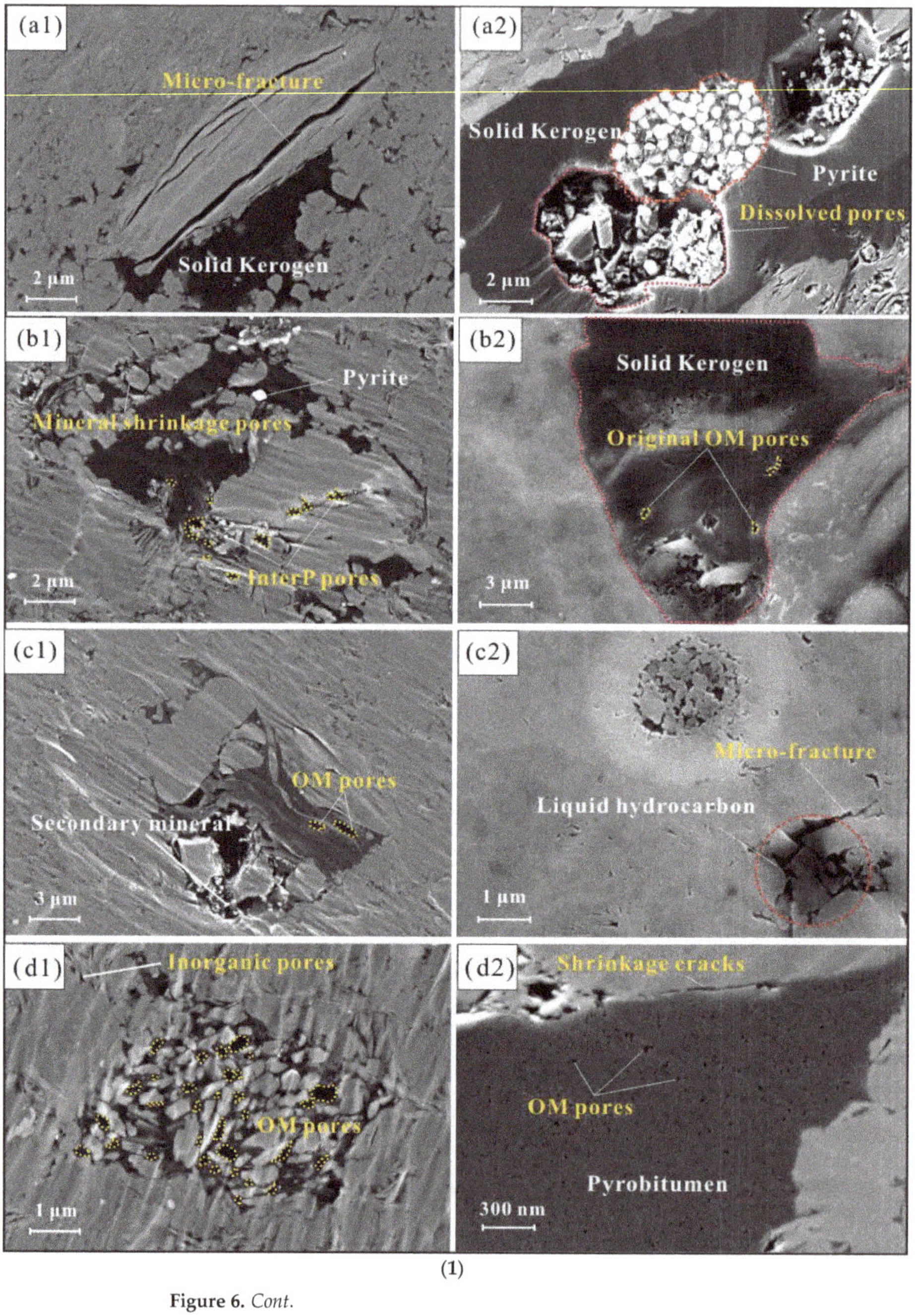

(1)

Figure 6. *Cont.*

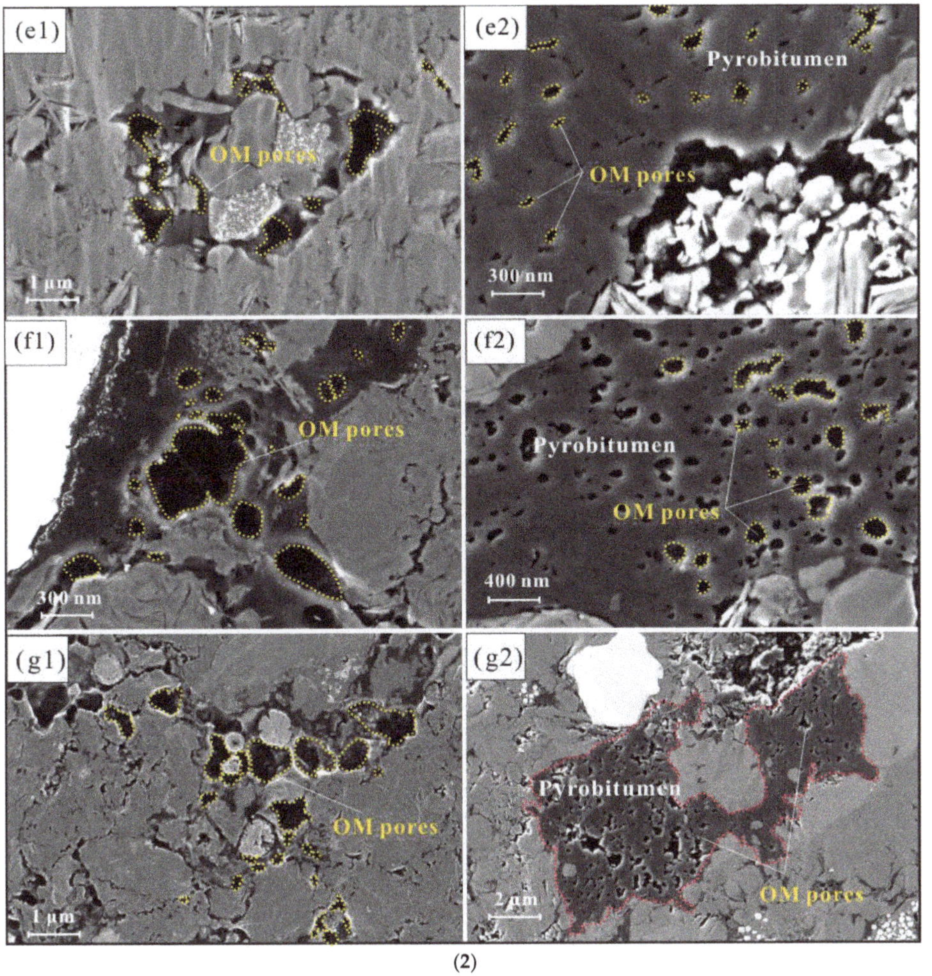

(2)

Figure 6. (**1**) SEM images of pyrolysis samples at temperatures from 300 to 450 °C. (**2**) SEM images of pyrolysis samples at temperatures from 500 to 600 °C. Notes: (**a1–g1**) samples based on open system pyrolysis: (**a1**) micro-fracture in inorganic mineral at 300 °C, R_o = 0.68%; (**b1**) mineral shrinkage pore and interpores at 350 °C, R_o = 0.86%; (**c1**) secondary mineral at 400 °C, R_o = 1.23%; (**d1**) OM pore at 450 °C, R_o = 1.61%; (**e1**) OM pore at 500 °C, R_o = 2.01%; (**f1**) OM pore at 550 °C, R_o = 2.58%; (**g1**) OM pore at 600 °C, R_o = 3.09%. (**a2–g2**) Samples based on semi-closed system pyrolysis: (**a2**) dissolved pores and pyrite at 300 °C, R_o = 0.69%; (**b2**) original OM pores developed at 350 °C, R_o = 0.89%; (**c2**) liquid hydrocarbon and microfracture at 400 °C, R_o = 1.25%; (**d2**) there are a large number of OM pores in the pyrobitumen of shale at 450 °C, R_o = 1.51%; (**e2**) OM pores at 500 °C, R_o = 1.95%; (**f2**) OM pores with good connectivity in shale at 550 °C, R_o = 2.60%; and (**g2**) large size OM pore with good connectivity in shale at 600 °C, R_o = 3.02%. (**a2–g2**) are cited from Fang et al. [21].

For shale samples based on semi-closed system pyrolysis experiments, organic pores begin to appear in organic matter on shale at the oil generation stage (350 °C–400 °C), and mineral-related pores are reduced by compaction and cementation [8]. At the wet gas stage (400 °C–500 °C), the solid bitumen and petroleum in pores cracking produce secondary organic pores. The presence of fluid pressure makes hydrocarbon expulsion easier, OM pores generated in pyrobitumen were isolatedly distributed with irregular shapes. At the dry gas stage (550 °C–600 °C), organic matter and liquid hydrocarbon were further cracked [21]. The relatively thin pore wall ruptures due to fluid pressure resulting in the formation of connected larger pores, thus improving the connectivity of shale pores [47]. It can be seen from the SEM images of the shale samples in two the thermal evolution experiments that the presence of fluid pressure causes part of the hydrocarbons originally stored in the pores to discharge, increasing the pore volume, especially those with small pore size. Meanwhile, formation pressure breaks down some of the thinner pore walls, increasing the pore connectivity.

The applicability of characterization technologies is variable. Several technologies should be integrated to get the full-scale pore parameter. In this work, CO_2 and N_2 adsorption/desorption and MIP were used to characterize the pores in shale. Many pieces of research have shown the applicability of this full-scale method [1,26,44]. The CO_2, N_2, and MIP curves of the shale samples from the two pyrolysis experiments are shown in Figure 7. According to the IUPAC classification, the N_2 adsorption/desorption isotherms of the shale samples based on open and semi-closed system pyrolysis are mainly of type IV owning to an obvious hysteresis loop [15]. The type of N_2 adsorption hysteresis loop in open system shale samples is Type H4, representing narrow slit pores. However, with the increase of thermal evolution maturity, the hysteresis loop type of shale samples in the semi-closed system gradually changes from Type H4 to Type H2, which represents the pore type gradually changing from narrow slit pore to ink bottle-shaped.

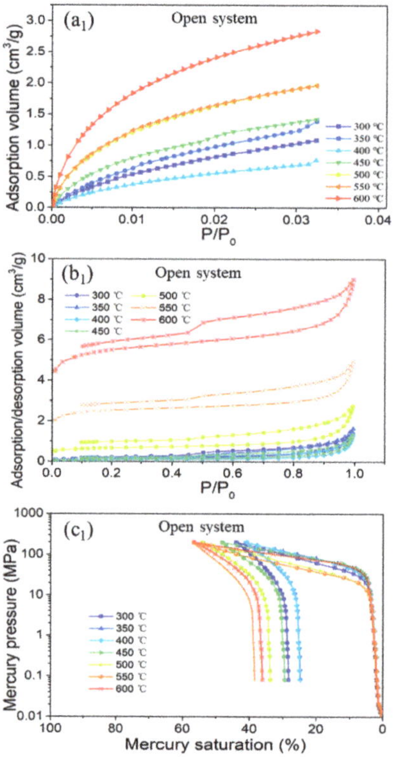

Figure 7. (**a1**,**a2**) CO_2 adsorption isotherms, (**b1**,**b2**) N_2 adsorption/desorption isotherms, and (**c1**,**c2**) MIP curves from the two pyrolysis experiments. Data of semi-closed pyrolysis are cited from Fang et al. [21].

The MIP curve type of the open system is relatively constant. The mercury injection curve has two stages, and the mercury withdrawal efficiency is between 30–40%. After the formation pressure is applied, the first half of the mercury injection curve slows down, indicating that the pore connectivity becomes better. To intuitively describe the evolution characteristics of pore structure, the pore volume of three pore size scales at each thermal evolution stage is shown in Figure 8. With the increase of R_o, the pore volume of the shale samples from the two pyrolysis experiment changes greatly. The volume difference between micropores, mesopores, and macropores of shale samples in the open system is significantly smaller than that in the semi-closed system. The mesopores content of shale samples in the semi-closed system obviously accounts for the largest proportion (Figure 8). The mesopore volumes of the two pyrolysis samples decrease first and then increase with the thermal evolution of shale maturity, however, the change of micropore volume is not obvious. The macropore volume of shales in open-system pyrolysis is increased with the increase of evolution degree, while the changing trend in shales of the semi-closed system pyrolysis is not obvious.

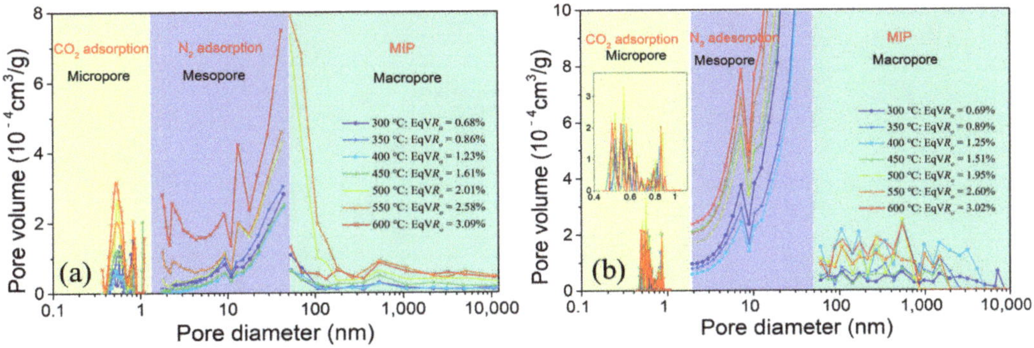

Figure 8. Full-scale pore size distribution of pyrolysis shales obtained by MIP, N_2 and CO_2 physisorption based on (**a**) open system and (**b**) semi-closed system pyrolysis experiments [21].

4. Discussion

4.1. Influence of Formation Pressure on Pore Structure

A full-scale pore size distribution (PSD) based on open system and semi-closed system pyrolysis experiments obtained by MIP, CO_2, and N_2 physisorption is shown in Figure 9. It is apparent that the mesopore volume of the shale samples at each temperature point increases significantly after formation pressure is applied, while the changes of macropore and micropore volume do not follow an obvious trend. At 400 °C–500 °C, part of the micropore size distribution curve of the semi-closed system pyrolysis is higher than that of the open system, indicating that there is a small segment of micropore volume increase due to the existence of formation pressure. At 500 °C and 550 °C, the macropore volume in the 50 nm–100 nm range is higher in the open system than in the semi-open system, indicating that the pore pressure reduces the pore volume at this stage. Therefore, it is inferred that the pressure drives the hydrocarbon flow and occupies the pores in the 50–100 nm pore range.

To determine the effect of formation pressure on the pores at different pore diameters, pore structure parameters of micropores, mesopores, and macropores from open and semi-closed system pyrolysis were calculated. The PSD-based two pyrolysis systems are shown in Table 3. Based on open system pyrolysis, the pore volume of micropores (<2 nm) varied from 7.83×10^{-4} cm^3/g to 44.62×10^{-4} cm^3/g with an average of 21.68×10^{-4} cm^3/g, the pore volume of mesopores changed from 9.80×10^{-4} cm^3/g to 57.37×10^{-4} cm^3/g with an average of 22.57×10^{-4} cm^3/g, and the pore volume of macropores were between 3.86×10^{-4} cm^3/g and 42.94×10^{-4} cm^3/g with an average of 16.25×10^{-4} cm^3/g. Based on semi-closed system pyrolysis, the pore volume of micropores varied from 5.52×10^{-4} cm^3/g to 13.81×10^{-4} cm^3/g with an average of 10.83×10^{-4} cm^3/g, the pore volume of mesopores changed from 37.30×10^{-4} cm^3/g to 166.49×10^{-4} cm^3/g with an average of 88.45×10^{-4} cm^3/g, and the PV of macropores were between 8.22×10^{-4} cm^3/g and 26.94×10^{-4} cm^3/g with an average of 14.81×10^{-4} cm^3/g. Based on open system pyrolysis, the total pore volume (TPV) of shale samples varied from 21.60 × 10^{-4} cm^3/g to 144.92 × 10^{-4} cm^3/g with an average of 60.49 × 10^{-4} cm^3/g, and the TPV of shales based on semi-closed system pyrolysis changed from 54.94 × 10^{-4} cm^3/g to 206.30 × 10^{-4} cm^3/g with an average of 114.09 × 10^{-4} cm^3/g [21].

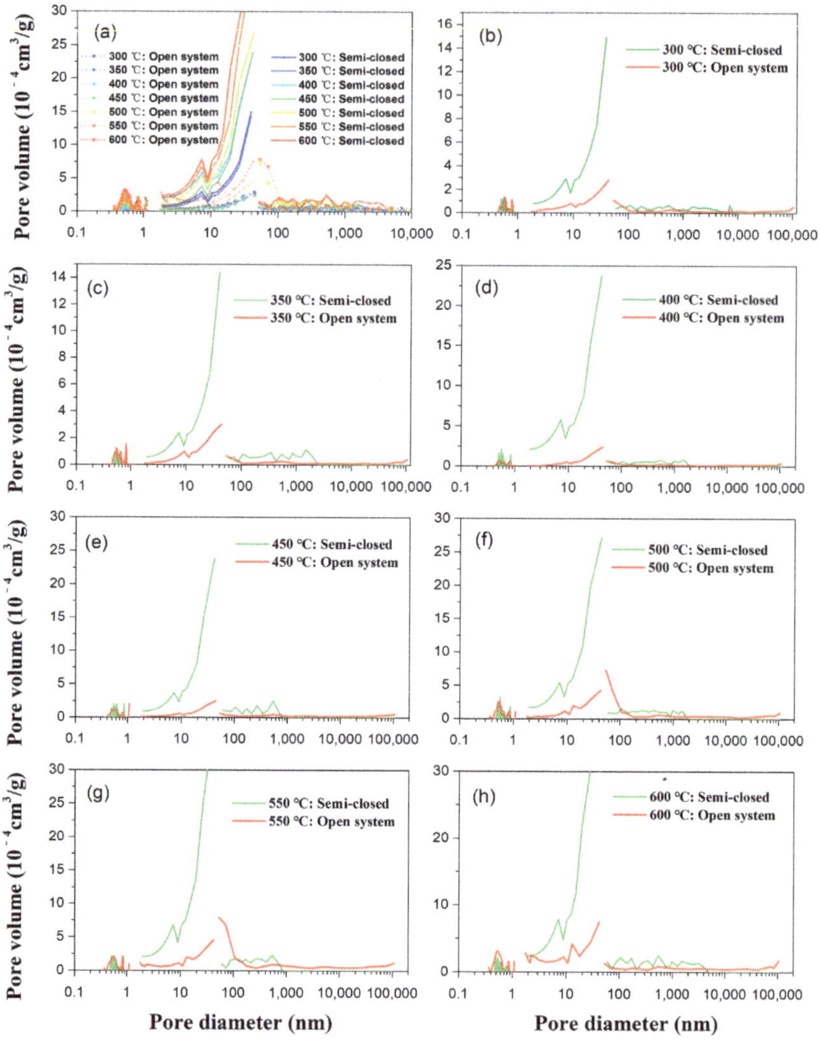

Figure 9. Full-scale pore size distribution based on open system and semi-closed system pyrolysis experiments by MIP, and CO_2 and N_2 physisorption. (**a**) Full-scale PSD of all samples; the dashed lines are data of open system pyrolysis, and the solid lines are semi-closed system pyrolysis. (**b**–**h**) Full-scale PSD of shale with temperature from 300 to 600 °C; the red curves are data of open system pyrolysis, the green curves are data of semi-closed system pyrolysis. Data of semi-closed pyrolysis are cited from Fang et al. [21].

Table 3. PSD and fractal dimension of shale samples based on open and semi-closed pyrolysis experiments.

Sample no.	PV_{-OSP} (10^{-4} cm^3/g)				$D1_{-OSP}$	$D2_{-OSP}$	Sample no.	PV_{-SSP} (10^{-4} cm^3/g)				$D1_{-SSP}$	$D2_{-SSP}$
	<2 nm	2–50 nm	>50 nm	TPV_{-OSP}				<2 nm	2–50 nm	>50 nm	TPV_{-SSP}		
XHY-1	10.39	14.37	4.91	29.67	2.45	2.62	XHY-01	7.08	51.70	9.28	68.06	2.37	2.55
XHY-2	12.86	15.50	3.86	32.22	2.41	2.56	XHY-02	5.52	45.62	11.14	62.28	2.36	2.48
XHY-3	7.83	9.80	3.97	21.60	2.50	2.49	XHY-03	9.42	37.30	8.22	54.94	2.39	2.63
XHY-4	17.38	11.63	5.98	34.99	2.44	2.56	XHY-04	13.48	79.14	13.81	106.44	2.43	2.53
XHY-5	28.98	21.38	20.20	70.55	2.78	2.72	XHY-05	13.81	104.19	16.92	134.92	2.49	2.61
XHY-6	29.68	27.94	31.87	89.49	2.86	2.87	XHY-06	13.63	134.68	17.39	165.71	2.55	2.60
XHY-7	44.62	57.37	42.94	144.92	2.86	2.91	XHY-07	12.87	166.49	26.94	206.30	2.58	2.59

Note: OSP = pore data based on open system pyrolysis; SSP = pore data based on semi-closed system pyrolysis; PV = pore volume; TPV = total pore volume; *D1* is the pore surface fractal dimension; *D2* is the pore structure fractal dimension. Data of semi-closed pyrolysis are cited from Fang et al. [21].

The pore volume initially decreased, but then increased during open-system pyrolysis (Figure 10a). The pore volume transition is basically consistent with the four stages of thermal maturation. At the maturity stage (300 °C–400 °C, 0.68% < $EqVR_o$ < 1.23%), the volume of the mesopores are the largest, followed by micropores and macropores. At the highly matured stage (400 °C–550 °C, 1.23% < $EqVR_o$ < 2.58%), before 550 °C, the micropores volume becomes the largest, mesopores follows and the macropores volume are the least. After 500 °C, the volume of the micropores are still the largest, however, they are now followed by the volume of macropores, and the mesopores volume are the smallest. At the overmatured stage (500 °C–600 °C, 2.58% < $EqVR_o$ < 3.09%), the mesopores volume are the largest, and the volume of the macropores are equivalent to that of the micropores. For the semi-closed system, the volume of mesopores decreases first and then increases significantly during the thermal evolution of shale, while the volume of micropores and macropores change insignificantly (Figure 10b). Mesopores account for most of the volume (nearly 70% of the total pore volume), which indicates the volume of mesopores is more significantly affected by formation pressure than micropore and macropore.

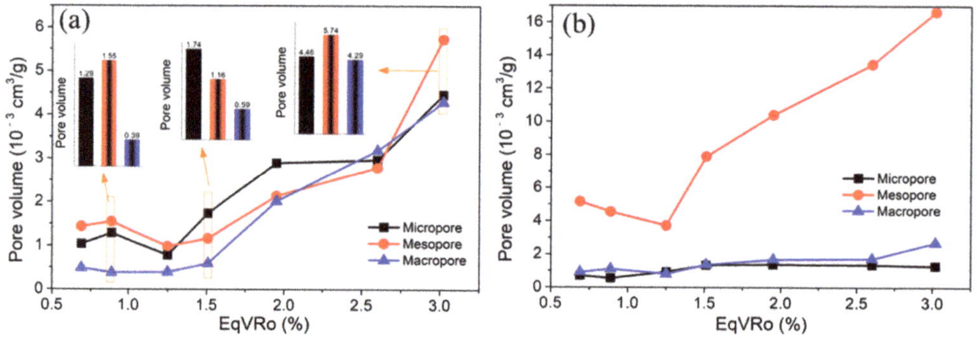

Figure 10. Changes of pore volume of shale during (**a**) open system and (**b**) semi-closed system pyrolysis experiments. Data of semi-closed pyrolysis are cited from Fang et al. [21].

To intuitively understand the difference in pore volume, the variations in the pore volumes between the two pyrolysis experiments are proposed in Figure 11. The micropore volumes of the shale samples from the semi-closed system pyrolysis is smaller than that of the open system. It shows that in the process of hydrocarbon expulsion driven by formation pressure, the migration of hydrocarbons will block some micropores and reduce the pore volume of micropores [5,48]. However, at the peak of oil generation ($EqVR_o$ = 1.23 %), the liquid hydrocarbons and solid bitumen produced by the open system would have filled the pores and reduced the micropore volume at this stage [35,47]. Thus, the micropore volumes of the samples from the two pyrolysis systems is similar at about 400 °C ($EqVR_o$ = 1.23 %) (Figure 11b). The mesopore volumes of shale from the semi-closed system pyrolysis greatly exceeds that of the open system (Figure 11c), which shows that the volumes of the mesopores increases significantly after the formation pressure applied. The reason should be that a large number of hydrocarbons are discharged from the pores and the application of pressure causes some thinner pore walls to break, which increases porosity [13,35]. At the same time, it also shows that the retained hydrocarbons are mainly stored in the mesopores of shale. The influence of formation pressure on the macropores of the shale samples presents two stages. Before 500 °C ($EqVR_o$ = 2.01%), the macropore volumes of the shale in the semi-closed system is larger than that of the open system, while after 500 °C, the trend is the opposite. Before 500 °C, the open system has no hydrocarbon expulsion pressure. Therefore, more retained hydrocarbons were blocked in the large pores, causing the pore volumes to be smaller than that in the semi-closed system. After 500 °C, solid bitumen and oil cracked to form gas, and the volume of macropores increased. Due to the existence of fluid pressure in the shale of the semi-closed system, the hydrocarbons discharged from the mesopores will occupy parts of the macropore volume or cracks [2,34], therefore, the macropore volumes of the semi-closed volume will be smaller.

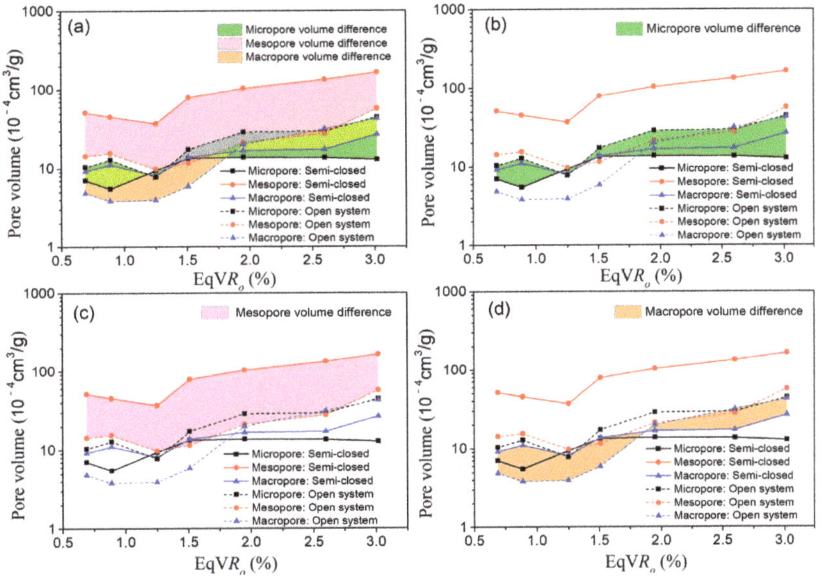

Figure 11. The difference in pore volume in the shale samples between the open and semi-closed system pyrolysis experiments. (**a**) Total comparison of pore volume; (**b**) micropore volume difference; (**c**) mesopore volume difference; (**d**) macropore volume difference. Data of semi-closed pyrolysis are cited from Fang et al. [21].

4.2. Influence of Formation Pressure on Pore Heterogeneity

The description of pore heterogeneity is difficult in shale gas reservoirs. The fractal theory is a common way to characterize the heterogeneity of shale pores, which are usually used to characterize the complexity of the pore surface and pore structure of porous material [28,49–51]. Through the data of N_2 physisorption, the fractal dimension (*D*) can be obtained to quantitatively describe the pore structure and surface heterogeneity. The relative pressure and adsorption capacity were used to quantitatively acquire the fractal characteristics of shale pores using the Frenkel–Halsey–Hill (FHH) method [28,49,51]:

$$\mathrm{lnV} = (\mathrm{D}-3)\ln\left[\ln\left(\frac{\mathrm{P_0}}{\mathrm{P}}\right)\right] + \mathrm{C} \tag{1}$$

where *D* is the fractal dimension of pore structure; P_0 is the vapor saturation pressure of gas (MPa); *C* is a constant; *P* is the equilibrium pressure of N_2 (MPa); *V* is the volume of adsorbed gas molecules at *P* (cm^3/g).

According to N_2 physisorption, there are two distinct segments on the plot of *lnV* vs *ln*[*ln*(*P0*/*P*)]. At low relative pressure ($P/P_0 < 0.5$), the N_2 physisorption is controlled by the van der Waals force. When $P/P_0 > 0.5$, N_2 physisorption is mainly dominated by surface tension [20,28,51,52]. The fractal dimension of the two segments can be calculated by using *ln*(*V*) and *ln*[*ln*(P_0/*P*)] from the N_2 adsorption/desorption experiment according to Equation (1). *D1* can represent the heterogeneity of pore surface, and *D2* can represent the heterogeneity of pore volume and structure [20,28,49–51].

Figure 12 shows the fractal dimension fitted curves of the shale samples in the two pyrolysis systems, which show that the adsorption data of the semi-closed system pyrolysis samples are higher than those of the open system except for the two points of 550 °C and

600 °C. The calculation results of the fractal dimensions are shown in Table 3, which shows that the heterogeneity of pore surface and pore volume decreases initially before increasing (creating a "U" shape) during the open system pyrolysis experiments (Figure 13a). Based on the semi-closed system pyrolysis experiment, *D1* increases with the increase of shale thermal evolution, while the pore volume heterogeneity (*D2*) did not show a correlation with shale maturity (Figure 13b). To make the difference of fractal dimension more intuitive, the variations in *D1* and *D2* between the two pyrolysis experiments are shown in Figure 13c,d. The existence of formation pressure significantly reduced the pore surface heterogeneity because the formation pressure promotes hydrocarbon expulsion, which makes the pore surface rougher and reduces the heterogeneity of the pore surface. The effect of formation pressure on pore structure heterogeneity is three-stage:

1. When EqVR_o < 1.03%, the formation pressure reduces the heterogeneity of pore volume in the other stages. The reason is that the amount of oil produced is low at this stage, the migration of hydrocarbons will block some micropores and reduce the pore volume of micropores (Figure 11b). Thus, the pore volume distribution is more concentrated, reducing the pore heterogeneity.
2. At the peak of oil generation (1.03% < EqVR_o < 1.51%), the formation pressure increases pore heterogeneity. The reason is that the retained hydrocarbons occupy the pores before the formation pressure is applied, which makes a small difference between micropore, mesopore, and macropore volumes (Figure 10a). After applying pressure, the difference between these pore volumes becomes larger due to hydrocarbon expulsion (Figure 10b), which increases the heterogeneity of the pore structure.
3. When EqVR_o > 1.51%, the formation pressure reduces the heterogeneity of pore volume. In this stage, the partial micropore volume begins to transform into mesopore volume [35], and the existence of formation pressure makes some thinner pore walls rupture, which increases pore connectivity [35,53] and decreases the heterogeneity of pore structure.

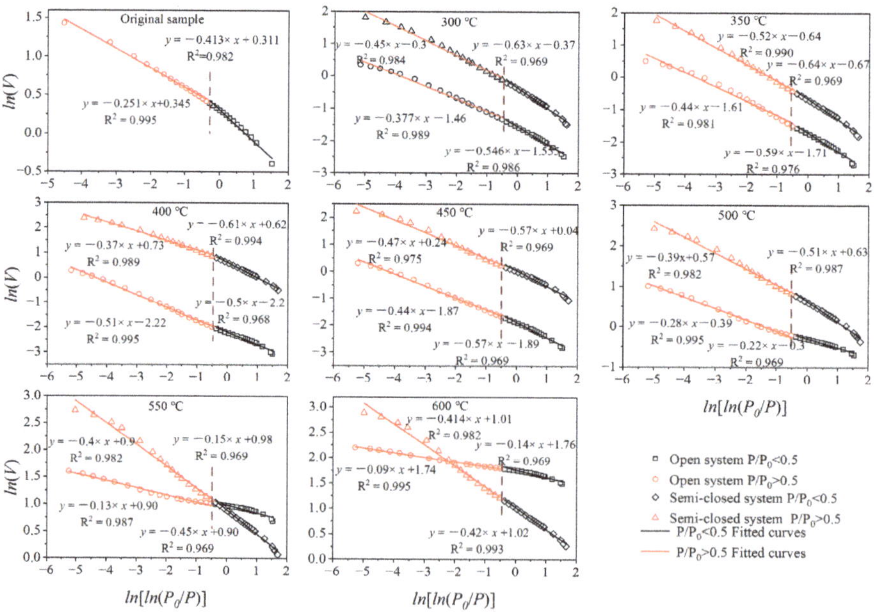

Figure 12. Plots of *ln*(*V*) versus *ln*[*ln*(*P0*/*P*)] from low-temperature nitrogen physisorption isotherm data using the open system and semi-closed system pyrolysis experiments. Data of semi-closed pyrolysis are cited from Fang et al. [21].

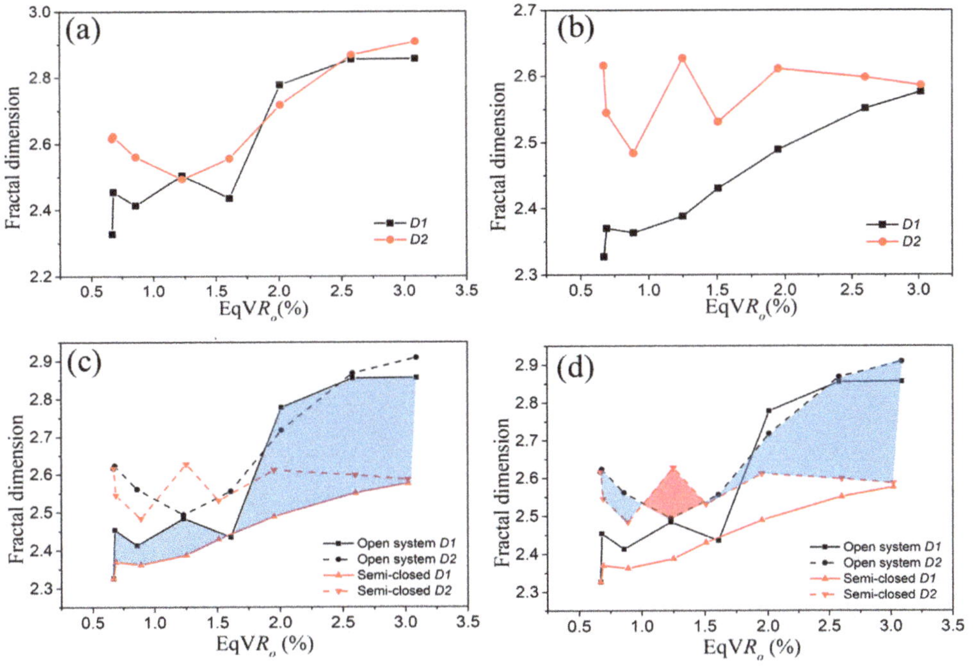

Figure 13. Relationships between fractal dimension and R_o based on (**a**) the open system and (**b**) semi-closed system pyrolysis experiments; (**c**) pore surface fractal dimension variation between the open and semi-closed system pyrolysis experiments during shale thermal evolution; (**d**) pore structure fractal dimension variation between the open and semi-closed system pyrolysis experiments during shale thermal evolution. Note: *D1* is the pore surface fractal dimension, and *D2* is the pore structure fractal dimension. Data of semi-closed pyrolysis are cited from Fang et al. [21].

4.3. Pore Evolutionary Pattern for Marine Shale

A shale pore evolutionary model based on fluid pressure was established by comparing the evolution process of pore structures and minerals, as well as the generation, expulsion, and retention of hydrocarbons in shale during the two pyrolysis experiments (Figure 14). The coupling relationships between the evolutionary characteristics of organic matter and the pore structure of Neoproterozoic marine shales can be roughly divided into four stages. In the first stage (EqVR_o < 0.69%) the evolution of organic matter is in the low mature period, and the shale has not yet entered the oil window. The pores are mostly mineral pores, and there are almost no primary organic pores in the FEM images. The transformation of clay minerals is not obvious. In the second stage (0.69% < EqVR_o < 1.25%), a large amount of the liquid hydrocarbons generated by kerogen pyrolysis are stored in pores, which reduces the pore volume of shale. After the formation pressure is applied, the micropore volume increases slightly and the mesopore volume increases significantly. Organic acids produced during hydrocarbon generation dissolves the organic matter and clay minerals, and secondary quartz begins to form [44,45]. The formation of shrinkage cracks is related to hydrocarbon generation (Figure 6b1). In the third stage (1.25% < EqVR_o < 2.60%), solid bitumen and crude oil began to pyrolyze to form wet gas, while kerogen is condensed to pyrobitumen. A large number of isolated and dispersed circular pores appear in pyrobitumen. The volume of the micropores in this stage is greatly affected by pressure. The transition from partial micropore volume to mesopore volume makes the micropore volume increase slowly [35]. At the same time, due to hydrocarbon expulsion, the pore morphology gradually changes from narrow slit pores to ink bottle-shaped pores, and the transformation of kaolinite and montmorillonite into illite is common [48]. The fourth stage (2.60% < $EqVR_o$ < 3.02%) is the dry gas stage, and light hydrocarbons are rapidly generated. The existence of formation pressure makes some thinner pore walls crack, and increases pore connectivity [35,53].

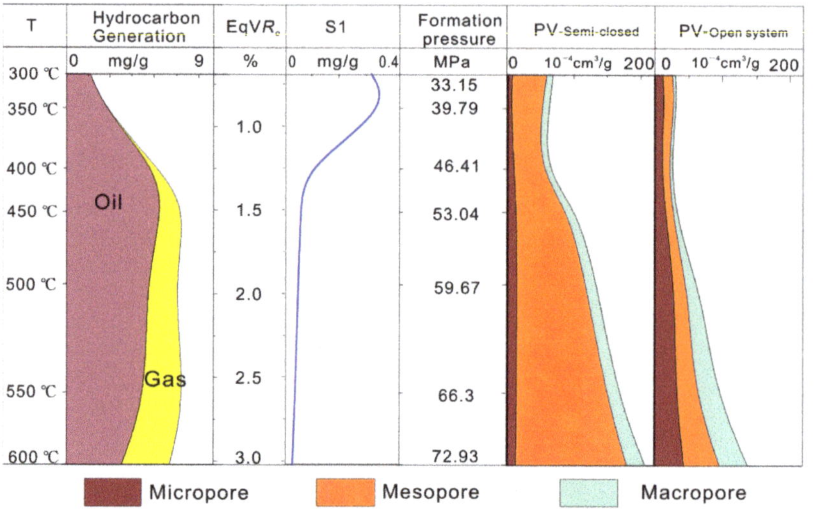

Figure 14. Pore evolution model of organic-rich marine shale based on formation pressure.

5. Conclusions

(1) With the thermal evolution process of marine shale, the quartz content increased, while the content of clay minerals decreased in both the open and semi-closed system pyrolysis experiments. The existence of formation pressure increased the percentage of quartz and decreased the content of clay minerals, and the change of formation pressure had no obvious influence on the maturity evolution of marine shales.

(2) The relationship between formation pressure and pore development is revealed. With the increase of formation pressure, the pore type of shale gradually changes from narrow slit pore to ink bottle-shaped. Formation pressure improves pore connectivity, especially in the high-overmature stage of shale. With the increase of formation pressure, the volume of micropores decreases slightly, the volume of mesopores increases significantly, and the volume of macropores changes in two stages.

(3) The relationship between formation pressure and pore heterogeneity is revealed. After applying formation pressure, the volume difference between the micropores, mesopores, and macropores of the shale samples became larger. Formation pressure makes the pore surface heterogeneity significantly reduced, while the effect of formation pressure on pore structure heterogeneity is three-stage.

(4) The pore size/volume results of the comparative thermal evolution experiment show that the retained hydrocarbons mainly occupy the mesopore space of shale, and the existence of formation pressure promotes hydrocarbon expulsion, especially the hydrocarbons in the mesopores.

Author Contributions: D.L. and Y.C. conceived and designed the experiments; X.F. performed the experiments and wrote the paper; X.F. and Y.C. analyzed the data; Q.H. and P.G. revised the paper and provided language support; Y.Q. provided technical support. All authors have read and agreed to the published version of the manuscript.

Funding: This research was funded by the National Natural Science Foundation of China (Grant nos.U19B6003-03-01 and 42030804), the Fundamental Research Funds for the Central Universities (2652022207), and the joint Ph.D. program from China Scholarship Council.

Institutional Review Board Statement: Not applicable.

Informed Consent Statement: Not applicable.

Data Availability Statement: The data used to support the findings of this study are available from the corresponding author upon request.

Conflicts of Interest: The authors declare no conflict of interest.

References

1. Wang, Y.; Liu, L.; Zheng, S.; Luo, Z.; Sheng, Y.; Wang, X. Full-scale pore structure and its controlling factors of the Wufeng-Longmaxi shale, southern Sichuan Basin, China: Implications for pore evolution of highly overmature marine shale. *J. Nat. Gas. Sci. Eng.* **2019**, *67*, 134–146. [CrossRef]
2. Wu, L.; Geng, A.; Wang, P. Oil expulsion in marine shale and its influence on the evolution of nanopores during semi-closed pyrolysis. *Int. J. Coal. Geol.* **2018**, *191*, 125–134. [CrossRef]
3. Han, Y.; Horsfield, B.; Wirth, R.; Mahlstedt, N.; Bernard, S. Oil retention and porosity evolution in organic-rich shales. *Am. Assoc. Pet. Geol. Bull.* **2017**, *101*, 807–827. [CrossRef]
4. Goodman, A.; Sanguinito, S.; Kutchko, B.; Natesakhawat, S.; Cvetic, P.; Allen, A.J. Shale pore alteration: Potential implications for hydrocarbon extraction and CO_2 storage. *Fuel* **2020**, *265*, 116930. [CrossRef] [PubMed]
5. Gao, Z.; Xuan, Q.; Hu, Q.; Jiang, Z.; Liu, X. Pore structure evolution characteristics of continental shale in China as indicated from thermal simulation experiments. *Am. Assoc. Pet. Geol. Bull.* **2021**, *105*, 2159–2180. [CrossRef]
6. Zhao, J.; Yang, L.; Yang, D.; Kang, Z.; Wang, L. Study on pore and fracture evolution characteristics of oil shale pyrolysed by high-temperature water vapour. *Oil Shale* **2022**, *39*, 79–95. [CrossRef]
7. Wang, P.; Zhangm, C.; Li, X.; Zhang, K.; Yuan, Y.; Zang, X. Organic matter pores structure and evolution in shales based on the he ion microscopy (HIM): A case study from the Triassic Yanchang, Lower Silurian Longmaxi and Lower Cambrian Niutitang shales in China. *J. Nat. Gas. Sci. Eng.* **2020**, *84*, 103682. [CrossRef]
8. Zhang, Y.; Yu, B.; Pan, Z.; Hou, C.; Zuo, Q.; Sun, M. Effect of thermal maturity on shale pore structure: A combined study using extracted organic matter and bulk shale from Sichuan Basin, China. *J. Nat. Gas. Sci. Eng.* **2020**, *74*, 103089. [CrossRef]
9. Liu, X.; Lai, J.; Fan, X.; Shu, H.; Wang, G.; Ma, X. Insights in the pore structure, fluid mobility and oiliness in oil shales of Paleogene Funing Formation in Subei Basin, China. *Mar. Pet. Geol.* **2020**, *114*, 104228. [CrossRef]
10. Wei, L.; Mastalerz, M.; Schimmelmann, A.; Chen, Y. Influence of Soxhlet-extractable bitumen and oil on porosity in thermally maturing organic-rich shales. *Int. J Coal. Geol.* **2014**, *132*, 38–50. [CrossRef]
11. Geng, Y.; Liang, W.; Liu, J.; Cao, M.; Kang, Z. Evolution of Pore and Fracture Structure of Oil Shale under High Temperature and High Pressure. *Energy Fuels* **2017**, *31*, 10404–10413. [CrossRef]
12. Li, G.; Li, G.; Luo, C.; Zhou, R.; Zhou, J.; Yang, J. Dynamic evolution of shale permeability under coupled temperature and effective stress conditions. *Energy* **2023**, *266*, 126320. [CrossRef]
13. Wang, T.; Wang, Q.; Lu, H.; Peng, P.; Zhan, X. Understanding pore evolution in a lacustrine calcareous shale reservoir in the oil window by pyrolyzing artificial samples in a semi-closed system. *J. Pet. Sci. Eng.* **2021**, *200*, 108230. [CrossRef]
14. Wang, E.; Feng, Y.; Guo, T.; Li, M. Oil content and resource quality evaluation methods for lacustrine shale: A review and a novel three-dimensional quality evaluation model. *Earth-Sci. Rev.* **2022**, *232*, 104134. [CrossRef]
15. Sing, K.S.W. Reporting physisorption data for gas/solid systems with special reference to the determination of surface area and porosity (Recommendations 1984). *Pure. Appl. Chem.* **1985**, *57*, 603–619. [CrossRef]
16. Cai, Y.; Li, Q.; Liu, D.; Zhou, Y.; Lv, D. Insights into matrix compressibility of coals by mercury intrusion porosimetry and N2 adsorption. *Int. J. Coal. Geol.* **2018**, *200*, 199–212. [CrossRef]
17. Bai, L.H.; Liu, B.; Du, Y.J.; Wang, B.Y.; Tian, S.S.; Wang, L. Distribution characteristics and oil mobility thresholds in lacustrine shale reservoir: Insights from N_2 adsorption experiments on samples prior to and following hydrocarbon extraction. *Pet. Sci.* **2022**, *19*, 486–497. [CrossRef]
18. Su, S.; Jiang, Z.; Xuanlong, S.; Zhang, C.; Zou, Q.; Li, Z. The effects of shale pore structure and mineral components on shale oil accumulation in the Zhanhua Sag, Jiyang Depression, Bohai Bay Basin, China. *J. Pet. Sci. Eng.* **2018**, *165*, 365–374. [CrossRef]
19. Clarkson, C.R.; Solano, N.; Bustin, R.M.; Bustin, A.M.M.; Chalmers, G.R.L.; He, L. Pore structure characterization of North American shale gas reservoirs using USANS/SANS, gas adsorption, and mercury intrusion. *Fuel* **2013**, *103*, 606–616. [CrossRef]
20. Wang, Q.; Hu, Q.; Ning, X.; Ilavsky, J.; Kuzmenko, I.; Tom, T. Spatial heterogeneity analyses of pore structure and mineral composition of Barnett Shale using X-ray scattering techniques. *Mar. Pet. Geol.* **2021**, *134*, 105354. [CrossRef]
21. Fang, X.; Cai, Y.; Hu, Q.; Liu, D.; Gao, P.; Qian, Y. Hydrocarbon Retention and Its Effect on Pore Structure Evolution of Marine Shale Based on Pyrolysis Simulation Experiments. *Energy Fuels* **2022**, *36*, 13556–13569. [CrossRef]
22. Graham, S.P.; Rouainia, M.; Aplin, A.C.; Cubillas, P.; Fender, T.D.; Armitage, P.J. Geomechanical characterisation of organic-rich calcareous shale using AFM and nanoindentation. *Rock. Mech. Rock. Eng.* **2021**, *54*, 303–320. [CrossRef]

23. Han, Y.; Horsfield, B.; Mahlstedt, N.; Wirth, R.; Curry, D.J.; LaReau, H. Factors controlling source and reservoir characteristics in the Niobrara shale oil system, Denver Basin. *Am. Assoc. Pet. Geol. Bull.* **2019**, *103*, 2045–2072. [CrossRef]
24. Sun, L.; Tuo, J.; Zhang, M.; Wu, C.; Wang, Z.; Zheng, Y. Formation and development of the pore structure in Chang 7 member oil-shale from Ordos Basin during organic matter evolution induced by hydrous pyrolysis. *Fuel* **2015**, *158*, 549–557. [CrossRef]
25. Tang, X.; Zhang, J.; Jin, Z.; Xiong, J.; Lin, L.; Yu, Y. Experimental investigation of thermal maturation on shale reservoir properties from hydrous pyrolysis of Chang 7 shale, Ordos Basin. *Mar. Pet. Geol.* **2015**, *64*, 165–172. [CrossRef]
26. Hu, H.; Hao, F.; Lin, J.; Lu, Y.; Ma, Y.; Li, Q. Organic matter-hosted pore system in the Wufeng-Longmaxi (O3w-S11) shale, Jiaoshiba area, Eastern Sichuan Basin, China. *Int. J. Coal. Geol.* **2017**, *173*, 40–50. [CrossRef]
27. Wu, S.; Yang, Z.; Zhai, X.; Cui, J.; Bai, L.; Pan, S. An experimental study of organic matter, minerals and porosity evolution in shales within high-temperature and high-pressure constraints. *Mar. Pet. Geol.* **2019**, *102*, 377–390. [CrossRef]
28. Xie, X.; Hu, L.; Deng, H.; Gao, J. Evolution of pore structure and fractal characteristics of marine shale during electromagnetic radiation. *PLoS ONE* **2020**, *15*, e0239662. [CrossRef]
29. Chandra, D.; Bakshi, T.; Vishal, V. Thermal effect on pore characteristics of shale under inert and oxic environments: Insights on pore evolution. *Microporous Mesoporous Mater.* **2021**, *316*, 110969. [CrossRef]
30. Shao, D.; Zhang, T.; Zhang, L.; Li, Y.; Meng, K. Effects of pressure on gas generation and pore evolution in thermally matured calcareous mudrock: Insights from gold-tube pyrolysis of the Eagle Ford Shale using miniature core plugs. *Int. J. Coal. Geol.* **2022**, *252*, 103936. [CrossRef]
31. Wang, L.; Yang, D.; Zhao, Y.; Wang, G. Evolution of pore characteristics in oil shale during pyrolysis under convection and conduction heating modes. *Oil Shale* **2020**, *37*, 224–241. [CrossRef]
32. Zhu, J.; Yang, Z.; Li, X.; Wang, N.; Jia, M. Evaluation of different microwave heating parameters on the pore structure of oil shale samples. *Energy Sci. Eng.* **2018**, *6*, 797–809. [CrossRef]
33. Yang, X.G.; Guo, S.B. Porosity model and pore evolution of transitional shales: An example from the Southern North China Basin. *Pet. Sci.* **2020**, *17*, 1512–1526. [CrossRef]
34. Zhang, Y.; Hu, S.; Shen, C.; Liao, Z.; Xu, J.; Zhang, X. Factors influencing the evolution of shale pores in enclosed and semi-enclosed thermal simulation experiments, Permian Lucaogou Formation, Santanghu Basin, China. *Mar. Pet. Geol.* **2022**, *135*, 105421. [CrossRef]
35. Song, D.; Tuo, J.; Zhang, M.; Wu, C.; Su, L.; Li, J. Hydrocarbon generation potential and evolution of pore characteristics of Mesoproterozoic shales in north China: Results from semi-closed pyrolysis experiments. *J. Nat. Gas. Sci. Eng.* **2019**, *62*, 171–183. [CrossRef]
36. Connan, J. Time-temperature relation in oil genesis: REPLY. *Am. Assoc. Pet. Geol. Bull.* **1976**, *60*, 2516–2521.
37. Gao, Z.; Fan, Y.; Xuan, Q.; Zheng, G. A review of shale pore structure evolution characteristics with increasing thermal maturities. *Adv. Geo-Energy Res.* **2020**, *4*, 247–259. [CrossRef]
38. Shi, M.; Yu, B.; Zhang, J.; Huang, H.; Yuan, Y.; Li, B. Evolution of organic pores in marine shales undergoing thermocompression: A simulation experiment using hydrocarbon generation and expulsion. *J. Nat. Gas. Sci. Eng.* **2018**, *59*, 406–413. [CrossRef]
39. Ma, Y.; Ardakani, O.H.; Zhong, N.; Liu, H.; Huang, H.; Larter, S. Possible pore structure deformation effects on the shale gas enrichment: An example from the Lower Cambrian shales of the Eastern Upper Yangtze Platform, South China. *Int. J. Coal. Geol.* **2020**, *217*, 103349. [CrossRef]
40. Li, Z.; Liu, D.; Cai, Y.; Ranjith, P.G.; Yao, Y. Multi-scale quantitative characterization of 3-D pore-fracture networks in bituminous and anthracite coals using FIB-SEM tomography and X-ray M-CT. *Fuel* **2017**, *209*, 43–53. [CrossRef]
41. Hu, Q.; Kalteyer, R.; Wang, J.; El-Sobky, H.F. Nanopetrophysical characterization of the Mancos Shale Formation in the San Juan Basin of northwestern New Mexico, USA. *Interpretation* **2019**, *7*, SJ45–SJ65. [CrossRef]
42. Wang, Y.; Qiu, N.; Ma, Z.; Ning, C.; Zheng, L.; Zhou, Y. Evaluation of equivalent relationship between vitrinite reflectance and solid bitumen reflectance. *J. China. Univ. Min. Technol.* **2020**, *49*, 563–575.
43. Baruch, E.T.; Kennedy, M.J.; Löhr, S.C.; Dewhurst, D.N. Feldspar dissolution-enhanced porosity in Paleoproterozoic shale reservoir facies from the Barney Creek Formation (McArthur Basin, Australia). *Am. Assoc. Pet. Geol. Bull.* **2015**, *99*, 1745–1770. [CrossRef]
44. Wang, F.; Guo, S. Influential factors and model of shale pore evolution: A case study of a continental shale from the Ordos Basin. *Mar. Pet. Geol.* **2019**, *102*, 271–282. [CrossRef]
45. Xu, L.; Yang, K.; Wei, H.; Liu, L.; Li, X.; Chen, L. Diagenetic evolution sequence and pore evolution model of Mesoproterozoic Xiamaling organic-rich shale in Zhangjiakou, Hebei, based on pyrolysis simulation experiments. *Mar. Pet. Geol.* **2021**, *132*, 105233. [CrossRef]
46. Zhang, S.; Yang, Y.; Zhang, L.; Cao, Y. Formation mechanism of secondary quartz and its influence on physical properties of xu 2 Member reservoir in western Sichuan Depression. *Xinjiang Pet. Geol.* **2023**, *44*, 25–32.
47. Kang, Z.; Zhao, J.; Yang, D.; Zhao, Y.; Hu, Y. Study of the evolution of micron-scale pore structure in oil shale at different temperatures. *Oil Shale* **2017**, *34*, 42–54. [CrossRef]
48. He, W.; Sun, Y.; Shan, X. Organic matter evolution in pyrolysis experiments of oil shale under high pressure: Guidance for in situ conversion of oil shale in the Songliao Basin. *J. Anal. Appl. Pyrolysis* **2021**, *155*, 105091. [CrossRef]
49. Ma, B.; Hu, Q.; Yang, S.; Zhang, T.; Qiao, H.; Meng, M. Pore structure typing and fractal characteristics of lacustrine shale from Kongdian Formation in East China. *J. Nat. Gas. Sci. Eng.* **2021**, *85*, 103709. [CrossRef]

50. Shan, C.; Zhao, W.; Wang, F.; Zhang, K.; Feng, Z.; Guo, L. Nanoscale pore structure heterogeneity and its quantitative characterization in Chang7 lacustrine shale of the southeastern Ordos Basin, China. *J. Pet. Sci. Eng* **2020**, *187*, 106754. [CrossRef]
51. Li, X.; Jiang, Z.; Jiang, S.; Li, Z.; Song, Y.; Jiang, H. Various controlling factors of matrix-related pores from differing depositional shales of the Yangtze Block in south China: Insight from organic matter isolation and fractal analysis. *Mar. Pet. Geol.* **2020**, *111*, 720–734. [CrossRef]
52. Liu, C.; Xu, X.; Liu, K.; Bai, J.; Liu, W.; Chen, S. Pore-scale oil distribution in shales of the Qingshankou formation in the Changling Sag, Songliao Basin, NE China. *Mar. Pet. Geol.* **2020**, *120*, 104553. [CrossRef]
53. Mastalerz, M.; Schimmelmann, A.; Drobniak, A.; Chen, Y. Porosity of Devonian and Mississippian New Albany Shale across a maturation gradient: Insights from organic petrology', gas adsorption, and mercury intrusion. *Am. Assoc. Pet. Geol. Bull.* **2013**, *97*, 1621–1643. [CrossRef]

processes

MDPI

Article

Pore Size Distribution Characterization by Joint Interpretation of MICP and NMR: A Case Study of Chang 7 Tight Sandstone in the Ordos Basin

Chaozheng Li [1], Xiangbai Liu [1,*], Fuliang You [2,3], Peng Wang [4], Xinluo Feng [5] and Zhazha Hu [6,*]

1 Research Institute of Petroleum Exploration and Development, PetroChina, Beijing 100083, China
2 State Key Laboratory of Petroleum Resources and Prospecting, China University of Petroleum, Beijing 102249, China
3 College of Geosciences, China University of Petroleum, Beijing 102249, China
4 No. 10 Oil Production Plant, PetroChina Changqing Oilfield Company, Qingcheng 745100, China
5 Exploration and Development Research Institute of Tarim Oilfield Company, PetroChina, Korla 841000, China
6 School of Energy Science and Engineering, Henan Polytechnic University, Jiaozuo 454003, China
* Correspondence: xiangbai_liu@163.com (X.L.); zhazha.hu@hpu.edu.cn (Z.H.)

Abstract: Pore size distribution characterization of unconventional tight reservoirs is extremely significant for an optimized extraction of petroleum from such reservoirs. In the present study, mercury injection capillary pressure (MICP) and low-field nuclear magnetic resonance (NMR) are integrated to evaluate the pore size distribution of the Chang 7 tight sandstone reservoir. The results show that the Chang 7 tight sandstones are characterized by high clay mineral content and fine grain size. They feature complex micro-nano-pore network system, mainly composed of regular primary intergranular pores, dissolved pores, inter-crystalline pores, and micro-fractures. Compared to the porosity obtained from MICP, the NMR porosity is closer to the gas-measured porosity (core analysis), and thus can more accurately describe the total pore space of the tight sandstone reservoirs. The pore throat distribution (PTD) from MICP presents a centralized distribution with high amplitude, while the pore size distribution (PSD) derived from NMR shows a unimodal distribution or bimodal distribution with low amplitude. It is notable that the difference between the PSD and the PTD is always related to the pore network composed of large pores connecting with narrow throats. The PSD always coincides very well with the PTD in the very tight non-reservoirs with a much lower porosity and permeability, probably due to the pore geometry that is dominated by the cylindrical pores. However, a significant inconsistency between the PSD and PTD in tight reservoirs of relatively high porosity and low permeability is usually associated with the pore network that is dominated by the sphere-cylindrical pores. Additionally, Euclidean distance between PSD and PTD shows a good positive correlation with pore throat ratio (PTR), further indicating that the greater difference of pore bodies and pore throats, the more obvious differentiation of two distributions. In summary, the MICP and NMR techniques imply the different profiles of pore structure, which has an important implication for deep insight into pore structure and accurate evaluation of petrophysical properties in the tight sandstone reservoir.

Keywords: pore size distribution; MICP; NMR; pore network model; tight sandstone; Ordos Basin

Citation: Li, C.; Liu, X.; You, F.; Wang, P.; Feng, X.; Hu, Z. Pore Size Distribution Characterization by Joint Interpretation of MICP and NMR: A Case Study of Chang 7 Tight Sandstone in the Ordos Basin. *Processes* **2022**, *10*, 1941. https://doi.org/10.3390/pr10101941

Academic Editors: Ping Gao, Yidong Cai, Yingfang Zhou and Quan Gan

Received: 7 August 2022
Accepted: 14 September 2022
Published: 26 September 2022

1. Introduction

With the growth in energy needs and advances in horizontal drilling and hydraulic fracturing techniques, tight oil reservoirs have become a focus for hydrocarbon exploration and exploitation in recent years [1,2]. In China, the Chang 7 of the Ordos Basin was regarded as one of the most typical lacustrine tight reservoirs, which possessed enormous potential for exploration and exploitation [3]. A large amount of hydrocarbon accumulates in the pore systems of these tight reservoirs, and these pore structures are one of the dominant factors controlling the properties of the reservoir rocks [4]. Therefore, a full understanding of pore geometry is critical for accurately evaluating the petrophysical properties and optimizing petroleum extraction from the tight reservoir.

It is a challenge to evaluate the pore structure of tight oil reservoirs due to the broad PSD, with a significant portion being nanoscale pores and substantially constricted pore throats [5–8]. However, the pore network system composed of pore bodies and their connecting throats controls the crucial storage space of hydrocarbon and the significant flowing pathway during petroleum accumulation and extraction. Therefore, it is highly important to effectively evaluate PSD and obtain pore throat distribution (PTD).

Mercury injection capillary pressure (MICP) has been a viable tool for reservoir rock characterization for over six decades [9,10] since Purcell [11] first introduced this technique into the petroleum industry. It was widely used to measure the porosity and PSD in porous materials [12], because it is fast and provides abundant information about pore structure and a wide range of pore sizes [13,14]. Washburn [15] mentioned a capillary tube model, which laid theoretical foundation for porous material pore structure determination using MICP method. This model gives an assumption that porous media is comprised of capillary tubes with different diameters. The PSD can be calculated according to a pressure–volume curve produced by mercury intrusion process [12].

Low-field nuclear magnetic resonance (NMR) is an important method for evaluating pore structure, fluid properties, and petrophysical properties of rocks [16–21]. This technique was widely used in the evaluation of various petroleum reservoirs, such as clastic rocks, carbonate rocks, and volcanic rocks, for nearly five decades [11,22–25]. In recent years, it has become an indispensable tool for characterizing reservoir properties, such as PSD, porosity, permeability, wettability, fluid movability, and saturation, due to the nondestructive feature, convenience in sample processing, and short test time [6,26–29].

Except for advantages described above, however, both methods also exhibit their own limitations. First, MICP allows calculation of the PTD, while NMR allows the PSD, due to the discrepancy of mercury intrusion method and fluid relaxation mechanism in porous materials [10,11,15,17]. In addition, MICP can usually identify pores down to about 3 nm due to the limitation of injection pressure. Excessive mercury injection pressure above about 70 MPa probably results in micro-fractures or some pore structure distortion occurring at the lower pore size limit (~3 nm) [12,30]. The NMR T_2 spectrum can be used to describe the entire pore distribution range, but the conversion of transverse relaxation time to pore size is constrained by many factors, such as lithology, paramagnetic minerals, and fluid types [17,20]. Consequently, it is necessary to combine these two methods to interpret pore size distribution of tight sandstone reservoir.

In this paper, based on the analysis of petrology and pore characteristics, we report the combination of MICP and NMR techniques that is performed to evaluate pore structure and calculate pore structure parameters and PSD. Additionally, we also make a comparison between the PSD derived from NMR and the PTD determined by MICP and give a detailed explanation of their similarities and differences. In short, a comprehensive understanding of the characteristics of pore structure of tight sandstone reservoir is highly important for accurate evaluation of the petrophysical properties of reservoir rocks and a successful exploitation.

2. Materials and Methods

2.1. Geological Setting and Samples

The Ordos Basin is located in the center of China (Figure 1), which has undergone multiple tectogeneses and was a multicycle superimposed basin, with the sediment changing from marine-continental transitional faces to continental faces in the Triassic [31,32]. The Chang 7 deposited a set of lacustrine detrital sediment rocks with a thickness of approximately 100 m and is regarded as one of the most typical lacustrine tight reservoirs in China, which mainly consists of fine sandstone and siltstone of gravity flow in delta front faces, as well as dark mudstone and black shale of semi-deep and deep lacustrine faces [33]. Due to a deep burial depth (>2000 m) and long-time evolution, the Chang 7 is significantly influenced by mechanical compaction, dissolution, and cementation, including clay minerals, carbonate, and silicate cements, and has formed a set of very tight sandstone reservoirs [34].

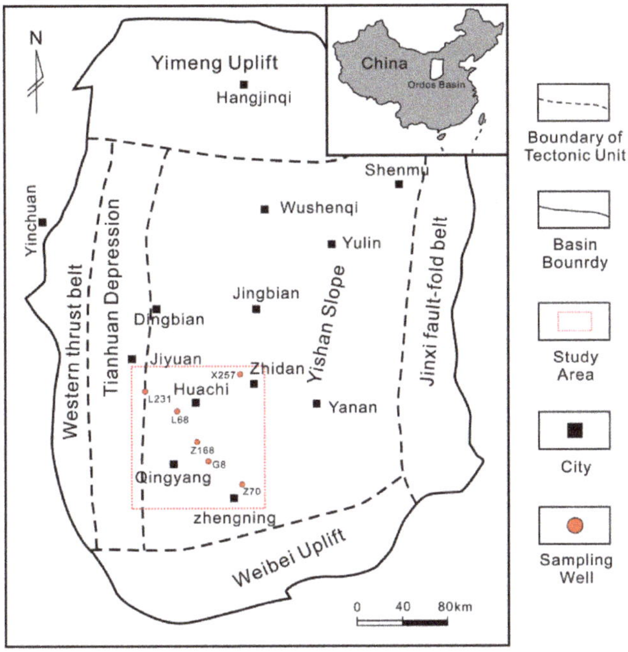

Figure 1. Tectonic division of the Ordos Basin and location of the study area and sampling wells.

Sixteen core samples were collected from six wells in the southwest area of the Ordos Basin (Figure 1). These samples are from the Chang 7 tight sandstone succession of the Upper Triassic Yanchang Formation, with a well depth interval of about 1720–2070 m. Cylindrical plugs drilled from those samples were prepared with about 2.5 cm in diameter and length, and were cleaned by hot extraction solvent. The porosity and permeability analysis, low-field NMR, and MICP analysis were then performed successively on these samples. In addition, thin section analysis, SEM, mineral composition, and grain size analysis were conducted to determine petrology and pore characteristics.

2.2. MICP

The MICP method is based on the fact that mercury behaves as a nonwetting liquid when in contact with most substances [11]. Consequently, mercury does not spontaneously penetrate into the openings and pores of these substances unless pressure is applied. With increasing pressure, smaller pore throats are invaded by mercury [35].

The capillary pressure (P_c) required for mercury to penetrate pores is a function of the contact angle (θ_{Hg}) between mercury and the porous material to be intruded, its gas/liquid surface tension (σ_{Hg}), and pore radius (r_p) [12]. This relationship was provided by the Young–Laplace equation with the assumption of cylindrical pores as the Washburn equation [36,37]:

$$P_c = \frac{2\sigma_{Hg} \cos \theta_{Hg}}{r_p} \tag{1}$$

Equation (1) indicates that, with increasing pressure, mercury progressively intrudes into narrower pores for constant values of σ_{Hg} and θ_{Hg}. The volume of mercury penetrating the pores is measured directly as the increasing applied pressure.

The mercury porosimetry experiments were performed with a Micromeritics AutoPore IV 9520 porosimeter. The results of mercury porosimetry are generally displayed as graphs of capillary pressure versus cumulative mercury intrusion. The equivalent pore radius was computed according to the capillary pressure using the Washburn equation with P_c ranging from approximately 0.003 to 413 MPa, using a contact angle of 130° [38] and surface tension of 485 dyne/cm [39]. Therefore, the corresponding pore radius ranges from 1.8 nm to 218 μm.

2.3. NMR

In porous rocks, the transverse relaxation T_2 of NMR measurement is composed of surface relaxation, bulk relaxation, and diffusion relaxation, which can be expressed as Equation (2) [17,40]:

$$\frac{1}{T_2} = \frac{1}{T_{2B}} + \frac{1}{T_{2S}} + \frac{1}{T_{2D}} = \frac{1}{T_{2B}} + \rho_2\frac{S}{V} + \frac{D\gamma^2G^2TE^2}{12} \tag{2}$$

where T_{2B} (ms) is the bulk relaxation time, T_{2S} (ms) is the surface relaxation time, T_{2D} (ms)is the diffusion-induced relaxation time, ρ_2 (μm/ms) is the surface relaxivity, S is the surface area of pore and V is the volume of the pore, D (μm^2/ms) is the molecular diffusion coefficient of the pore fluid, G (G/cm) is the field-strength gradient, γ is the gyromagnetic ratio and T_E (ms) is the inter-echo spacing used in the CPMG sequence.

The bulk relaxation is produced by interactions of hydrogen nuclei of fluid, which can be negligible due to its long relaxation time compared to the surface relaxation [17]. Diffusion relaxation occurs in the inhomogeneous (gradient) magnetic field and it is almost close to zero in the homogeneous magnetic field [41]. So, the transverse relaxation T_2 can be approximated as the surface relaxation generated by interactions between fluid nuclei and solid interface (pore walls). That is, Equation (2) is converted to Equation (3):

$$\frac{1}{T_2} \approx \frac{1}{T_{2S}} = \rho_2\frac{S}{V} \tag{3}$$

This relationship is based on an important assumption that diffusion within the pore is in the fast diffusion, which is valid when $\rho2\gamma/D << 1$ [40]. In the condition of the fast diffusion, the limit diffusion across the pore is much more rapid than the relaxation at the pore surface [42] and is met in most reservoir rocks [43].

The ρ_2 is close to a constant coefficient when the measured samples come from the same region, same formation, and similar lithology [20]. Therefore, the surface relaxation is a function of the surface-to-volume ratio (S/V) of the pores, which means that the small pores have short T_2 times and the large ones have long T_2 times. Additionally, the S/V depends on pore geometry. Consequently, Equation (3) is converted to Equation (4):

$$\frac{1}{T_2} = \rho_2\frac{F_s}{R} \tag{4}$$

where F_s is the shape factor, which is a constant with values of 3 and 2 for spherical and cylindrical pores [44], respectively. R is pore radius (μm). Therefore, there is a linear relationship between the value of T_2 and pore size for the same area and formation and negligible lithological difference in clastic rocks.

The NMR measurements were conducted on a SPEC-023 instrument. It has a constant magnetic field strength of 1200 Gauss and a resonance frequency of 2.38 MHz. The measurement parameters were set as follows: echo spacing, 0.2 ms; waiting time, 3 s; echo numbers, 2048; and number of scans, 64 [27]. After the measurements, transverse relaxation time (T_2) distributions were computed by multi-exponential inversion of the echo data with 64 preset decay times logarithmically spaced from 0.1 ms to 10 s [17]. The samples were fully saturated in water for 48 h under an ambient pressure of 30 MPa. In addition, the centrifugation was performed on the 100% water-saturated core plugs under a centrifugal force of 2.76 MPa (400 psi) for 1.5 h to obtain irreducible water condition. The centrifuge pressure used in the centrifuge experiment, which is an empirical value derived from the repeated experiment. The higher centrifugal force may induce a change in the pore structure [45]. The T_2 spectrum at the water-saturated and irreducible conditions are obtained successively.

3. Results

3.1. Petrology and Pore Characteristics

3.1.1. Petrology Characteristics

The Chang 7 tight sandstones in the study area are mainly made up of lithic arkose and feldspathic litharenite [12,34]. The content of quartz mainly ranges from about 50% to 60%, except for one sample with a lower quartz content of about 35% (Figure 2a). As a sedimentary debris, feldspar generally have lower content of about 10–15% compared with quartz. The content of carbonate minerals shows large fluctuations, ranging from 5% to 20%, while that of clay minerals is generally high and relatively stable from 15% to 20% (Figure 2a). In addition, skeleton particles are mainly composed of quartz, feldspar, and lithic fragment (Figure 3a,b), while the interstitial material is mainly composed of high content clastic clay matrix, as well as cements dominated by carbonate, silicate, illite, and chlorite minerals (Figure 3c,j–l). Laser grain size analysis shows that the median grain size of most samples mainly ranges from 80 to 100 μm (Figure 2b), which reveals a predominantly fine size fraction in the Chang7 tight sandstones (Figure 3a–f) [46]. Therefore, the Chang 7 tight sandstones are composed of silt fine clastic grains and are characterized by high content of clay minerals or carbonate carbon minerals. Moreover, fine grain structure and high clay content lead to strong compaction and the precipitation of carbonate cements in the pores, which finally leads to complex pore structure and micro-nano pore throat system in the Chang 7 tight sandstones (Figure 3).

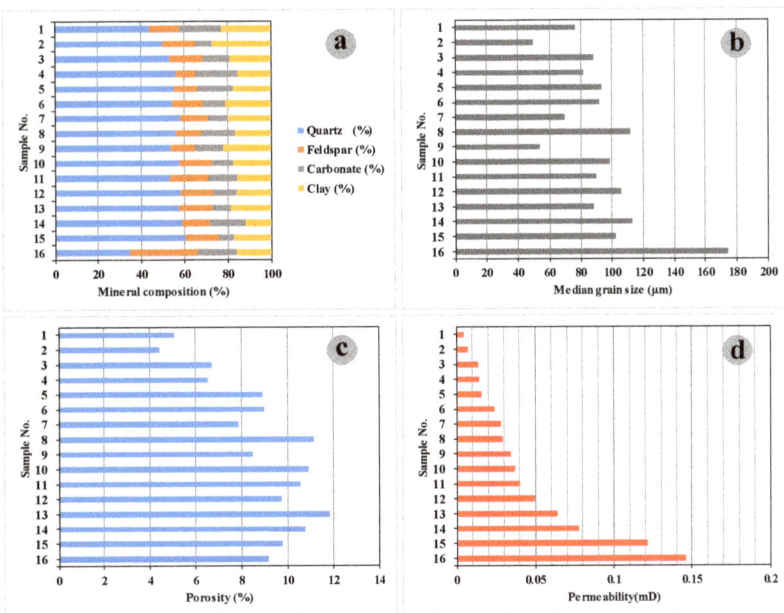

Figure 2. Mineral composition (**a**), median grain size (**b**), porosity (**c**) and permeability (**d**) of the Chang 7 tight sandstones.

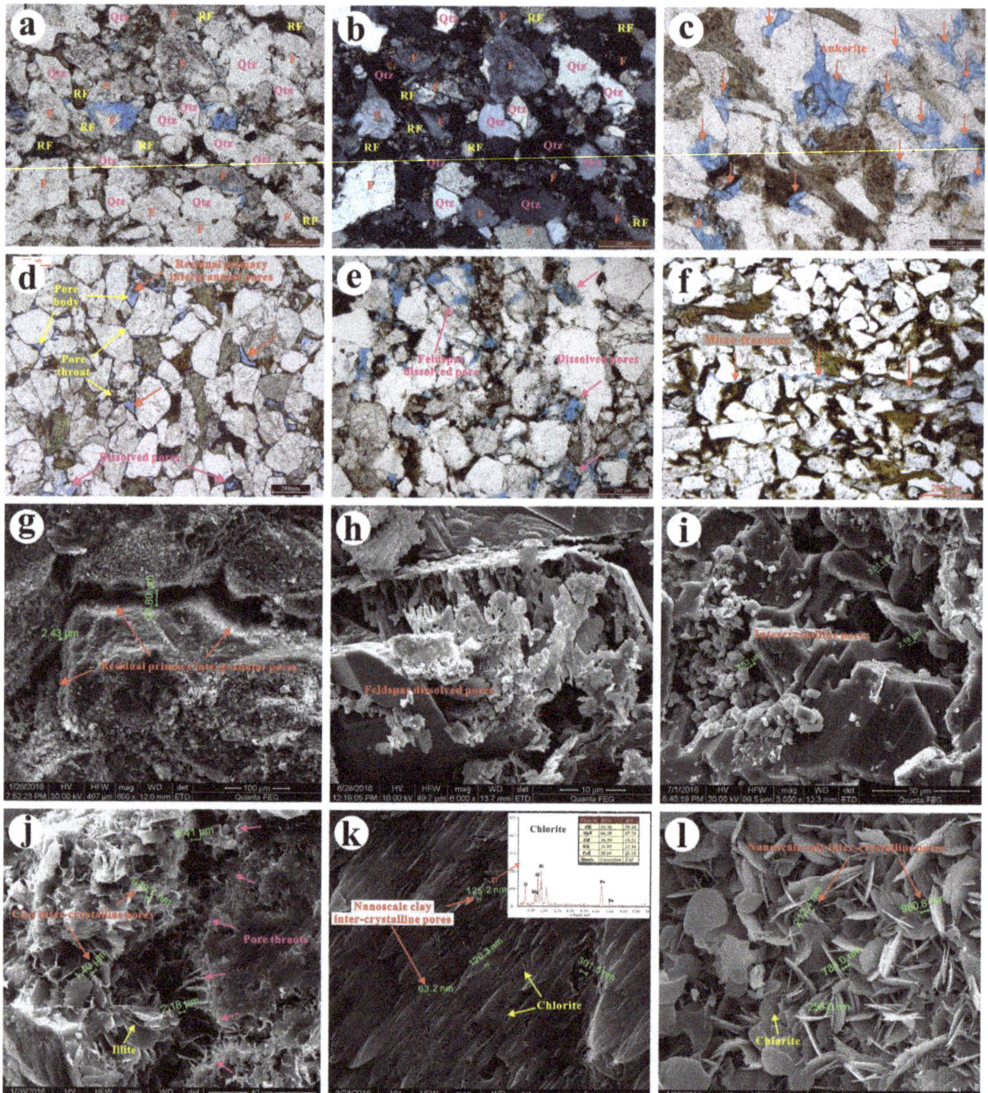

Figure 3. Petrology and pore characteristics of the Chang 7 tight sandstone reservoirs in the Ordos Basin. (**a**,**b**) Z168, 1754.56 m, fine-grained, subrounded to subangular, poorly sorted lithic arkose. Qtz—quartz; F—feldspar; RF—rock fragment. (**c**) H144, 2554.10 m, ankerite cement (blue) is the most common pore-filling constituents, thin section colored by the mixture of sodium alizarinsulfonate and potassium ferricyanide. (**d**) X257 1927.10 m, micro-scale regular residual primary intergranular pores. (**e**) Z388, 2080.77 m, large amounts of dissolved pores, including feldspar dissolved pores and lithic fragment dissolved pores. (**f**) W100, 1968.00 m, micro-fractures. (**g**) X257, 1927.10 m, microscale regular residual primary intergranular pores, SEM. (**h**) Z70, 1617.10 m, feldspar dissolved pores, SEM. (**i**) G8, 1795.60 m, authigenic albite inter-crystalline pores. (**j**) Z168, 1725.40 m, clay inter-crystallite pores developed in rolled sheet illite, SEM. (**k**) X257, 1927.10 m, nanoscale triangular or planar clay inter-crystallite pores developed in foliated chlorite aggregates, argon ion polishing, SEM. (**l**) X257, 1927.10 m, nanoscale clay inter-crystalline pores developed in foliated chlorite aggregates, SEM.

The porosity varies from 4% to 12% and can be divides into two parts, mainly ranging from 4% to 6% and from 8% to 10% (Figure 2c). The permeability shows significant change, mainly ranging from about 0.01 mD to 0.10 mD (Table 1, Figure 2d). Additionally, the lower part of porosity (4–6%) always corresponds to the ultra-low permeability, less than 0.02 mD, indicating that such tight sandstones are generally not effective in the tight oil accumulation [47] and are non-reservoir rocks, such as samples of No.1 to No.4.

Table 1. The pore structure parameters from MICP.

Sample No.	So	P_{c50} (MPa)	P_d (MPa)	r_{max} (μm)	S_{max} (%)	We (%)	PTR	φ_{Hg} (%)	$r_{2.5}$ (μm)	r_{25} (μm)	r_{75} (μm)
1	1.69	46.11	6.82	0.108	71.83	32.42	2.08	3.24	0.104	0.034	0.002
2	1.87	59.51	4.39	0.167	77.35	44.04	1.27	2.85	0.173	0.043	0.003
3	1.59	19.37	3.56	0.206	86.70	36.99	1.70	5.74	0.203	0.084	0.010
4	1.43	18.97	3.56	0.206	85.21	24.22	3.13	6.06	0.195	0.075	0.012
5	1.51	26.88	6.02	0.122	89.33	11.85	7.44	7.07	0.213	0.048	0.009
6	1.49	11.02	2.87	0.256	86.99	35.04	1.85	8.45	0.212	0.118	0.016
7	1.69	11.42	1.84	0.400	90.31	32.78	2.05	8.18	0.415	0.151	0.019
8	1.53	11.30	2.60	0.283	91.92	41.67	1.40	9.39	0.258	0.129	0.021
9	1.65	12.21	1.83	0.401	91.20	20.41	3.90	5.97	0.471	0.117	0.020
10	1.23	10.06	4.40	0.167	87.87	27.73	2.61	8.96	0.214	0.110	0.028
11	1.54	5.90	2.25	0.326	91.59	19.83	4.04	9.14	0.293	0.180	0.033
12	0.98	7.11	2.26	0.326	89.88	26.09	2.83	9.20	0.287	0.155	0.051
13	1.49	7.06	2.88	0.256	83.71	22.73	3.40	8.99	0.318	0.167	0.017
14	0.81	3.94	0.94	0.782	91.23	13.58	6.36	8.90	0.645	0.254	0.135
15	1.59	7.86	4.39	0.167	92.46	13.13	6.61	9.37	0.834	0.139	0.032
16	2.03	2.59	1.19	0.619	89.70	25.49	2.92	8.03	0.917	0.460	0.032

So: sorting coefficient. P_d: displacement pressure. P_{c50}: median capillary pressure. r_{max}: maximum pore throat radius. S_{max}: maximum mercury saturation. We: mercury withdrawal efficiency. PTR: the pore throat ratio, the average ratio of the pore volume and the throat volume derived from the mercury injection curve and ejection curve, respectively; $PTR = \frac{S_R}{S_{max}-S_R}$, S_R is the residual mercury saturation in the pores after completely mercury withdrawal. φ_{Hg}: the porosity derived from MICP. $r_{2.5}$, r_{25}, and r_{75} (μm) are the pore-throat radius corresponding to 2.5%, 25% and 75% of mercury injection cumulative saturation, respectively.

3.1.2. Pore Characteristics

Four types of pores are recognized in the Chang 7 tight sandstone reservoirs, including residual intergranular pores, dissolved pores, inter-crystalline pores, and micro-fractures (Figure 3). Residual primary intergranular pores are relatively larger in size but fewer in number, with pore diameter mainly ranging from 10 to 100 μm. The shape of these pores is usually triangular or polygonal with smooth and straight edges (Figure 3d,e,g) [12]. Fine throats (nanoscale) always occur in the position of grain contacts due to a fierce mechanical compaction and presence of grain coating chlorite (Figure 3d,e,g) [2]. These large residual intergranular pores connected with fine throats can form a special pore network with large pore bodies and small pore throats. Dissolved pores are produced by the dissolution of unstable minerals, such as feldspar and carbonate cement (Figure 3e,h) [48], which are an important component to the pore space of the Chang 7 tight sandstones. These pores are characterized by rather irregular pore surfaces and large pore sizes, generally ranging from several hundred nanometers to 100 μm. These pores are usually connected by fine throats due to a heterogeneity of the grain dissolution. Inter-crystalline pores are widely distributed and generated from the precipitation of authigenic quartz, albite, ankerite, and clay minerals (Figure 3i–l), including illite, chlorite, and mixed-layer illite/smectite, in intergranular pores and dissolved pores. The dimensions of these pores are generally smaller than several microns, and the inter-crystalline pores within authigenic quartz or albite are larger than those within clay minerals ranging within the nanoscale (Figure 3i–l). These pores often work as throats connecting other pores, such as residual intergranular pores, dissolution pores, and relatively large inter-crystalline pores.

Micro-fractures are also observed in some samples, with widths ranging from 2 μm to 10 μm, while their lengths were up to 10 mm (Figure 3f). These are significant for the percolation capacity of the tight reservoir due to the connection to large of micro pores.

3.2. MICP Curves and Parameters

MICP is a useful tool to obtain the information of pore structure and pore throat distribution (PTD) [9,10,49,50]. MICP curves can directly and visually reflect pore geometry characteristics and are easily used to calculate pore structure parameters and PTD. The main mercury injection capillary pressure of the Chang 7 tight sandstone distributes from 1 MPa to 100 MPa (Figure 4a), indicating that the dominant portion of pore throats is in the nanoscale (0.00735–0.735 μm). The displacement pressure (P_d) is between 0.94 MPa and 6.82 MPa, with an average value of 2.92 MPa, while the median capillary pressure (P_{c50}) ranges from 2.9 MPa to 59.51 MPa, with an average value of 16.33 MPa, suggesting that hydrocarbon cannot easily enter into these tight reservoirs without overpressure (Table 1). In addition, the PTD has a wide distribution, mainly ranging from 0.01 μm to 1 μm (Figure 4b). The maximum pore throat radius (r_{max}) is between 0.94 μm and 6.82 μm, with an average value of 2.92 μm. The median pore throat radius (r_{50}) ranges from 0.012 μm to 0.284 μm, with an average value of 0.085 μm [47]. The line section slope of the MICP curves is a visual indication of pore throat sorting characteristics that the smaller is the line slope, and the better is the sorting quality, which can be quantified by sorting coefficients (So) of approximately 0.81 to 2.03, with an average So of 1.51 (Table 1). In short, Chang 7 tight sandstone reservoir is characterized by high capillary pressure, a dominant nanoscale pore throat, and poorer sorting.

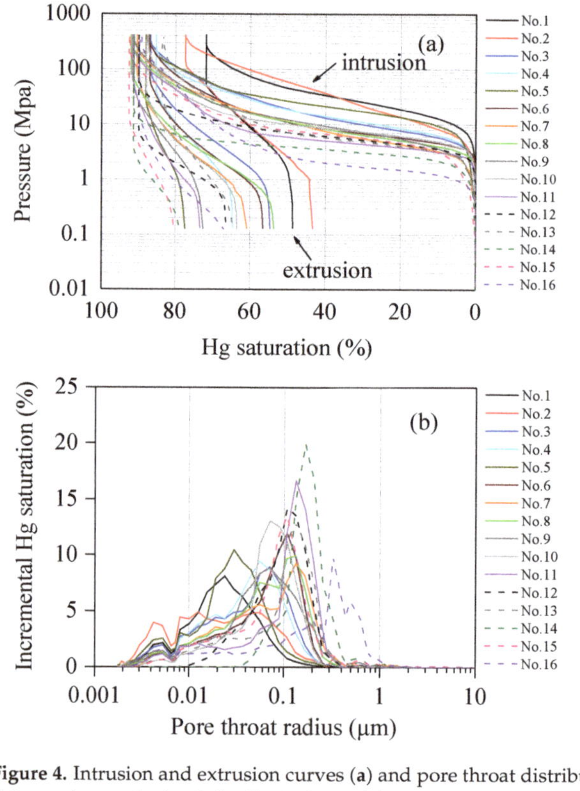

Figure 4. Intrusion and extrusion curves (**a**) and pore throat distribution curves (**b**) of the Chang 7 tight sandstones in the Ordos Basin derived from MICP.

The mercury intrusion and extrusion curves exhibit significant hysteresis (Figure 4a), indicating that a large amount of injected mercury remains in the pore network after complete mercury withdrawal. This phenomenon is intimately related to the crucial role of the connectivity of the pore network system in filling the pore space with mercury [51]. During the extrusion process, mercury remains trapped in the pore bodies of the ink-bottle pores due to the difference in capillary pressure between the pore bodies and pore throats [26,52]. This difference can be roughly defined by the parameter of the pore throat ratio (PTR). The PTR is the average ratio of the pore volume and the throat volume derived from the mercury injection curve and ejection curve, respectively. The PTR ranges from 1.27 to 7.44, with an average of 3.35 (Table 1). Additionally, a large amount of mercury trapped in the pores can also be quantified by the pore structure parameters, e.g., the maximum mercury saturation (S_{max}) and mercury withdrawal efficiency (We). The maximum mercury saturation (S_{max}) ranges from 71.83% to 92.46% and its mean is 87.33%. In addition, mercury withdrawal efficiency (We) is between 11.85% and 44.04%, with an average value of 26.75%. It is apparent that about 40% to 80% of mercury remains trapped in the pore network (Figure 4a).

3.3. NMR T_2 Spectrum

The NMR T_2 spectra of the Chang 7 tight sandstones mainly exhibit the unimodal or bimodal distributions (Figure 5), which generally occur in sandstone and shale rocks [27,53]. The T_2 spectrum mainly distributes from 0.1 to 100 ms, showing a unimodal distribution, unimodal distribution with a positive skewness, or bimodal distribution (Figure 5). According to Equation (4), the T_2 spectra of saturated water implies the PSD of rocks, and the long T_2 relaxation time indicates the large pores and the short T_2 relaxation time indicates the small pores. Therefore, the samples with unimodal distribution would have larger percentage composition of small pores than those with bimodal distribution.

As is shown in Figure 5, the left peak (or the peak of the unimodal distribution) and right peak of the T_2 distribution present at 1 to 10 ms and 10 to 100 ms, respectively. The peak number and location of the T_2 distribution can reflect the pore type of rocks. The left peak is representative of the micro pores mainly composed of clay inter-crystallite pores and micro dissolved pores (Figure 3), while the right peak is indicative of residual primary intergranular pores and some macro pores generated from particle dissolution, which are the dominant contribution to permeability. In addition, the T_2 geometric mean (T_{2gm}) ranges from 0.78 to 5.94 ms, with an average value of 3.46 ms [27], which further indicates a tight pore system has been developed in the Chang 7 tight sandstones.

Fluid flow in a porous media is mainly controlled by PSD, interfacial tension, and wettability [6,28,36]. For porous reservoirs, the interfacial tension and wettability changes would significantly affect oil migration and recovery [54–57]. However, it is generally believed that the interfacial tension and wettability of the core plugs are consistent, considering that they are from the same stratum and contain similar lithology. Therefore, the fluid flow is mainly controlled by pore throat size during centrifugation.

After centrifugation, the right peak is almost completely absent except in sample No. 5 and No. 11, while the left peak remains, indicating that the fluid in the small pores is not easy to flow (Figure 5). The fluid trapped in the large pores (>10 ms) for No. 5 and No. 11 is probably due to occurrence of a large amount of the complex pore geometry that is composed of large pores connected with fine throats (or ink bottle pores) [27,58]. The irreducible fluid is mainly around 0.1 to 10 ms (Figure 5), consisting of clay-bound water and capillary-bound water that are limited in clay micro pores and small capillary pores, respectively. This phenomenon is mainly related with the sufficient molecular adsorption force and capillary pressure generated from the pore fluid interaction with the clay mineral surface or pore walls of the rocks. The $T_{2cutoff}$ is a relaxation time threshold that divides the T_2 distribution into the movable fluid and irreducible fluid, ranging from 0.87 to 7.73 ms, with an average value of 3.00 ms (Table 2) [27]. The movable water saturation (S_m) of Chang 7 tight sandstone ranges from 32.01% to 84.84%, with an average value of 50.53%, while the irreducible water saturation (S_{ir}) ranges from 15.06% to 67.99%, with an average value of 49.47%

(Table 2). Consequently, the movable water volume is about a half of the total pore volume. This result suggests that these micro pores contribute a significant portion of the storage space but may not be important for oil or gas percolation in tight sandstone reservoirs.

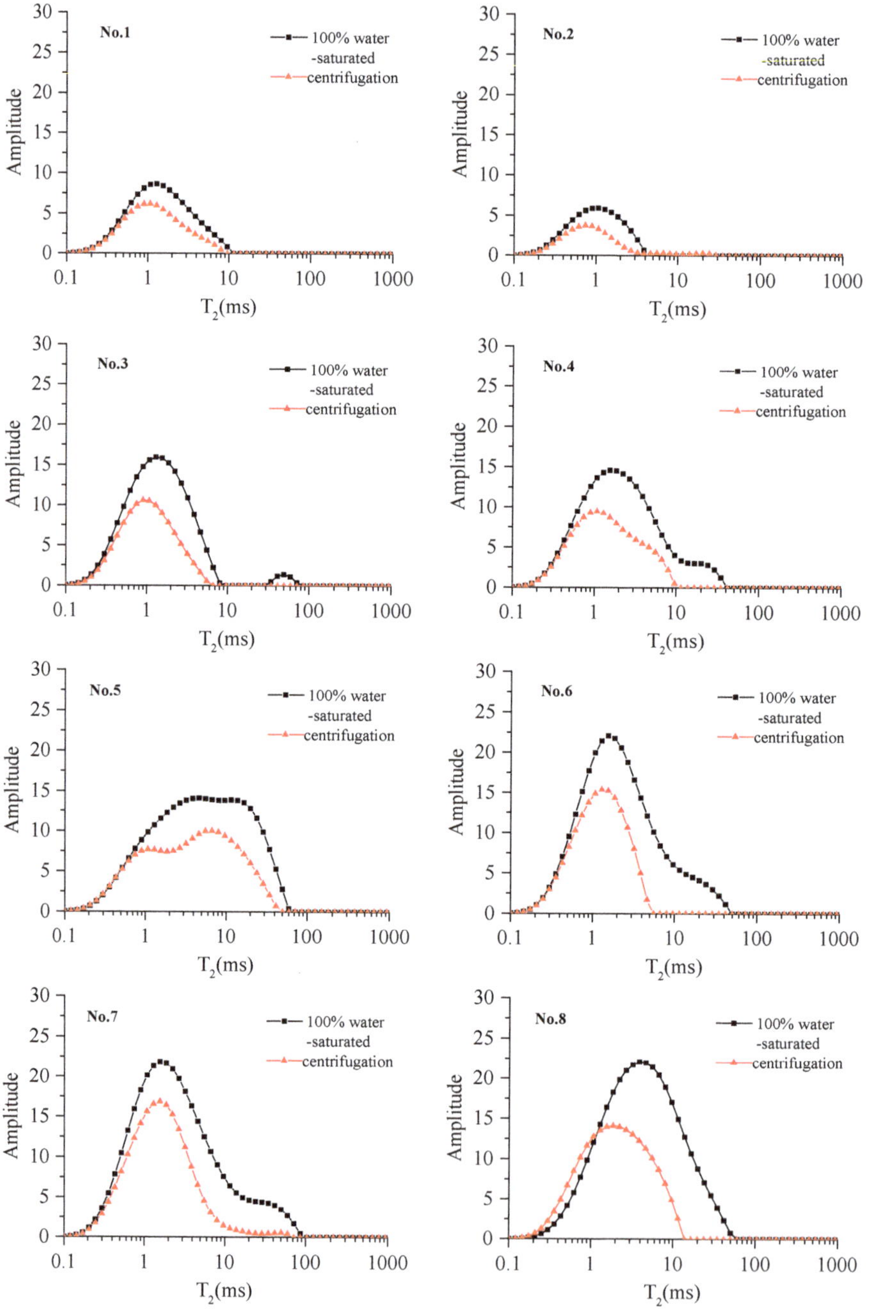

Figure 5. *Cont.*

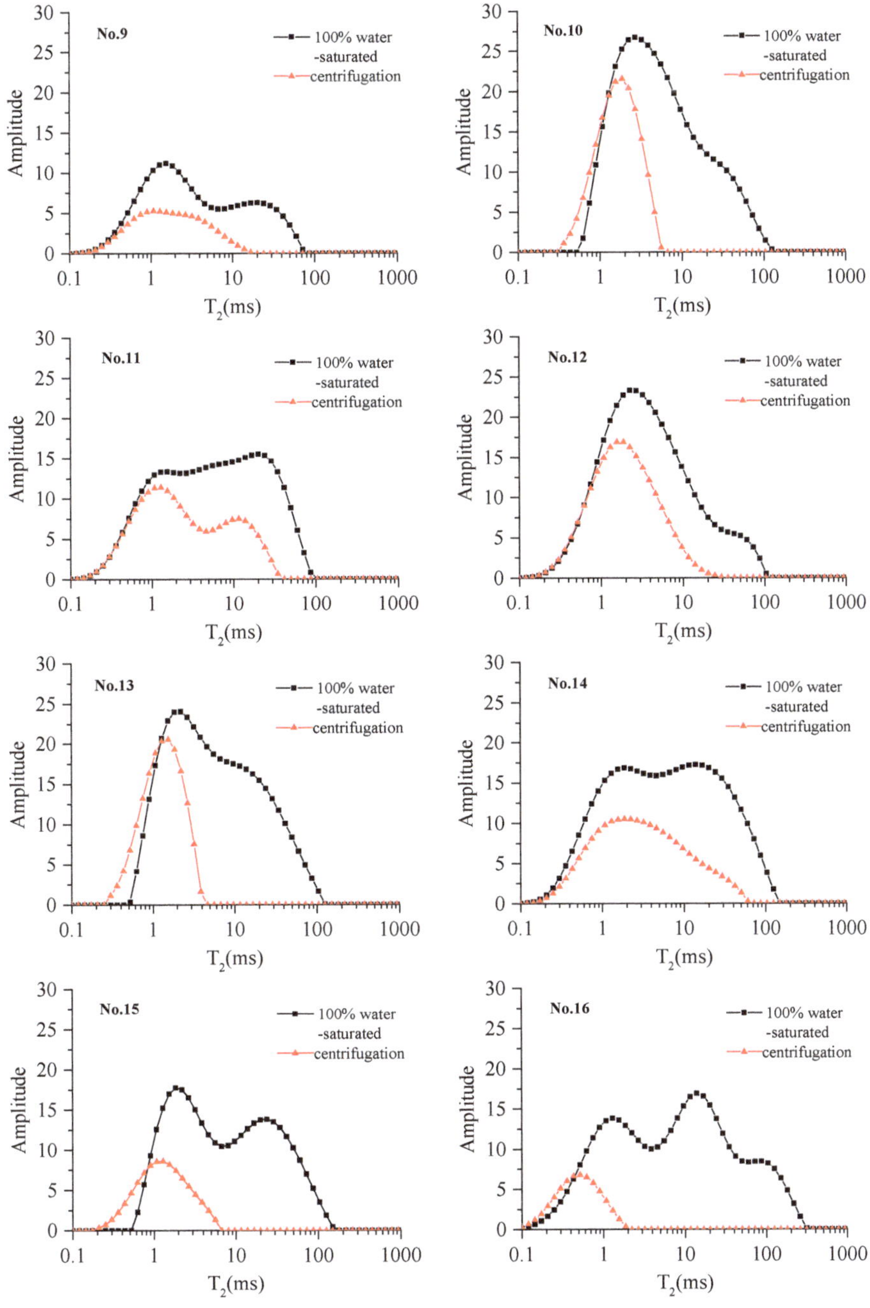

Figure 5. The 100% water-saturated and irreducible T_2 spectrum distribution of the Chang 7 tight sandstones in the Ordos Basin.

Table 2. The petrophysical parameters from NMR and centrifuge experiment.

Sample No.	S_{ir}* (%)	S_w* (%)	C (μm/ms)	$R_{2.5}$ (μm)	R_{25} (μm)	R_{50} (μm)	R_{75} (μm)
1	67.99	32.01	0.027	0.151	0.037	0.022	0.013
2	58.57	41.43	0.047	0.068	0.025	0.016	0.010
3	56.80	43.20	0.034	0.153	0.053	0.029	0.017
4	56.47	43.53	0.024	0.491	0.096	0.040	0.021
5	63.85	36.15	0.006	0.826	0.261	0.090	0.034
6	53.02	46.98	0.027	0.525	0.077	0.038	0.019
7	58.87	41.13	0.019	0.954	0.123	0.046	0.023
8	58.37	41.63	0.011	0.630	0.221	0.084	0.030
9	40.80	59.20	0.027	1.019	0.175	0.066	0.026
10	42.26	57.74	0.015	1.318	0.177	0.064	0.028
11	52.47	47.53	0.009	1.184	0.308	0.092	0.032
12	55.57	44.43	0.014	1.224	0.229	0.085	0.038
13	39.60	60.40	0.017	1.347	0.310	0.083	0.034
14	44.94	55.06	0.012	1.680	0.381	0.122	0.038
15	26.83	73.17	0.024	0.954	0.306	0.069	0.024
16	15.06	84.94	0.057	3.414	0.419	0.128	0.032

* Data from Li et al. [27]. S_{ir}: irreducible water saturation. S_w: movable water saturation. C is a constant conversion coefficient representative of the shape factor (Fs) and surface relaxivity (ρ_2). $R_{2.5}$, R_{25}, R_{50}, and R_{75} (μm) are the pore radius corresponding to 2.5%, 25%, 50%, and 75% of cumulative pore volume percentage from NMR, respectively.

4. Discussion

4.1. Comparison of Pore Volume and Size from MICP and NMR

Petrophysical parameters obtained from MICP and NMR always present disagreement due to the differences in test principles and corresponding petrophysical properties [10,11,17,19,50]. Therefore, the porosity and pore or pore throat size obtained from both MICP and NMR are compared with each other to clarify the difference between two methods in pore space and size characterization.

4.1.1. Porosity

As is shown in Figure 6, there are good correlations between gas-measured porosity (φ), MICP porosity (φ_{MICP}), and NMR porosity (φ_{NMR}) despite some differences. φ from core analysis is usually considered as the total porosity due to an assumption of gas molecules moving into almost all connected pore spaces. The porosity from MICP is always lower than the total porosity because of incomplete mercury injection, which is mainly caused by limited intruded mercury pressure. The deviation between φ and φ_{NMR} may be related to several reasons: the presence of paramagnetic minerals [59], gas molecules being smaller than water molecules [19], or external surface water of core plugs. In short, φ_{NMR} is much closer to φ, and thus can more accurately describe the total pore space of tight sandstone reservoirs when compared to mercury intrusion porosimetry.

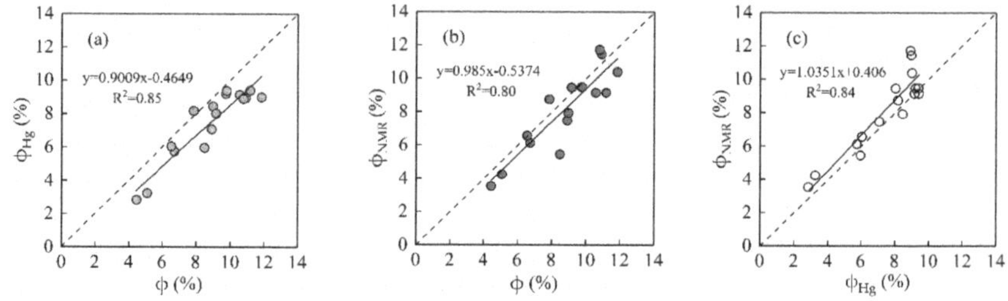

Figure 6. Cross-plots of φ and φ_{Hg} (**a**), φ and φ_{NMR} (**b**), φ_{Hg} and φ_{NMR} (**c**). φ: Gas-measured porosity. φ_{Hg}: MICP porosity. φ_{NMR}: NMR porosity. Data of φ_{NMR} is from Li et al. [27].

4.1.2. Pore and Pore Throat Size

Figure 7 shows the method of obtaining the pore radius and pore throat radius corresponding to the different pore volume percentages of the cumulative distribution curves of MICP and NMR. The results are shown in Table 2.

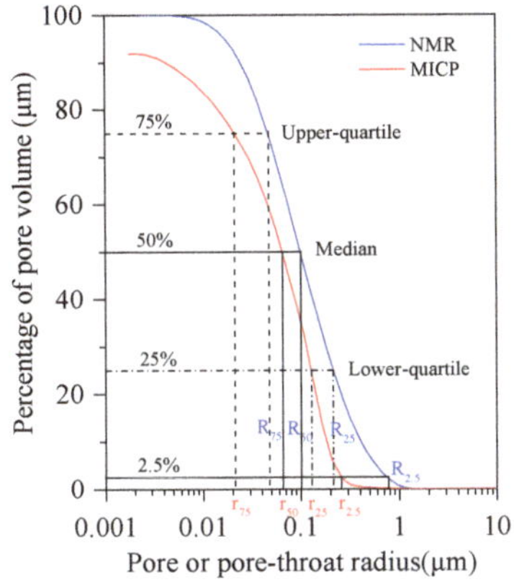

Figure 7. The method to calculate pore radius and pore throat radius corresponding to 2.5%, 25%, 50%, and 75% of the cumulative pore volume percentage from NMR and MICP.

The result shows that the pore radius (R_{75}) and pore throat radius (r_{75}) corresponding to the upper quartile provide a good match, except for a slight deviation in some samples (Figure 8a). It is notable that the median pore radius (R_{50}) almost coincides with the median pore throat radius (r_{50}), which is important for accurately predicting permeability by use of R_{50} (Figure 8b). The pore radius corresponding to the low quartile (R_{25}) is slightly larger than the pore radius corresponding to the low quartile (r_{25}) (Figure 8c), while the pore radius $R_{2.5}$ significantly exceeds the pore throat radius $r_{2.5}$ on the location of cumulative percentage of 2.5% (Figure 8d). This indicates that pore size and throat size differ greatly from each other in the large-size pore system. Although there is significant difference between pore and throat in tight sandstones, the medians of the pore and pore throat radius show greatly consistent scalar values, which is highly valuable for the accurate and indirect evaluation of permeability and capillary pressure by use of R_{50}.

4.2. Pore Size Distribution

4.2.1. Calibration of PSD

In NMR relaxation, the hydrogen atoms in smaller pores experience a greater surface relaxation, and thus relax and decay faster than that in the larger pores. So, the T_2 distribution of rocks with water saturation corresponds to the PSD: large pores correspond to a long relaxation time and small pores to a short relaxation time. There was a linear relationship between the T_2 value and pore size in single pores of clastic rocks, as Equation (4) verified. Therefore, the T_2 distribution can be calibrated to the PSD. Previously, many researchers directly overlapped the PTD derived from MICP to the T_2 distribution to obtain the PSD, based on the similarity of PTD and PSD in conventional reservoirs [23,60–62]. However, the PSD and PTD tend to present obvious differences due to the complex pore structure in tight sandstone reservoirs, which makes the reliability of calibrated PSD for this method low or even incorrect. Li et al. [63] found that there was not a noticeable difference within 0.05 µm between pore size and throat size when considering the similarity of pore volumes derived from MICP and nitrogen adsorption. Thus, the T_2 distribution can be converted into the PSD through the "$T_{2cutoff}$" method proposed by Yao et al. [19]. Although the method is based on a centrifuge experiment controlled by the Washburn equation, the effect of the significant difference between the pore and pore throat in the large pore system is avoidable to a great extent. The "$T_{2cutoff}$" method is as follows:

For any relaxation time (T_{2i}) in a T_2 relaxation distribution, the corresponding pore size (R_i) can be determined by Equation (4).

$$R_i = \rho_2 F_s \times T_{2i} = C \times T_{2i} \quad (5)$$

where C is a constant conversion coefficient representative of the shape factor (F_s) and surface relaxivity (ρ_2).

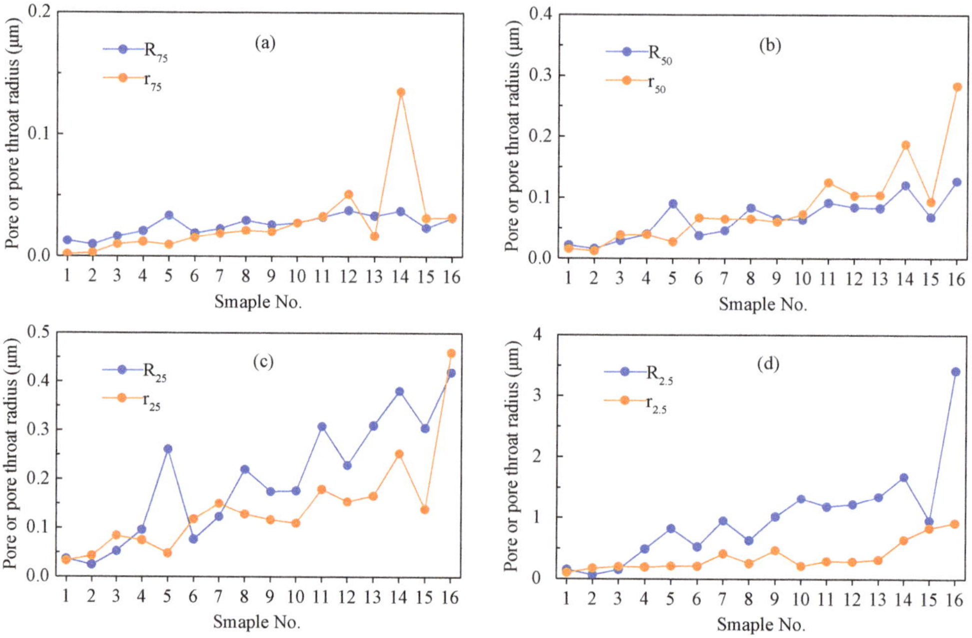

Figure 8. Plots of pore radius and pore throat radius. (**a**) The plot of r_{75} and R_{75} show a good match except for sample 14. (**b**) The plot of r_{50} from Li et al. [47] and R_{50}. r_{50} are closely equal to R_{50}, indicating R_{50} can be used in place of r_{50} to predict permeability [27]. The plot of r_{25} and R_{25} (**c**), $r_{2.5}$ and $R_{2.5}$ (**d**) show that pore radius is much larger than pore throat radius in macropore interval.

As a specific T_{2i}, the $T_{2cutoff}$ can be used to calculate the conversion coefficient C. Thus, the following formula is converted from Equation (5):

$$R_c = C \times T_{2cutoff} \quad (6)$$

where R_c is the cut-off pore radius, i.e., the minimal pore radius (μm) for water to discharge at the centrifuge pressure.

In the centrifuge experiment, R_c can be obtained from the following formula based on the Washburn equation:

$$R_c = \frac{2\sigma_{Wr} \cos\theta_{wr}}{P_{centr}} \quad (7)$$

where P_{centr} is the centrifuge pressure in MPa; R_c is the cut-off pore radius, the minimal pore radius (μm) for water to discharge at P_{centr}; θ_{wr} is the contact angle between water molecule and pore surface; and σ_{wr} is the interfacial tension of rocks and water.

The σ_{wr} and θ_{wr} of rock to water vary with samples due to the mineral composition difference. They are assigned with values of 0.072 N/m and 0° according to the fact that the rock is completely water-wet after cleaning [64]. Therefore, the R_c that corresponds to $T_{2cutoff}$ under the centrifuge pressure of 2.76 MPa is 0.05 μm.

The $T_{2cutoff}$ can be obtained from centrifuge experiment and NMR test. The detailed method to obtain a $T_{2cutoff}$ is based on Li et al. [27]. Therefore, the PSD from NMR T_2 distribution can be determined according to Equations (6) and (7).

4.2.2. The PSD and PTD

An average conversion coefficient (C_{av} = 0.023 μm/ms) is used when calibrating the PSD from the T_2 distribution. The PSD of the 16 tight sandstone samples is shown in Figure 9. The result shows that the constructive PSD is not always a good match with the PTD determined by MICP, with some of the samples presenting significant difference between the two distributions.

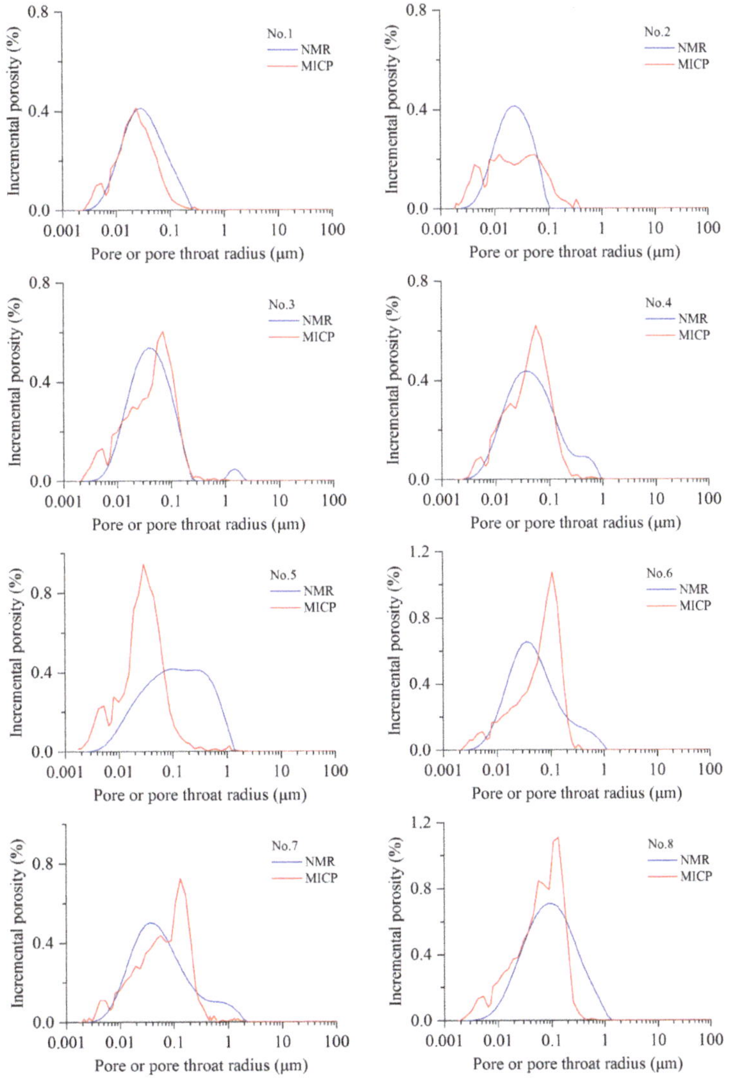

Figure 9. *Cont.*

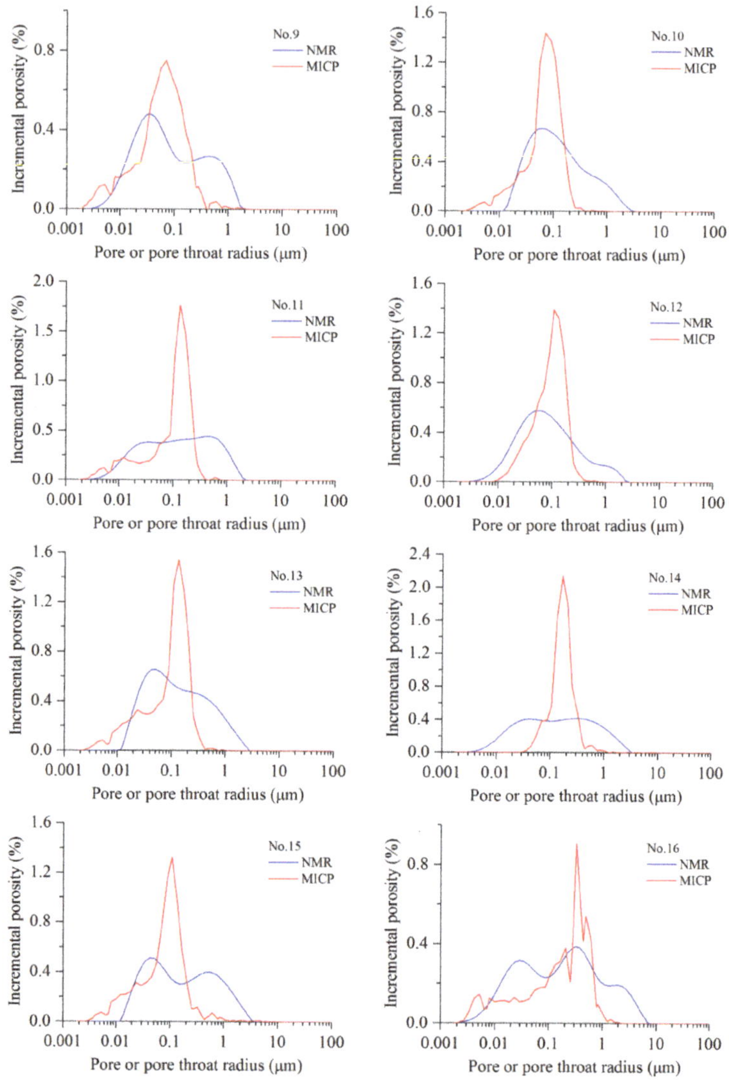

Figure 9. The NMR pore size distribution (PSD) and MICP pore throat distribution (PTD) of the Chang 7 tight sandstones in the Ordos Basin. An average conversion coefficient (C_{av} = 0.023 μm/ms) is used when the PSD is derived from the T_2 distribution. When comparing to the PTD and PSD, a discrepancy is found. This inconsistency is mainly because MICP quantifies pore throat size distribution, while NMR reveals the pore body size distribution.

The NMR-derived PSD presents a broad distribution with pore radius spanning three orders of magnitude from nanoscale to several microns. The PSD can be mainly divided into three groups based on the distribution shape and petrophysical properties.

The first group (No. 1, No. 2, No. 3, and No. 4) is a unimodal distribution similar to a logarithmic normal distribution, and some of them present a slightly positive skewness, including a weak tail due to the presence of a few large pores (Figure 9). The peak of the PSD is mainly located in the range of 0.02 to 0.04 μm, while the peak of the PTD ranges within 0.02 to 0.06 μm. It is notable that the PSD of this group always coincides very well with the PTD, probably due to the similar pore geometry. The pores of this group are mainly composed of intergranular triangular or sheet-like micro pores (generally less than 0.2 μm). These pores are related to the fierce compaction and/or tight cementation because petrography composition of these tight sandstones is mainly characterized by very-fine grain and high clay, plastic grain, or carbonate content (Figures 2, 3 and 10). Additionally, samples of this group are the extremely tight non-reservoir rock, which are mainly characterized by a very low porosity and an extremely low permeability, generally with a porosity less than 7% and permeability less than 0.015 mD (Table 1, Figure 2).

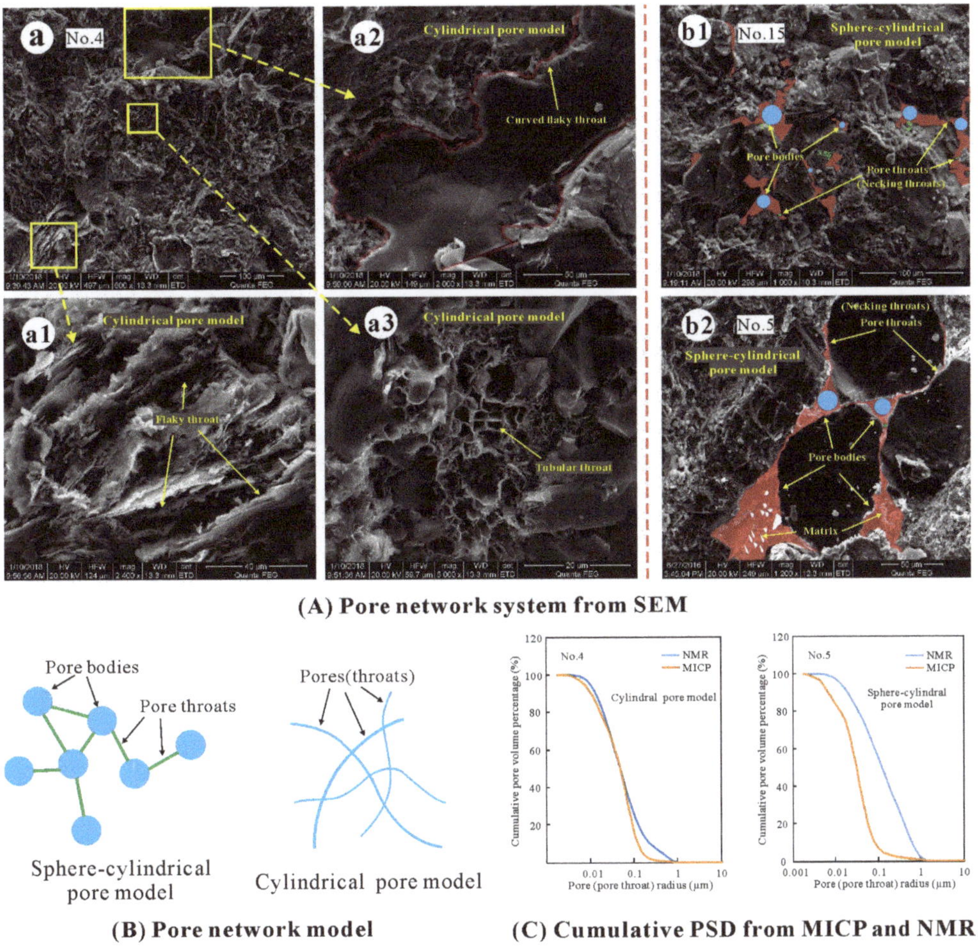

Figure 10. The pore network model of the Chang 7 tight sandstones in the Ordos Basin. (**A**) (**a**) No. 4, L231, 2027.90 m, large amounts of tubular/cylindrical throats or long strip throats are developed in the very tight sandstones, including: (**a1**) the curved lamellar throat, (**a2**) the laminated throat; (**a3**) the tubular throat. (**b**) Necking throats are fine throats connecting large-size pore bodies and small-size pore throats, origin from the close contact of grains caused by fierce compaction and grain coating chlorite. (**b1**) No. 15, L231, 2022.25 m; (**b2**) No. 5, L231, 2108.13 m. (**B**) Pore network model. (**C**) The cumulative PSD (or PTD) from MICP and NMR. No. 4, the cylindrical pore model (CPM), the PSD and PTD match well with each other. No. 5, the sphere-cylindrical pore model (SPM), a significant difference between the PSD and PTD is presented due to large amounts of necking throats.

The second group (No. 6, No. 7, No. 10, and No. 12) is a unimodal distribution with a positive skewness (Figure 9). They include an obvious tail or minor peak which is mainly related to large pores (Figure 9). The dominant peak is located at 0.04 μm, while its right tail is approximately 0.2 to 3 μm and reveals a second minor population accounting for approximately 20% to 30% of the total pore volume. The pores of this group are composed of a large number of small pores and a small number of large pores. These large pores are probably generated from the limited dissolution. Additionally, the porosity and permeability of this sandstone is generally about 7% to 11% and 0.02 to 0.03 mD, respectively. However, the sample No.8 presents a logarithmic normal distribution (unimodal), with a maximum pore radius exceeding 1 μm and a peak value of approximately 0.1 μm. A large number of clay inter-crystalline pores are observed under SEM (Figure 3j) and the macro petrophysical properties are closer to the second group.

The third group (No. 5, No. 9, No. 11, No. 13, No. 14, No. 15, and No. 16) presents a typical bimodal distribution (Figure 9). The left peak ranges within 0.1 μm with the peak located in 0.02 μm. The right peak distributes from 0.1 to 10 μm with the peak location mainly ranging from 0.2 to 1 μm. Obviously, the pore system of this group possesses two modes, with one corresponding to clay inter-crystalline pores dominated by micro pores and another corresponding to macro pores mainly composed of residual primary intergranular pores and large dissolved pores or inter-crystalline pores (Figures 3 and 10). The two pore types may form the pore geometry of large pores connecting with narrow throats, which further expounds the main reason of the difference between PSD and PTD. The porosity and permeability of the third group is generally more than 9% and 0.04 mD, respectively.

The PTD from MICP always presents a unimodal distribution with a negative skewness including a left tail (Figure 9). In particular, the sample of No.16 presents a second population with small pore throats (<0.1 μm). The PTD is characterized by leptokurtosis and high amplitude, which indicates that the main pore throat presents a concentrated distribution, significantly different from the low-amplitude and platykurtic bimodal distribution of the PSD from NMR. In fact, it is the special pore structure where the finer throats with centralized distribution control much of the larger pores that results in the main difference between the PSD and PTD in tight sandstones.

4.3. The Difference between MICP-PTD and NMR-Derived PSD

4.3.1. The Pore Network Model

The pore network, which is composed of large pore bodies and fine pore throats, is a system of storage space and percolation pathways for fluids. Macro pores of residual intergranular pores or dissolution pores comprise pore bodies that are connected by the fine passages composed of necking throats and clay inter-crystalline micro pores.

It is noteworthy that two pore network models are identified in the Chang 7 tight reservoir of the study area (Figure 10). One is the cylindrical pore model (CPM) composed of the flaky throat, curved flaky throat, and tubular throat, while the other is the sphere-cylindrical pore model (SPM), also known as ink-bottle pores [58], composed of large pore bodies connecting necking throats.

The flaky throat or curved flaky throat originates from the close contact of grains (line contact or even concavo-convex contact) under fierce compaction, and tubular throats are widely distributed in clay inter-crystalline pores and micro dissolved pores (Figure 10A). The tight sandstones of these throat types tend to have a very fine throat (generally, radius <0.2 μm). Therefore, the samples that are dominated by the CPM are always related to the very tight non-reservoir rock in the study area (e.g., No. 1 to No. 4). However, the pore network of the SPM dominant is composed of large amounts of necking throats, characterized by large pores connecting with narrow throats (Figure 10A). It is usually connected with the tight reservoirs of relatively high porosity and low permeability (e.g., No. 5, No. 11, and No. 13 to No. 16). Necking throats are generated from the spot contact of grains caused by fierce compaction or grain coating chlorite. Rocks of this type are characterized by interstitial material of low content or dissolution of material and grains, which is favorable to residual intergranular pores, dissolution pores, and relatively large inter-crystalline pores. Additionally, clay inter-crystalline pores are also one of the most important components in the samples of SPM dominant. These pores can act as throats, form ink-bottle pores in combination with macro pores and then control the flow percolation. The abundant micro pores in clay minerals are a main contributor to the left peak of NMR-derived PSD and the fine tail of the PTD from MICP.

These two pore network models make a vast difference between the PSD and PTD. As is discussed above, the PSD and PTD are in a good agreement in the CPM dominant, whereas they show a large discrepancy in the SPM dominant (Figures 9 and 10C).

4.3.2. Difference of Pores and Throats

The CPM is assumed to study pore geometry of porous material in mercury porosimetry, but in fact, as discussed above, the network of pores and throats in rocks is more similar to the SPM [10,65], also referred to as ink-bottle pores [58] (Figure 10). The ink-bottle pores, which comprise a serial of narrow throats (small pores) connecting to the expanded pore bodies (large pores), are widely distributed in reservoir rocks, especially in tight sandstone reservoirs. Therefore, the real pore geometry is quite different from the cylinder pore assumption, and this assumption can lead to significant differences between the MICP analysis and reality. In the MICP experiment, mercury gradually breaks through the narrow throats with increasing injection pressure, and then enters into the pore bodies connected by these throats. The injected mercury volume under a certain pressure is equal to the total volume of these pores and throats. However, the pore throat radius controlled by this injected pressure equates to the radius of the largest entrance of a pore void and is smaller than the real pore body radius (Figure 10).

The connectivity of the pore system of ink-bottle pores deteriorates with increasing the PTR in reservoir rocks, and thus the similarity of the PTD and PSD decreases. That is, an increase in PTR reflects a decrease in the volume of cylindrical pores or an increase in the volume of ink-bottle pores (sphere-cylindrical pores), which makes reservoir rocks a transformation from the CPM dominant to the SPM dominant and further presents more significant differences between the two distribution curves.

Euclidean distance is introduced to define the similarity of the two curves of the PTD and PSD. Euclidean distance is a numerical description of the total distance between points or vectors in space [66] and is widely applied for quantification of similarity or for classification in many research fields [67–69]. Therefore, it can be used to characterize the similarity of the two curves: the smaller the Euclidean distance is, the higher the similarity between the curves is [70]. Euclidean distance can be expressed as follows [66]:

$$D_{edu} = \sqrt{\sum_{x=x_i,i=1}^{n} \left(y_i^{NMR} - y_i^{MICP}\right)^2} \tag{8}$$

where D_{edu} is the Euclidean distance, x_i is the pore radius or pore throat radius, and y_i^{NMR} and y_i^{MICP} are the incremental porosity of NMR-derived PSD and MICP-PTD in fraction, respectively.

The result shows that the Euclidean distance increases with the increasing PTR of the Chang 7 tight sandstones (Figure 11), which indicates that the difference in pore body and throat size controls the similarity between the PTD and PSD and further reflects the phenomenon of "large pores connecting with narrow throats" in tight sandstones. In the MICP experiment, this phenomenon makes the volume of large pores connected by narrow throats attached to the corresponding throat volume, with the PTD visually shifting the smaller pore size direction compared to the authentic pore size distribution [58]. As discussed above, it is not rigorous or even wrong to directly overlap the curve of the T_2 distribution to the curve of PTD to obtain the PSD of NMR. In addition, Euclidean distance can effectively evaluate the similarity of the PTD obtained from MICP and NMR-derived PSD and indicate pore network characteristics (CPM and SPM) in tight sandstones.

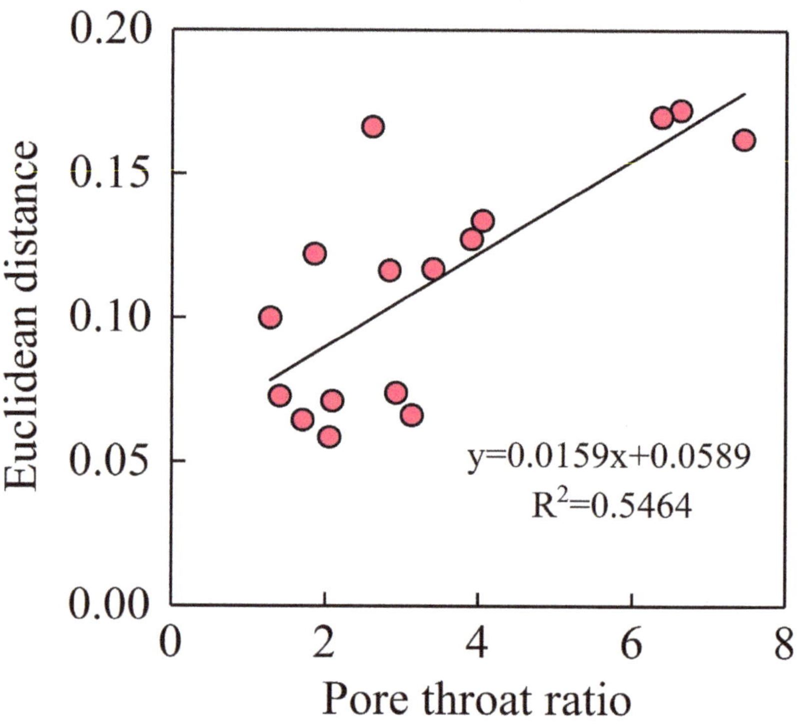

Figure 11. The relationship between the Euclidean distance and pore throat ratio (PTR).

5. Conclusions

In this study, based on the analysis of petrology and pore characteristics, the MICP and NMR techniques were combined to evaluate the PSD and PTD in the Chang 7 tight sandstone reservoirs, and the CPM and SPM (pore network models) were proposed to reveal their differences. The conclusions are as follows:

High capillary pressure, dominated nanoscale pore throat, and low mercury withdrawal efficiency indicate that ink-bottle pores and necking throats are developed in the Chang 7 tight sandstones. The micro pores filled with irreducible fluid contribute a significant portion of the storage space but may not be important for oil or gas percolation in tight sandstone reservoirs.

The comparison of pore and throat radius corresponding to different pore volume percentage suggests that pore and throat size dominantly differ in the large-size pore compartment. The constructive PSD does not always match well with the PTD determined by MICP. The main pore throat presents a centralized distribution, while the PSD of NMR is a low-amplitude and platykurtic bimodal distribution. In very tight core samples, the PSD always coincides very well with PTD probably due to the pore network dominated by cylindrical pores, while the samples of high porosity and permeability present a significant difference between the two distributions due to the pore network dominated by the sphere-cylindrical pores.

The CPM and SPM effectively explain the difference between PSD and PTD of the Chang 7 tight sandstones, providing us a good insight into the heterogeneity of pore geometry (pore and throat). This is of great significance for the application of MICP and NMR techniques to evaluate the pore structure and petrophysical properties of tight sandstone reservoirs.

Author Contributions: Conceptualization, C.L.; methodology, C.L., F.Y. and Z.H.; formal analysis, C.L. and X.L.; investigation, C.L.; resources, P.W.; writing—original draft preparation, C.L.; writing—review and editing, C.L., X.F. and Z.H.; visualization, C.L. and F.Y. All authors have read and agreed to the published version of the manuscript.

Funding: This research was funded by National Natural Science Foundation of China and grant number [NO. 41472114].

Data Availability Statement: The data presented in this study are available on request from the corresponding author. The data are not publicly available due to patent protection.

Acknowledgments: Support from the National Natural Science Foundation of China is highly acknowledged.

Conflicts of Interest: The authors declare no conflict of interest.

References

1. Zou, C.; Tao, S.; Yang, Z.; Yuan, X.; Zhu, R.; Hou, L.; Jia, J.; Wang, F.; Wu, S.; Bai, B.; et al. New advance in unconventional petroleum exploration and research in China. *Bull. Mineral. Petrol. Geochem.* **2012**, *31*, 312–322.
2. Zhou, Y.; Ji, Y.; Xu, L.; Che, S.; Niu, X.; Wan, L.; Zhou, Y.; Li, Z.; You, Y. Controls on reservoir heterogeneity of tight sand oil reservoirs in Upper Triassic Yanchang Formation in Longdong Area, southwest Ordos Basin, China: Implications for reservoir quality prediction and oil accumulation. *Mar. Pet. Geol.* **2016**, *78*, 110–135. [CrossRef]
3. Li, C.; Liu, G.; Cao, Z.; Niu, Z.; Wang, P.; Zhang, M.; Zhang, K. The study of Chang 7 tight sandstone micro pore characteristics in Longdong area, Ordos Basin. *Nat. Gas Geosci.* **2016**, *27*, 1236–1247.
4. Hu, Z.; Gaus, G.; Seemann, T.; Zhang, Q.; Littke, R.; Fink, R. Pore structure and sorption capacity investigations of Ediacaran and Lower Silurian gas shales from the Upper Yangtze platform, China. *Geomech. Geophys. Geo-Energy Geo-Resour.* **2021**, *7*, 1–26. [CrossRef]
5. Nelson, P.H. Pore-throat sizes in sandstones, tight sandstones, and shales. *AAPG Bull.* **2009**, *93*, 329–340. [CrossRef]
6. Rezaee, R.; Saeedi, A.; Clennell, B. Tight gas sands permeability estimation from mercury injection capillary pressure and nuclear magnetic resonance data. *J. Pet. Sci. Eng.* **2012**, *s88–s89*, 92–99. [CrossRef]
7. Zhao, H.; Ning, Z.; Wang, Q.; Zhang, R.; Zhao, T.; Niu, T.; Zeng, Y. Petrophysical characterization of tight oil reservoirs using pressure-controlled porosimetry combined with rate-controlled porosimetry. *Fuel* **2015**, *154*, 233–242. [CrossRef]
8. Hu, Z.; Lu, S.; Klaver, J.; Dewanckele, J.; Amann-Hildenbrand, A.; Gaus, G.; Littke, R. An Integrated Imaging Study of the Pore Structure of the Cobourg Limestone—A Potential Nuclear Waste Host Rock in Canada. *Minerals* **2021**, *11*, 1042. [CrossRef]
9. Comisky, J.; Santiago, M.; Mccollom, B.; Buddhala, A. Sample Size Effects on the Application of Mercury Injection Capillary Pressure for Determining the Storage Capacity of Tight Gas and Oil Shales. In Proceedings of the Canadian Unconventional Resources Conference, Society of Petroleum Engineers, Calgary, AB, Canada, 15–17 November 2011.
10. Schmitt, M.; Fernandes, C.P.; Wolf, F.G.; Neto, J.A.B.D.C.; Rahner, C.P.; Santos, V.S.S.D. Characterization of Brazilian tight gas sandstones relating permeability and angstrom-to micron-scale pore structures. *J. Nat. Gas Sci. Eng.* **2015**, *27*, 785–807. [CrossRef]
11. Purcell, W.R. Capillary pressures-their measurement using mercury and the calculation of permeability therefrom. *J. Pet. Technol.* **1949**, *1*, 39–48. [CrossRef]
12. Cao, Z.; Liu, G.; Zhan, H.; Li, C.; Yuan, Y.; Yang, C.; Jiang, H. Pore structure characterization of chang-7 tight sandstone using MICP combined with N2GA techniques and its geological control factors. *Sci. Rep.* **2016**, *6*, 36919. [CrossRef] [PubMed]
13. Bustin, R.M.; Bustin, A.M.M.; Cui, A.; Ross, D.; Pathi, V.M. Impact of shale properties on pore structure and storage characteristics. In Proceedings of the Society of Petroleum Engineers Shale Gas Production Conference, Fort Worth, TX, USA, 16–18 November 2008; p. SPE119892.
14. Zhang, N.; He, M.; Zhang, B.; Qiao, F.; Sheng, H.; Hu, Q. Pore structure characteristics and permeability of deep sedimentary rocks determined by mercury intrusion porosimetry. *J. Earth Sci.* **2016**, *27*, 670–676. [CrossRef]
15. Washburn, E.W. Note on the method of determining the distribution of pore sizes in a porous material. *Proc. Natl. Acad. Sci. USA* **1921**, *7*, 115–116. [CrossRef] [PubMed]
16. Timur, A. Pulsed nuclear magnetic resonance studies of porosity, movable fluid, and permeability of sandstones. *J. Pet. Technol.* **1969**, *21*, 775–786. [CrossRef]
17. Coates, G.; Xiao, L.; Prammer, M. *NMR Logging Principles and Applications*; Halliburton Energy Services: Houston, TX, USA, 1999; pp. 2–72.
18. Al-Mahrooqi, S.H.; Grattoni, C.A.; Moss, A.K.; Jing, X.D. An investigation of the effect of wettability on NMR characteristics of sandstone rock and fluid systems. *J. Pet. Sci. Eng.* **2003**, *39*, 389–398. [CrossRef]
19. Yao, Y.; Liu, D.; Che, Y.; Tang, D.; Tang, S.; Huang, W. Petrophysical characterization of coals by low-field nuclear magnetic resonance (NMR). *Fuel* **2010**, *89*, 1371–1380. [CrossRef]
20. Xiao, L.; Xie, R.; Liao, G. *NMR Logging Principles and Applications of Complex Hydrocarbon Reservoirs in China*; Science Press: Beijing, China, 2012; pp. 2–6.

21. Wang, F.; Tang, T.; Liu, T.; Zhang, H. Evaluation of the pore structure of reservoirs based on NMR T2 Spectrum decomposition. *Appl. Magn. Reson.* **2016**, *47*, 361–373. [CrossRef]
22. Cohen, M.H.; Mendelson, K.S. Nuclear magnetic relaxation and the internal geometry of sedimentary rocks. *J. Appl. Phys.* **1982**, *53*, 1127–1135. [CrossRef]
23. Hodgkins, M.A.; Howards, J.J. Application of NMR logging to reservoir characterization of low-resistivity sands in the Gulf of Mexico. *AAPG Bull.* **1999**, *83*, 114–127.
24. Westphal, H.; Surholt, I.; Kiesl, C.; Thern, H.; Kruspe, T. NMR Measurements in carbonate rocks: Problems and an approach to a solution. *Pure Appl. Geophys.* **2005**, *162*, 549–570. [CrossRef]
25. Xu, H.; Tang, D.; Zhao, J.; Li, S. A precise measurement method for shale porosity with low-field nuclear magnetic resonance: A case study of the Carboniferous–Permian strata in the Linxing area, eastern Ordos Basin, China. *Fuel* **2015**, *143*, 47–54. [CrossRef]
26. Hinai, A.A.; Rezaee, R.; Esteban, L.; Labani, M. Comparisons of pore size distribution: A case from the western Australian gas shale formations. *J. Unconv. Oil Gas Resour.* **2014**, *8*, 1–13. [CrossRef]
27. Li, C.; Liu, G.; Cao, Z.; Yuan, W.; Wang, P.; You, Y. Analysis of petrophysical characteristics and water movability of tight sandstone using low-field nuclear magnetic resonance. *Nat. Resour. Res.* **2020**, *29*, 2547–2573. [CrossRef]
28. Soete, J.; Claes, S.; Claes, H.; Erthal, M.; Hamaekers, H.; De Boever, E.; Foubert, A.; Klitzsch, N.; Swennen, R. Unravelling the pore network and its behaviour: An integrated NMR, MICP, XCT and petrographical study of continental spring carbonates from the Ballık area, SW Turkey. *Depos. Recor* **2020**, *8*, 292–316. [CrossRef]
29. Zhang, N.; Wang, S.; Zhao, F.; Sun, X.; He, M. Characterization of the pore structure and fluid movability of coal-measure sedimentary rocks by nuclear magnetic resonance (NMR). *ACS Omega* **2021**, *6*, 22831–22839. [CrossRef]
30. Clarkson, C.R.; Solano, N.; Bustin, R.M.; Bustin, A.M.M.; Chalmers, G.R.L.; He, L.; Melnichenko, Y.B.; Radliński, A.P.; Blach, T.P. Pore structure characterization of north american shale gas reservoirs using USANS/SANS, gas adsorption, and mercury intrusion. *Fuel* **2013**, *103*, 606–616. [CrossRef]
31. Yang, J. *Tectonic Evolution and Oil-gas Reservoirs Distribution in Ordos Basin*; Petroleum Industry Press: Beijing, China, 2002; pp. 36–37.
32. Yuan, W.; Liu, G.; Xu, L.; Niu, X.; Li, C. Petrographic and geochemical characteristics of organic-rich shale and tuff of the upper triassic yanchang formation, ordos basin, China: Implications for lacustrine fertilization by volcanic ash. *Can. J. Earth Sci.* **2019**, *56*, 47–59. [CrossRef]
33. Yang, H.; Dou, W.; Liu, X.; Zhang, L. Analysis on sedimentary facies of member 7 in Yanchang Formation of Triassic in Ordos Basin. *Acta Sedimentol. Sin.* **2010**, *28*, 254–263.
34. Zhu, H.; Zhong, D.; Yao, J.; Sun, H.; Niu, X.; Liang, X.; You, Y.; Li, X. Alkaline diagenesis and its effects on reservoir porosity: A case study of Upper Triassic Chang 7 Member tight sandstone in Ordos Basin, NW China. *Pet. Explor. Dev.* **2015**, *42*, 56–65. [CrossRef]
35. Wardlaw, N.C. Pore Geometry of carbonate rocks as revealed by pore casts and capillary pressure. *AAPG Bull.* **1976**, *60*, 245–257.
36. Rootare, H.M.; Prenzlow, C.F. Surface area from mercury porosimetry measurements. *J. Phys. Chem.* **1967**, *71*, 2733–2735. [CrossRef]
37. Giesche, H. Mercury porosimetry: A general (practical) overview. *Part. Part. Syst. Charact.* **2006**, *23*, 9–19. [CrossRef]
38. Gan, H.; Nandi, S.P.; Walker, P.L. Nature of the porosity in American coals. *Fuel* **1972**, *51*, 272–277. [CrossRef]
39. Mosher, K.; He, J.; Liu, Y.; Rupp, E.; Wilcox, J. Molecular simulation of methane adsorption in micro- and mesoporous carbons with applications to coal and gas shale systems. *Int. J. Coal Geol.* **2013**, *109*, 36–44. [CrossRef]
40. Kenyon, W.E. Petrophysical principles of applications of NMR logging. *Log Analyst.* **1997**, *38*, 21–40.
41. Daigle, H.; Johnson, A.; Thomas, B. Determining fractal dimension from nuclear magnetic resonance data in rocks with internal magnetic field gradients. *Geophysics* **2014**, *79*, 425–431. [CrossRef]
42. Brownstein, K.R.; Tarr, C.E. Importance of classical diffusion in NMR studies of water in biological cells. *Phys. Rev. A* **1979**, *19*, 2446–2453. [CrossRef]
43. Kleinberg, R.L.; Horsfield, M.A. Transverse relaxation processes in porous sedimentary rock. *J. Magn. Reson.* **1990**, *88*, 9–19. [CrossRef]
44. Sigal, R.F. Pore-size distributions for organic-shale-reservoir rocks from nuclear-magnetic-resonance spectra combined with adsorption measurements. *SPE J.* **2015**, *20*, 824–830. [CrossRef]
45. Yan, W.; Sun, J.; Cheng, Z.; Li, J.; Sun, Y.; Shao, W.; Shao, Y. Petrophysical characterization of tight oil formations using 1D and 2D NMR. *Fuel* **2017**, *206*, 89–98. [CrossRef]
46. Folk, R.L.; Ward, W.C. Brazos River bar: A study in the significance of grain size parameters. *J. Sediment. Res.* **1957**, *27*, 3–16. [CrossRef]
47. Li, C.; Liu, G.; Cao, Z.; Sun, M.; You, Y.; Liu, N. Oil charging pore throat threshold and accumulation effectiveness of tight sandstone reservoir using the physical simulation experiments combined with NMR. *J. Pet. Sci. Eng.* **2022**, *208*, 109338. [CrossRef]
48. Giles, M.R.; Boer, R.B.D. Origin and significance of redistributional secondary porosity. *Mar. Pet. Geol.* **1990**, *7*, 378–397. [CrossRef]
49. Mastalerz, M.; Drobniak, A.; Hower, J. Controls on reservoir properties in organic-matter-rich shales: Insights from MICP analysis. *J. Pet. Sci. Eng.* **2021**, *196*, 107775. [CrossRef]

50. Zhou, Y.; You, L.; Zi, H.; Lan, Y.; Cui, Y.; Xu, J.; Fan, Q.; Wang, G. Determination of pore size distribution in tight gas sandstones based on Bayesian regularization neural network with MICP, NMR and petrophysical logs. *J. Nat. Gas Sci. Eng.* **2022**, *100*, 104468. [CrossRef]
51. Kaufmann, J.; Loser, R.; Leemann, A. Analysis of cement-bonded materials by multi-cycle mercury intrusion and nitrogen sorption. *J. Colloid Interface Sci.* **2009**, *336*, 730–737. [CrossRef]
52. Wardlaw, N.C.; Malcolm, M.; Li, Y. Pore and throat size distributions determined by mercury porosimetry and by direct observation. *Carbonates Evaporites* **1988**, *3*, 1–16. [CrossRef]
53. Rylander, E.; Philip, M.S.; Jiang, T.; Lewis, R. NMR T_2 distributions in the Eagle Ford shale: Reflections on pore size. In Proceedings of the SPE Unconventional Resources Conference, The Woodlands, TX, USA, 10–12 April 2013; p. SPE164554.
54. Schowalter, T.T. Mechanics of secondary hydrocarbon migration and entrapment. *AAPG Bull.* **1979**, *63*, 723–760.
55. Hognesen, E.J.; Strand, S.; Austad, T. Waterflooding of Preferential Oil-wet Carbonates: Oil Recovery Related to Reservoir Temperature and Brine Composition. In Proceedings of the SPE Europec/EAGE Annual Conference, Madrid, Spain, 13–16 June 2005; p. SPE94166.
56. Meng, W.; Haroun, M.R.; Sarma, H.K.; Adeoye, J.T.; Aras, P.; Punjabi, S.; Rahman, M.M.; Al Kobaisi, M. A novel approach of using phosphate-spiked smart brines to alter wettability in mixed oil-wet carbonate reservoirs. In Proceedings of the Abu Dhabi International Petroleum Exhibition and Conference, Abu Dhabi, United Arab Emirates, 9–12 November 2015; p. SPE17755.
57. Dai, C.; Cheng, R.; Sun, X.; Liu, Y.; Zhou, H.; Wu, Y.; You, Q.; Zhang, Y.; Sun, Y. Oil migration in nanometer to micrometer sized pores of tight oil sandstone during dynamic surfactant imbibition with online NMR. *Fuel* **2019**, *245*, 544–553. [CrossRef]
58. Wardlaw, N.C.; Mckellar, M. Mercury porosimetry and the interpretation of pore geometry in sedimentary rocks and artificial modelsy. *Powder Technol.* **1981**, *29*, 127–143. [CrossRef]
59. Lala, A.M.S.; El-Sayed, N.A.A. Effect of pore framework and radius of pore throats on permeability estimation. *J. Afr. Earth Sci.* **2015**, *110*, 64–74. [CrossRef]
60. Volokitin, Y.; Looyestijn, W.J.; Slijkerman, W.F.J.; Hofman, J.P. A Practical approach to obtain primary drainage capillary pressure curves from NMR core and log Data. *Petrophysics* **2001**, *42*, 334–343.
61. Liu, T.; Wang, S.; Fu, R.; Zhou, M.; Li, Y.; Luo, M. Analysis of rock pore throat structure with NMR spectra. *Oil Geophys. Prospect.* **2003**, *18*, 737–742.
62. Li, H.; Zhu, J.; Guo, H. Methods for calculating pore radius distribution in rock from NMR T2 spectra. *Chin. J. Magn. Reson.* **2008**, *25*, 273–280.
63. Li, H.; Guo, H.; Li, H.; Liu, W.; Jiang, B.; Hua, J. Thickness analysis of bound water film in tight reservoir. *Nat. Gas Geosci.* **2015**, *26*, 186–192.
64. Yang, S.; Wei, J. *Reservoir Physics*; Petroleum Industry Press: Beijing, China, 2004; pp. 188–208.
65. Liu, P.; Chen, X. *Foam Metel*; Centrol South University Press: Changsha, China, 2012; pp. 298–309.
66. Mesquita, D.P.P.; Gomes, J.P.P.; Junior, A.H.S.; Nobre, J.S. Euclidean distance estimation in incomplete datasets. *Neurocomputing* **2017**, *248*, 11–18. [CrossRef]
67. Wang, C.; Jiang, M.; Liu, Z. A new method to distinguish sedimentary environment based on the data of the grain size frequency distribution. *Bull. Tianjing Inst. Geol. Min. Res.* **1989**, *23*, 61–69.
68. Yang, T.; Li, T.; Zhang, J.; Li, J.; Liu, H.; Wang, Y. Rapid identification of bolete mushrooms by UV spectroscopy combined with Euclidean distance and principal component analysis. *Food Sci.* **2014**, *35*, 105–109.
69. Ghosh, A.; Barman, S. Application of Euclidean distance measurement and principal component analysis for gene identification. *Gene* **2016**, *583*, 112–120. [CrossRef]
70. Wang, X.; Wang, W.; Zhang, K.; Bi, K. Approaching the study on the similarity analysis of HPLC fingerprints spectra for traditional Chinese medicines. *J. Shenyang Pharm. Univ.* **2003**, *20*, 360–362.

processes MDPI

Article

Geochemical Characteristics and Development Model of the Coal-Measure Source Rock in the Kuqa Depression of Tarim Basin

Xiujian Ding [1,*], Tianze Gao [1,2], Xianzhang Yang [3], Zhenping Xu [3], Changchao Chen [3], Keyu Liu [1] and Xueqi Zhang [3]

1 School of Geosciences, China University of Petroleum, Qingdao 266580, China; z20010040@s.upc.edu.cn (T.G.)
2 China National Offshore Oil Corporation International Limited Research Institute, Beijing 100028, China
3 Research Institute of Exploration and Development, PetroChina Tarim Oilfield Company, Korla 841000, China
* Correspondence: dingxj@upc.edu.cn

Abstract: The development model of the coal-measure source rock may be different from that of the lacustrine source rock. The depositional environment of the coal-measure source rock is dominated by weak oxidation and weak reduction, and the majority of the organic material originates from terrestrial higher plants. Taking the Jurassic coal-measure source rock in the Kuqa Depression as the research object, the geochemical characteristics of the source rock are comprehensively analyzed, the primary controlling elements of source rock development are made clear, and the development model of the coal-measure source rock is established. This study contributes to the field of source rock prediction and oil and gas exploration. The lithology of the coal-measure source rock in the Kuqa Depression is mainly mudstone, carbonaceous mudstone, and coal, which are medium- to good-quality source rocks, and the organic matter type is mainly II_2 and III. Terrestrial organic matter is a key factor in controlling the formation of coal-measure source rocks, and the sedimentation rate also has a certain influence. The redox degree of the depositional environment, water salinity, and clay mineral content has little influence on the development of coal-measure source rocks. By integrating the main control factors, the development model of the coal-measure source rock is established. It is considered that the development model and distribution characteristics of the coal-measure source rock are different from the traditional understanding of lacustrine source rocks, and it is pointed out that the coal-measure source rock in the gentle slope zone is more developed than the sag area.

Keywords: coal-measure source rock; input of terrestrial organic matter; depositional environment; sedimentary rate; Jurassic system; Kuqa Depression

Citation: Ding, X.; Gao, T.; Yang, X.; Xu, Z.; Chen, C.; Liu, K.; Zhang, X. Geochemical Characteristics and Development Model of the Coal-Measure Source Rock in the Kuqa Depression of Tarim Basin. *Processes* **2023**, *11*, 1777. https://doi.org/10.3390/pr11061777

Academic Editor: Carlos Sierra Fernández

Received: 5 May 2023
Revised: 1 June 2023
Accepted: 9 June 2023
Published: 11 June 2023

1. Introduction

The study on the development of source rocks is of great significance to predict the distribution of source rocks and hydrocarbon accumulation. With regard to the distribution and development of lacustrine source rocks, the source rock of the large depression-type lake basin is mainly distributed in the deep and semi-deep lacustrine facies in the center of the lake basin [1], whereas the source rock of the faulted lake basin is mainly distributed in the steep side belt with a larger thickness and a bigger water depth [2,3].

As for the development of source rocks, researchers have studied the controlling effects of different lake types on the formation of source rocks under macroscopic conditions. Carroll and Bohacs (1999, 2001), through the comparative study of the sediments of modern lakes and ancient lakes, divided lakes into three categories: underfilled, balanced fill, and overfilled, and concluded that balanced fill lakes are most conducive to the development of source rock, underfilled lakes are in the middle, and overfilled lakes are the least conducive to the development of source rocks [4,5]. These three simple types cannot solve the problem of source rock heterogeneity in the basin, which is a critical problem in petroleum exploration.

Most scholars believe that productivity and preservation conditions are the main factors controlling the development of source rocks. Some scholars believe that the input of rich nutrients causes productivity to flourish, thus promoting the development of source rocks [6,7]. Some scholars also pointed out that the proportion of organic matter preserved in modern marine sediments in the world is less than 0.5% of the original organic matter, and most of the organic matter has been degraded [8,9]. The strongly reducing depositional environment of the underwater environment is the key factor in promoting the preservation of organic matter and the development of the source rock [10]. Coal-measure source rocks are mainly deposited in delta swamps [11], delta plains and tidal flat lakes [12], river delta systems, and other sedimentary environments with shallow water, gentle terrain, and high input of terrestrial organic matter [13–16]. Compared with marine source rock and lacustrine source rock, the coal-measure source rock lacks a strong reduction environment, and the depositional environment is dominated by weak oxidation and weak reduction. The main source of organic matter is terrestrial higher plants, and its development mechanism is different from the traditional understanding of the marine source rock and lacustrine source rock. Taking the Jurassic coal-measure source rock in the Kuqa Depression as the research object, the geochemical characteristics of the coal-measure source rock are comprehensively analyzed. The control of terrigenous organic matter input, depositional environment, and sedimentation rate on the development of coal-measure source rocks is discussed. The primary controlling factors of the coal-measure source rock development are identified, a coal-measure source rock development model is developed, and the source rock development mechanism is improved. In addition, relevant research can guide the distribution prediction of coal-measure source rocks and facilitate oil and gas exploration. This study may be useful in the analysis of hydrogeological processes [17–19].

2. Geological Setting

The Kuqa Depression is located at the northern edge of Tarim Basin and is usually divided into four tectonic zones and three sags. From north to south, the four structural belts are the Northern Structural Belt, Kelasu Structural Belt, Qiulitag Structural Belt, and Front Uplift Belt. From west to east, the three sags are Wushi Sag, Baicheng Sag, and Yangxia Sag, as shown in Figure 1. In recent years, the Dibei gas reservoir, Tuzi gas reservoir, and other tight sandstone gas reservoirs have been successively discovered in the Jurassic system in the northern structural belt of the Kuqa Depression, which has become an important area for natural gas exploration of the Tarim Basin [20,21].

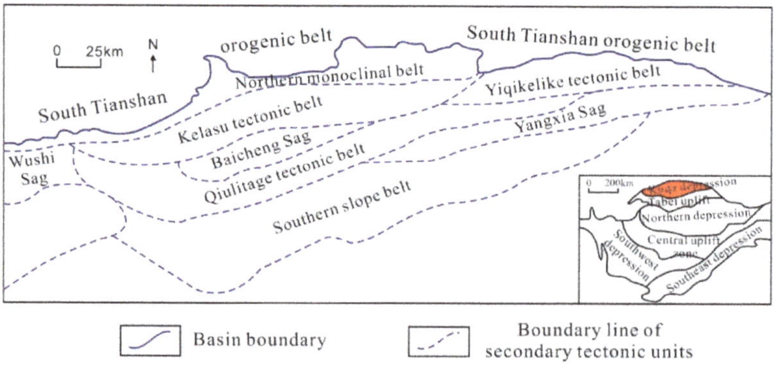

Figure 1. Geographical location of the northern tectonic belt in the Kuqa Depression (Wang et al., 2021 [18]).

The source rock in the Kuqa Depression is mainly developed in the Triassic and Jurassic systems. Five sets of hydrocarbon source rocks are developed from bottom to top, including the Upper Triassic Huangshanjie Formation (T_3h) and Tariqike Formation (T_3t), the Lower Jurassic Yangxia Formation (J_1y), the Middle Jurassic Kezilenuer Formation (J_2kz), and the Qiakemake Formation (J_2q). The Triassic source rock is mainly lacustrine mudstone, and the Jurassic source rock is mainly coal-measure mudstone, as shown in Figure 2.

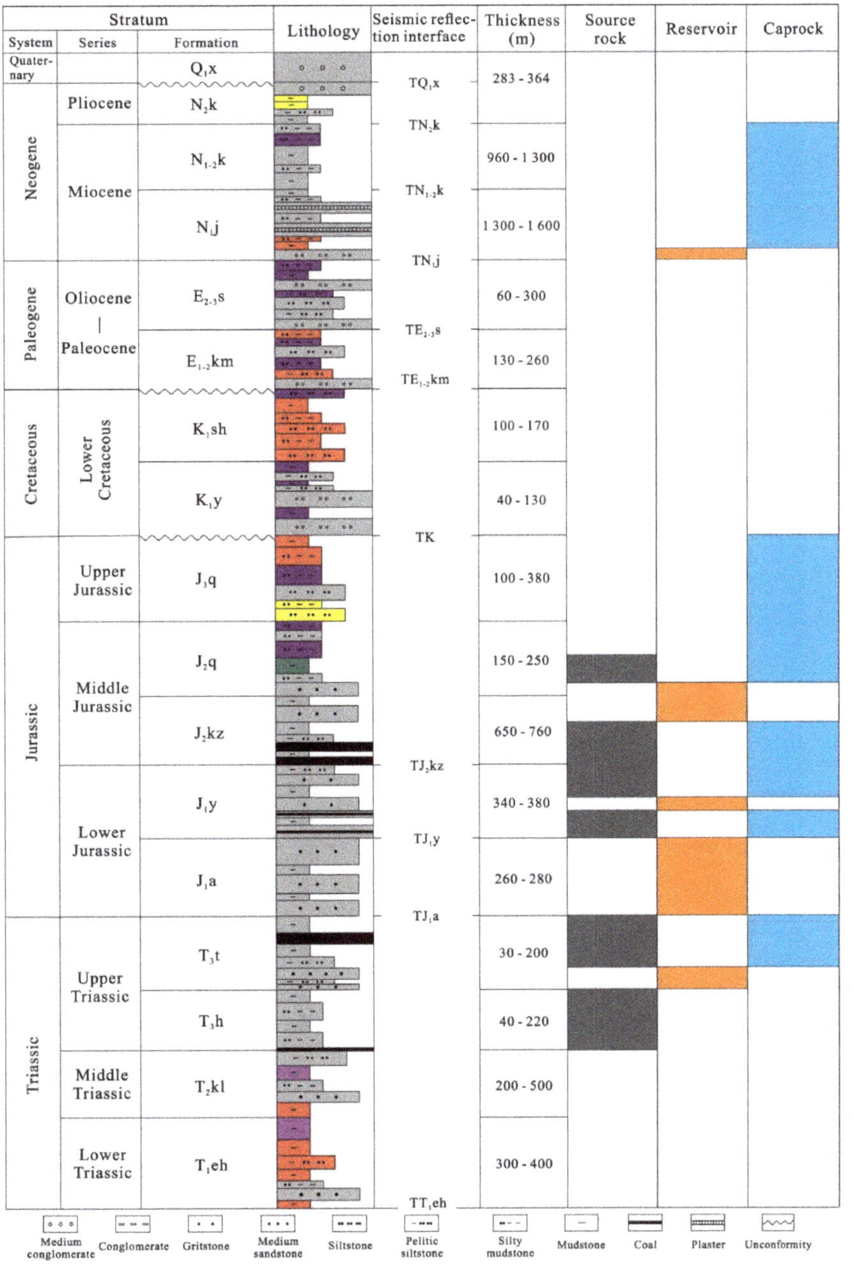

Figure 2. Mesozoic-Cenozoic stratigraphic system in the northern tectonic belt of the Kuqa Depression (Wang et al., 2021 [18]).

3. Samples and Analytical Methods

3.1. Samples

The Jurassic source rock samples, including mudstone samples, carbonaceous mudstone samples, and coal samples, were collected from J_2kz and J_1y strata in the northern tectonic belt.

3.2. Analytical Methods

The total organic carbon content (TOC), total hydrocarbon content (HC), rock pyrolysis hydrocarbon generation potential (S_1 + S_2), hydrogen index (HI), and pyrolysis hydrocarbon peak temperature (T_{max}) samples from the Kuqa Depression were analyzed at the Key Laboratory of Deep Oil and Gas of the China University of Petroleum (East China) using a Rock-Eval 7 analyzer. The vitrinite reflectance (R_O) was determined using a Leica DM4500P polarizing microscope with an MPS200 photometer at the State Key Laboratory of Heavy Oil, China University of Petroleum (Beijing).

The samples of source rocks used the Soxhlet extraction method to extract chloroform asphalt, and the chloroform asphalt was divided into saturated and aromatic hydrocarbons (the separation standard was SY/T 5119-2008). The saturated and aromatic hydrocarbons were finally tested by GC-MS. GC-MS uses the gas chromatography–mass spectrometer manufactured by Agilent, and the GC model is Agilent 9000.

4. Results and Discussion

4.1. Geochemical Characteristics

4.1.1. Abundance of Organic Matter

The abundance of organic matter reflects the relative content of organic matter in the source rocks and is one of the main factors controlling hydrocarbon generation potential. At present, the commonly used indicators of organic matter abundance mainly include total organic carbon content (TOC), chloroform asphalt "A", total hydrocarbon content (HC), and rock pyrolysis hydrocarbon generation potential (S_1 + S_2). Among them, organic carbon content (TOC) and pyrolysis hydrocarbon generation potential (S_1 + S_2) are commonly used indicators to evaluate organic matter abundance.

The coal-measure source rock in the Kuqa Depression is mainly divided into three lithologies: mudstone, carbonaceous mudstone, and coal. The ratings for organic matter abundance in mudstone, carbonaceous mudstone, and coal vary significantly. The TOC of mudstone is mostly less than 6%, whereas the TOC distribution range of carbonaceous mudstone is mainly between 6% and 40%, while that of coal is generally greater than 40%. The TOC distribution range of coal-measure mudstone is about 0.1~8%, and the S_1 + S_2 distribution range is about 0.2~20 mg/g. The distribution range of TOC in carbonaceous mudstone is about 6~40%, and the distribution range of S_1 + S_2 is about 10–90 mg/g. The distribution range of coal TOC is about 40~90%, and the distribution range of S_1 + S_2 is about 30~200 mg/g, as shown in Figure 3. According to the evaluation criteria for organic matter abundance of coal-measure source rock [22], the organic matter abundance of coal-measure mudstone in the Kuqa Depression varies greatly, and the non-source rock (TOC < 0.75%, S_1 + S_2 < 0.5 mg), poor source rock (0.75% < TOC < 1.5%, 0.5 mg/g < S_1 + S_2 < 2.0 mg/g), medium source rock (1.5% < TOC < 3.0%, 2.0 mg/g < S_1 + S_2 < 6.0 mg/g) and good source rock (TOC > 3.0%, S_1 + S_2 > 6.0 mg/g) are all developed. Carbonaceous mudstone is mainly a medium source rock (10% < TOC < 18%, 35 mg/g < S_1 + S_2 < 70 mg/g) or a good source rock (18% < TOC < 35%, 70 mg/g < S_1 + S_2 < 120 mg/g). Coal is mainly a poor source rock (100 mg/g < S_1 + S_2 < 200 mg/g) or a non-source rock (S_1 + S_2 < 100 mg/g). It should be pointed out that although coal and carbonaceous mudstone may not meet the standard of a good source rock according to the previous standards, the hydrocarbon generation potential (S_1 + S_2) of coal and carbonaceous mudstone is significantly greater than that of the mudstone, with a relatively higher hydrocarbon generation potential.

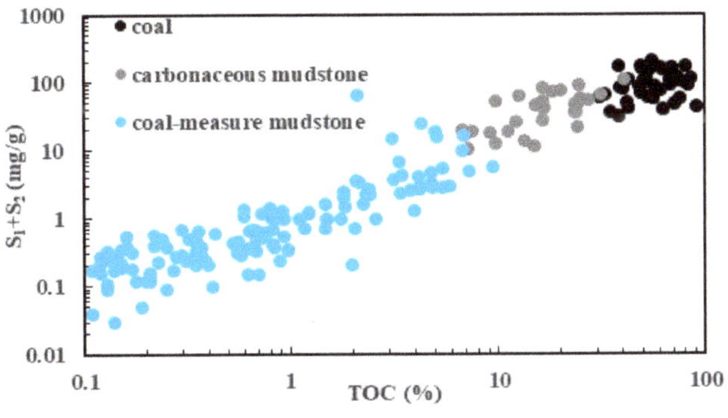

Figure 3. Crossplot of TOC and $S_1 + S_2$ of the Jurassic coal-measure source rocks.

4.1.2. Type of Organic Matter

The $S_1 + S_2$ is not only affected by the organic matter abundance but is also related to the organic matter type. The most popular technique for identifying the type of organic matter is the graphical analysis of kerogen elements. Type I kerogen typically has the largest potential for oil production, a high original hydrogen content, and a low oxygen content. The original hydrogen content of type II kerogen is relatively high, but slightly lower than that of type I kerogen, with a moderate oil generation potential. Type III kerogen has a low original hydrogen content and a high oxygen content, and it does not have the ability to generate oil, relying mainly on gas generation.

The hydrogen index (HI) distribution range of Jurassic coal-measure mudstone is about 50~400 mg/g·TOC, with most of it being less than 200 mg/g·TOC. The organic matter types are mainly II_2 and III types. The difference between HI in the distribution range of carbonaceous mudstone and coal is relatively small, mainly ranging from 100 to 500, and the organic matter type is mainly type II, as shown in Figure 4.

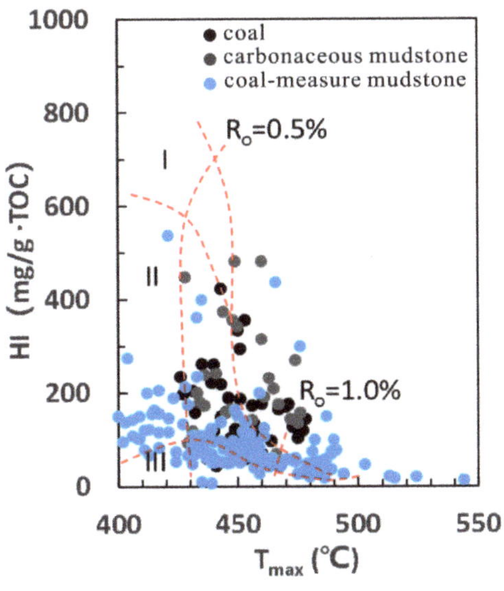

Figure 4. Organic matter types of the Jurassic coal-measure source rocks.

4.1.3. Maturity of Organic Matter

In addition to the abundance and type of organic matter, the degree of thermal evolution of organic matter is also closely related to the hydrocarbon generation capacity of source rocks. There are currently many indicators for determining the maturity of organic matter, such as vitrinite reflectance (R_o), rock pyrolysis peak temperature (T_{max}), kerogen H/C atomic ratio, and biomarker parameters. Among them, R_o is currently the most common method for determining the maturity of organic matter.

The R_o of Jurassic coal-measure source rock is 0.4~1.3%, and the organic matter evolution is mainly in the mature stage. R_o gradually increases with the increase in depth and enters the mature stage when R_o is more than 0.5% at a depth of about 1000 m, as shown in Figure 5.

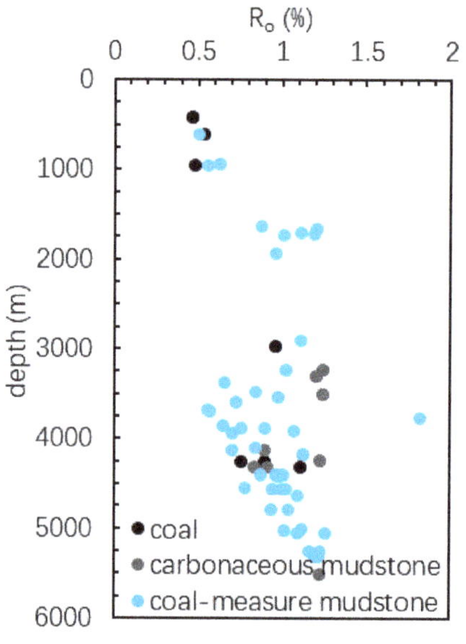

Figure 5. Kerogen types of the Jurassic coal-measure source rocks.

4.2. *Main Control Factors of Coal-Measure Source Rock Development*

The development of source rock is a process of organic matter enrichment. The organic matter in the lake basin mainly comes from aquatic organisms and terrestrial higher plants. Organic matter slowly settles to the bottom of the lake and is degraded during the long process of burial. The remaining organic matter enters the sediment and is preserved in the rock via diagenesis. Researchers have carried out systematic research on the effects of productivity, redox environment, sedimentation rate, and other factors on the development of source rocks [23–26], and believe that organic matter supply, organic matter preservation, and organic matter dilution are the main factors controlling the development of source rocks [27–29]. The input of terrestrial organic matter and the paleoproductivity of water bodies determine the supply of organic matter. It is generally believed that the greater the supply of organic matter, the better the development of source rocks. Organic matter preservation refers to the process of preserving the organic matter at the bottom of a lake basin via degradation. When the organic matter is well preserved, it suffers less degradation, and the proportion of preserved organic matter is high, which is conducive to the development of source rocks. The dilution of organic matter means that the input of non-organic terrigenous detritus will dilute the organic matter and reduce the abundance of organic matter, which is not conducive to the development of source rocks.

4.2.1. Organic Matter Supply

Biomarkers are commonly used parameters for studying the source of organic matter, the depositional environment (redox degree and salinity), the maturity of organic matter, and biodegradation [30]. The relative abundance of regular steroids is commonly used to identify the source of the organic matter. It is generally believed that C_{27} regular steroids mainly come from aquatic organic matter, whereas C_{29} regular steroids mainly come from terrestrial organic matter [31]. The content of C_{27} regular steranes in the coal-measure source rock in the Kuqa Depression is about 5~45%, mainly distributed in 5~30%, and the content of coal and carbonaceous mudstone is lower than that of the mudstone. There is a negative correlation between C_{27} regular sterane and TOC. As the content of C_{27} regular sterane increases, TOC shows a significant downward trend, as shown in Figure 6a. The content of C_{29} regular sterane in the coal-measure source rock is generally greater than 35%, and the content of C_{29} regular sterane in carbonaceous mudstone and coal is significantly higher than that of the mudstone. The C_{29} regular sterane content shows a positive correlation with TOC. As the C_{29} regular sterane content increases, TOC shows a gradually increasing trend, as shown in Figure 6b. Tricyclic terpenes are ubiquitous in source rock extracts, and the carbon number can extend from C_{19} to C_{45} [32,33]. Usually, lacustrine terrestrial organic matter is characterized by a high content of C_{21} tricyclic terpenoids (C_{21}TT), nearshore sediments are mainly characterized by C_{19} tricyclic terpenoids (C_{19}TT) and C_{20} tricyclic terpenoids (C_{20T}T), whereas the shale facies is characterized by a high content of C_{23} tricyclic terpenoids (C_{23}TT) [34]. It is believed that the higher the C_{19}TT/C_{23}TT and C_{21}TT/C_{23}TT ratios, the more terrestrial organic matter input [35,36]. The distribution range of C_{19}TT/C_{23}TT of coal-measure source rocks in the Kuqa Depression is about 0~15. Among them, the ratio of C_{19}TT/C_{23}TT of coals is the highest, that of carbonaceous mudstone is in the middle, and that of mudstone is the lowest. The TOC of the source rock shows an obvious positive correlation with C_{19}TT/C_{23}TT, as shown in Figure 6c. The distribution range of C_{21}TT/C_{23}TT in the coal-measure source rock is about 0~5, which also shows the characteristics of the highest ratio of coal, the middle ratio of carbonaceous mudstone, and the lowest ratio of mudstone. The positive correlation between TOC and C_{21}TT/C_{23}TT in the source rock is good, as shown in Figure 6d. These indicate that the greater the input of terrestrial organic matter, the higher the TOC of the source rock.

In conclusion, the input of terrigenous organic matter has obvious control over the development of coal-measure source rocks. The proportion of terrigenous organic matter in coal is the highest, followed by carbonaceous mudstone, and mudstone is the lowest. The more terrestrial organic matter input, the higher the organic matter abundance of coal-measure source rocks.

4.2.2. Organic Matter Preservation

There are many factors affecting organic matter preservation. It is generally believed that the redox degree [9], water salinity [37], and clay minerals [38] have obvious effects on organic matter preservation. Strong reducing environments have low oxygen concentrations and poor oxidative destruction of organic matter, which favors the preservation of organic matter and the formation of source rock [39–41]. High water salinity promotes the development of stable stratification, hypoxia in the bottom water body, and the preservation of organic matter by reducing the rate at which organic matter is oxidized by oxygen [42,43]. Clay minerals are highly effective in absorbing organic materials. Some organic matter can also enter the clay minerals to prevent oxygen degradation. A high concentration of clay minerals is thought to favor the preservation of organic materials and the formation of source rocks [44].

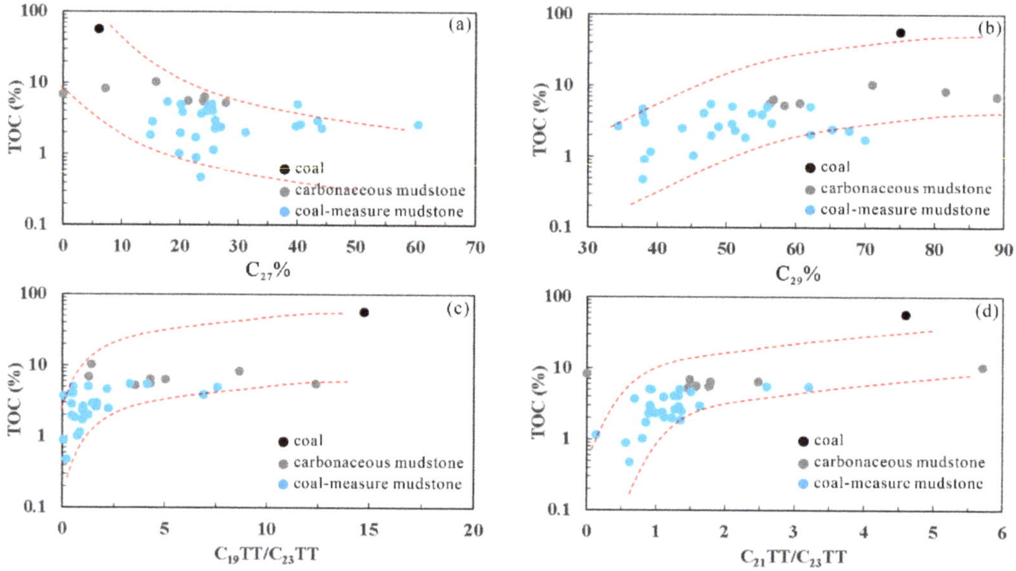

Figure 6. Cross plot of organic matter supply and TOC of the Jurassic coal-measure source rocks. (**a**) Cross plot of C_{27} regular sterane and TOC in coal-measure source rock; (**b**) cross plot of C_{29} regular sterane and TOC in coal-measure source rock; (**c**) cross plot between $C_{19}TT/C_{23}TT$ and TOC of coal-measure source rock; (**d**) cross plot of $C_{21}TT/C_{23}TT$ and TOC of coal-measure source rock.

To determine the redox degree of the depositional environment, it is frequently utilized to cross-plot the source rock's Ph/nC_{18} and Pr/nC_{17} data [45]. The distribution range of Ph/nC_{18} and Pr/nC_{17} of the coal-measure source rock in the Kuqa Depression is about 0.1~0.7 and 0.1~2, respectively. Partial oxidation characterizes the depositional environment, and the majority of the organic material comes from terrestrial sources rather than aquatic sources (Figure 7).

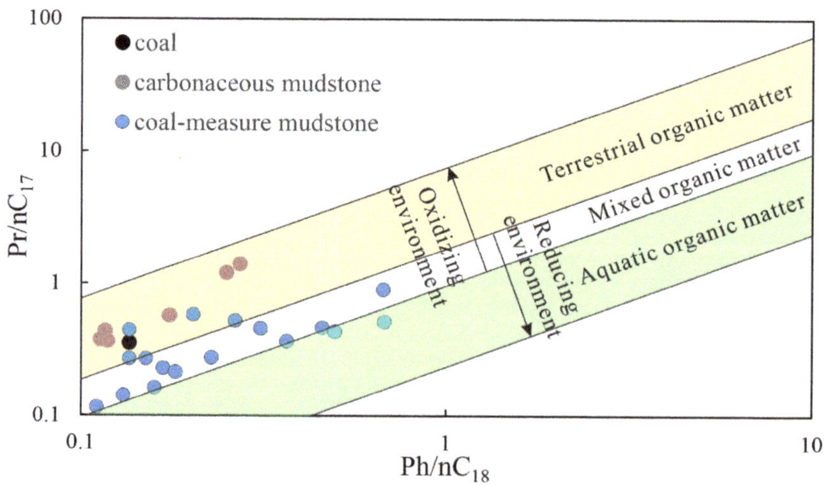

Figure 7. Cross plot of organic matter supply and TOC of the Jurassic coal-measure source rocks.

The primary metric that describes the level of redox reactions in depositional environments is the ratio of pristane to phytane (Pr/Ph). According to conventional wisdom, an environment is said to be reducing when the Pr/Ph ratio is less than 1, and oxidizing when the Pr/Ph ratio is greater [46]. The ratio of gammacerane to C30 hopane (Ga./C30H) is a commonly used parameter to evaluate water salinity [47]. Normally, Pr/Ph and Ga./C_{30}H show a negative correlation, and when the salinity of the water is high, the reducibility of the depositional environment is strong. The Pr/Ph of coal-measure source rocks in the Kuqa Depression is mostly greater than 1, with a distribution range of 0.8~4, indicating weak reducing and oxidizing environments. Ga./C_{30}H is usually less than 0.8, and the salinity of the water is relatively low. Pr/Ph and Ga./C_{30}H show a significant negative correlation, as shown in Figure 8a. While Ga./C_{30}H is positively correlated with TOC, the Pr/Ph of source rocks is typically adversely correlated with TOC. There is a weak positive association between Pr/Ph and TOC of coal-measure source rocks, which shows that the TOC of source rocks is high in an oxidizing environment (Figure 8b). There is no obvious positive correlation between Ga./C_{30}H and TOC, but a weak negative correlation exists, indicating that the TOC of source rocks in the freshwater environment is higher (Figure 8c). The clay mineral content and TOC in coal-measure source rocks have a modest negative connection rather than a positive association, indicating that the TOC in the source rocks is higher when the clay mineral content is lower (Figure 8d).

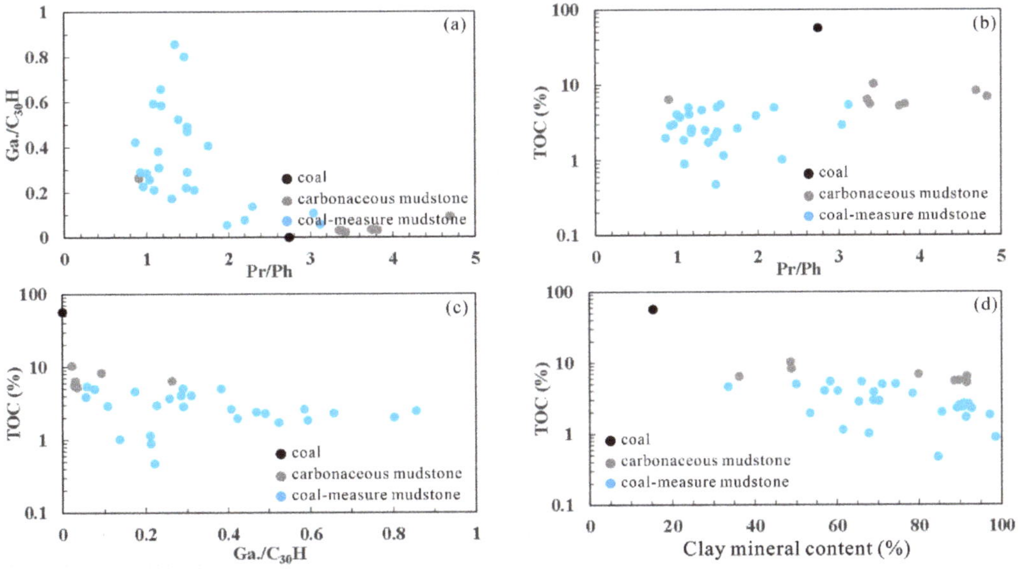

Figure 8. Organic matter preservation and TOC of the Jurassic coal-measure source rocks. (**a**) Cross plot of Pr/Ph and Ga./C_{30}H of coal-measure source rocks; (**b**) cross plot of Pr/Ph and TOC of coal-measure source rocks; (**c**) cross plot of Ga./C_{30}H and TOC in coal-measure source rocks; (**d**) cross plot of clay minerals content and TOC of coal-measure source rocks.

In conclusion, the coal-measure source rock's development is not strongly influenced by the organic matter preservation circumstances, as shown by the weak connections between the redox degree, water salinity, and clay mineral concentration of the depositional environment and TOC. This is in agreement with the sedimentary background of coal-measure source rocks with a significant input of terrigenous organic materials and partial oxidation of the depositional environment. At the two international seminars on "marine source rocks" (1983) and "lacustrine source rocks" (1985) organized by the Geological Society of London, there was debate about which is more important for the development of source rocks: organic matter supply or organic matter preservation. The conclusion is that the former is the more essential factor because, as long as the supply of organic matter is high enough, some organic matter will take too long to be degraded and enriched to form source rocks at the bottom of the oxygen-bearing water [48].

4.2.3. Organic Matter Dilution

The absolute amount of organic matter in the basin is influenced by the supply and preservation of organic matter, and whether the organic matter can be enriched to create source rocks depends on the amount of non-organic matter. The development of the source rock is hindered by the intake of a large amount of terrigenous detrital, which dilutes the organic matter and reduces its abundance. Due to the difficulty in quantitatively restoring the source quantity, the sedimentation rate is generally used to study the dilution effect of organic matter.

There are mainly two opposing views on the role of organic matter dilution in the development of source rocks. One theory holds that the dilution of organic matter is not obvious and that it is buried quickly at a high sedimentation rate, which reduces the time it takes for organic matter to degrade and is advantageous for its enrichment [49,50]. The enrichment of organic matter is at its best when the sedimentation rate is less than 1 cm/ka, according to another theory, which claims that the dilution of organic matter is obvious and only beneficial to the development of source rock when the dilution of organic matter is small [51,52]. In addition, some scholars believe that dilution only occurs when the sedimentation rate is high, which is not conducive to the development of source rocks. At the same time, it is also pointed out that the degradation of organic matter is serious when the sedimentation rate is too low, which is not conducive to the development of source rocks [53,54].

Due to the fact that compaction does not affect the relative relationship between sedimentation rate and organic carbon content, compaction correction is not required for the thickness of the formation in the study, and the sedimentation rate can be directly calculated based on the thickness and sedimentation time of the formation [50]. The average TOC value is calculated for a single well with more TOC data. The relationship between the average deposition rate of the source rock and the average TOC is shown in Figure 9. The sedimentation rates of coal and carbonaceous mudstone are generally low, with a corresponding sedimentation rate range of approximately 1~4 cm/ka. The range of mudstone sedimentation rates is wide, with sedimentation rates ranging from 0.5 cm/ka to 7.5 cm/ka. The trend of TOC changes in the coal, carbonaceous mudstone, and mudstone with the sedimentation rate is mostly consistent. Overall, TOC shows a trend of first increasing and then decreasing as the deposition rate increases. Because an increase in the deposition rate shortens the time it takes for organic matter to degrade, which is advantageous for the preservation of organic matter, TOC is positively linked with the deposition rate when it is less than 2 cm/ka. When the sedimentation rate is greater than 2 cm/ka, the excessive sedimentation rate leads to the obvious dilution of organic matter. The higher the sedimentation rate, the lower the organic matter abundance, and the TOC of the coal-measure source rock decreases with an increase in the sedimentation rate.

The sedimentary environment of the coal-measure source rock is often shallow water, and the association between the Jurassic coal-measure source rock's sedimentary rate and TOC in the Kuqa Depression is consistent with the characteristics of shallow water sedimentation. An increase in the sedimentation rate can greatly reduce the time it takes for organic matter to degrade in shallow water sedimentation, where the depositional environment is primarily one of weak oxidation, helping to preserve organic matter. A high rate of sedimentation results in a large dilution of organic matter and a decrease in the quantity of organic materials. As a result, it typically demonstrates a trend where TOC first rises and then falls as the deposition rate rises. The association between the source rock sedimentation rate and TOC in Erlian Basin's shallow lake basin is similar [55,56].

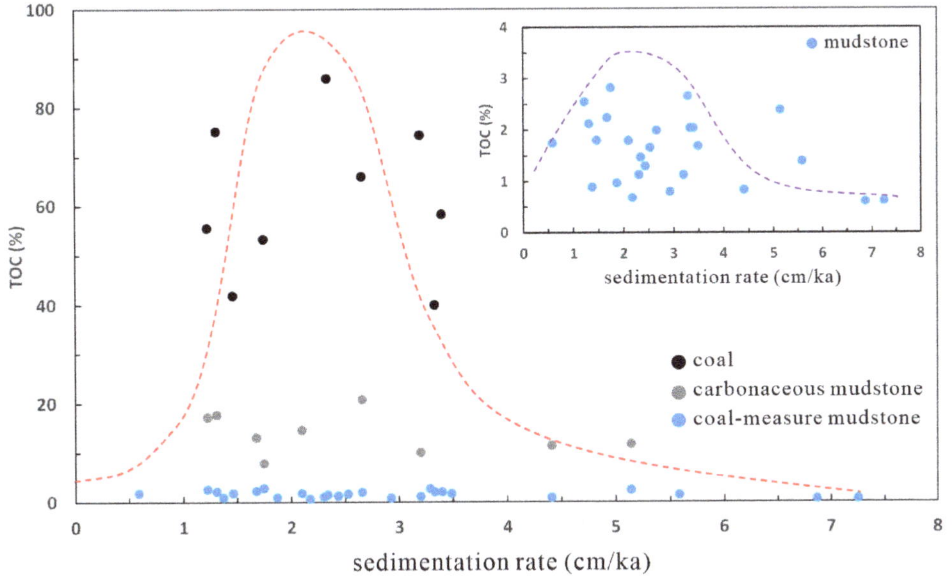

Figure 9. Relationship between the sedimentation rates and TOC of the Jurassic coal-measure source rocks.

4.3. Development Model of Coal-Measure Source Rock

The development model of the source rock has been the subject of extensive inquiry by forerunners [57], such as the marine model [58], marine upwelling current model and anticyclone current model [59], terrestrial marine source rock development model [60], green river shale model [61,62], deep lake hypoxia model [10,63,64], salt lake model [65], alkali lake model [66–68], and small faulted lake basin model [69,70]. There are many factors affecting the development of source rocks, including physical factors such as water depth, temperature, and humidity, chemical factors such as redox degree and salinity, and even disturbance of aquatic organisms. Although there are many factors influencing the development of source rocks, researchers have generally established a development model of the hydrocarbon source rock through the supply, preservation, and dilution of organic matter. It is believed that most development models belong to the "preservation model" or "high productivity model" [23,24,71]. As it has been discussed above, the development of coal-measure source rocks in the Kuqa Depression is mainly controlled by the input of terrestrial organic matter, and the sedimentation rate also has a certain effect but is less affected by the redox degree, salinity, and clay minerals.

The depositional environment of coal-measure source rocks in the Kuqa Depression is mainly delta and shallow lakes [21]. The input of terrestrial organic matter into the delta system is controlled by source and transport conditions. With an increase in transport distance, the input of terrestrial organic matter typically exhibits a characteristic of first increasing and then declining, and it typically reaches its highest value in the zone of a gentle slope [72]. Based on the analysis of major controlling factors, such as terrestrial organic matter input, sedimentation rate, and depositional environment, the development model of coal-measure source rocks in the Kuqa Depression is established, as shown in Figure 10.

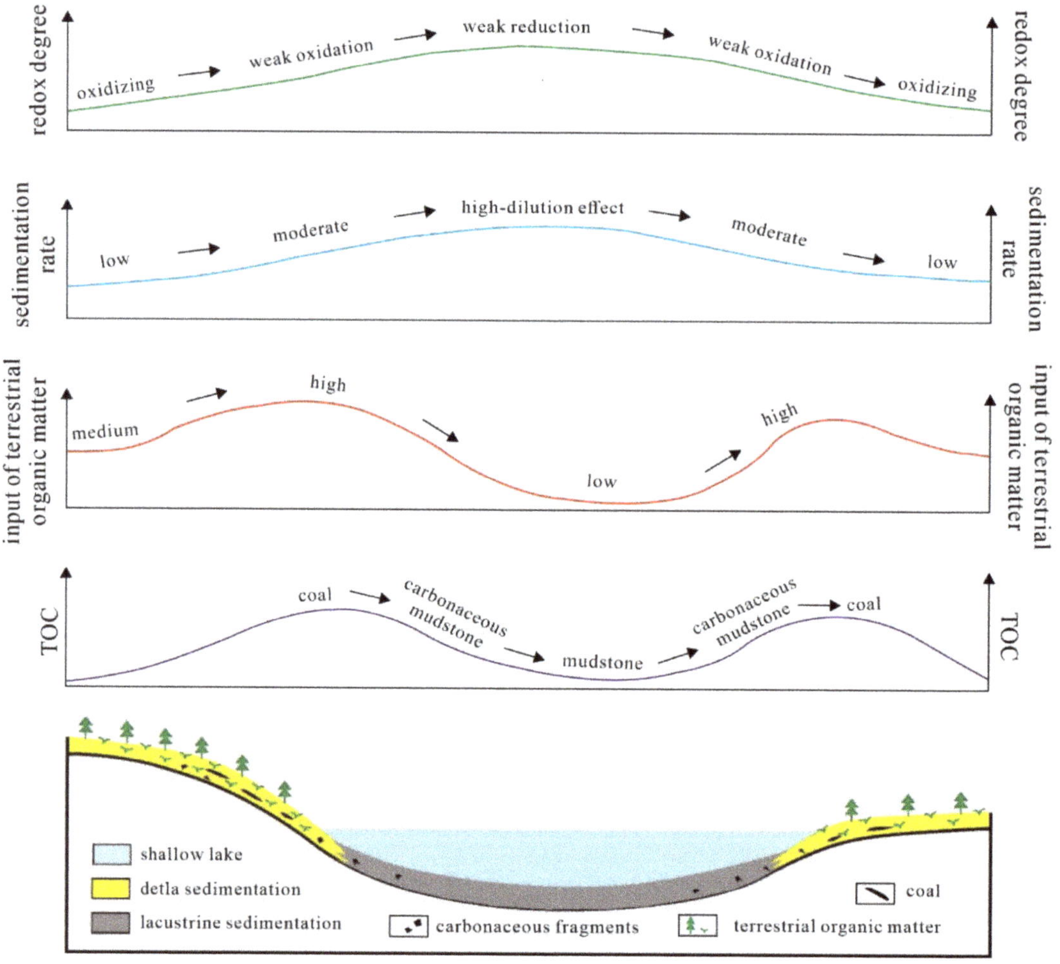

Figure 10. Development model of the Jurassic coal-measure source rocks in the Kuqa Depression.

The depositional environment of the delta is oxidizing, whereas that of the depositional environment of the shallow lake is weakly oxidizing. The central lake has the highest degree of reduction and is a weak reduction environment, with generally no strong reduction developing in the environment. The sedimentation rate of the coal-measure source rock gradually increases from the delta to the semi-deep to deep lake, and the dilution of organic matter gradually becomes obvious. The climate of the coal-measure source rock deposition period was mainly humid and hot, and the delta developed many higher plants, providing sufficient terrestrial organic matter. The intake of terrestrial organic matter first rises and then falls as sedimentation moves from the delta to the lacustrine zone, with the lake's center receiving the least amount of this material. Terrigenous organic matter input and sedimentation rate are the key determinants of how coal-measure source rock develops. The gentle slope zone has a large input of terrestrial organic matter, and a moderate sedimentation rate, and can develop coal, carbonaceous mudstone, and mudstone with the highest abundance of organic matter. The central lake basin has low terrestrial input, a high sedimentation rate, strong organic matter dilution, and mainly develops mudstone with a low organic matter abundance. The gentle slope has a higher richness of organic matter than the central lake basin, and the gentle slope is the primary development zone of coal-measure source rocks, as per the coal-measure source rock development model.

5. Conclusions

(1) The lithology of coal-measure source rocks in the Kuqa Depression is mainly mudstones, carbonaceous mudstones, and coal, which are medium- to good-quality source rocks, and the organic matter type is mainly II_2 and III.

(2) Combined with the depositional environment of isoprenoids and regular steranes, it shows that the depositional environment of coal-measure source rocks is mainly a shallow, freshwater continental environment with partial oxidation, and the source of organic matter is mainly terrestrial higher plants.

(3) The development of coal-measure source rocks is mainly controlled by the input of terrigenous organic matter and the sedimentation rate. Depositional environmental factors, such as redox degree, water salinity, and clay mineral content, have little influence on the development of coal-measure source rocks.

(4) Considering the main control factors of coal-measure source rocks, a development model of the coal-measure source rock is established. It is believed that the coal-measure source rock is more developed in the gentle slope zone than in the depression area. The development model and distribution characteristics of coal-measure source rocks are different from the traditional understanding of lacustrine source rocks.

Author Contributions: Writing—review and editing, X.D.; writing—original draft preparation, T.G.; investigation, X.Y. and C.C.; re-sources, Z.X.; data curation, K.L.; validation, X.Z. All authors have read and agreed to the published version of the manuscript.

Funding: This research was funded by the Major Science and Technology Project of PetroChina (Grant No. ZD2019-183-01-04).

Data Availability Statement: The data presented in this study are available on request from the corresponding author.

Conflicts of Interest: The authors declare no conflict of interest.

References

1. Liu, G.D. *Petroleum Geology*; Petroleum Industry Press: Beijing, China, 2009.
2. Qin, J.Z. *Hydrocarbon Source Rock in China*; Science Press: Beijing, China, 2005.
3. Hou, D.J.; Zhang, S.W.; Xiao, J.X. *Formation Mechanism of High Quality Source Rock in Faulted Continental Basin—A Case of Jiyang Depression*; Geology Press: Beijing, China, 2008; pp. 57–66.
4. Carroll, A.R.; Bohacs, K.M. Stratigraphic classification of ancient lakes: Balancing tectonic and climatic controls. *Geology* **1999**, *27*, 99–102. [CrossRef]
5. Carroll, A.R.; Bohacs, K.M. Lake type controls on petroleum source rock potential in nonmarine basins. *AAPG Bull.* **2001**, *85*, 1033–1053.
6. Pedersen, T.F.; Calvert, S.E. Anoxic vs. productivity: What controls the formation of organic-carbon-rich sediments and sedimentary rocks? *Am. Assoc. Pet. Geol. Bull.* **1990**, *74*, 454–466.
7. Katz, B.J. Lacustrine basin hydrocarbon exploration-current thoughts. *J. Paleolimnol.* **2001**, *26*, 161–179. [CrossRef]
8. Arthur, M.A.; Dean, W.E. Organic matter production and preservation and evolution of anoxia in the Holocene Black Sea. *Paleoceanography* **1998**, *13*, 395–411. [CrossRef]
9. Hedges, J.I.; Keil, R.G. Sedimentary organic matter preservation: An assessment and speculative synthesis. *Mar. Chem.* **1995**, *49*, 81–115. [CrossRef]
10. Demaison, G.J.; Moore, G.T. Anoxic environments and oil source bed genesis. *Org. Geochem.* **1980**, *2*, 9–31. [CrossRef]
11. Sang, S.X.; Chen, S.Y.; Liu, H.J. On diversity of late Paleozoic coal-forming environment and models in North China. *Chin. J. Geol.* **2001**, *36*, 212–221.
12. Lu, Y.B. On the Permian coal accumulating law in south China. *J. Huainan Min. Insitute* **1992**, *12*, 23–34.
13. Jiao, Y.Q.; Li, S.T.; Yang, S.G. Delta-lacustrine depositional system and coal accumulation research. *Earth Sci. J. China Univ. Geosci.* **1992**, *17*, 113–120.
14. Cao, Z.; Jiang, H.; Zeng, J.; Saibi, H.; Lu, T.; Xie, X.; Zhang, Y.; Zhou, G.; Wu, K.; Guo, J. Nanoscale liquid hydrocarbon adsorption on clay minerals: A molecular dynamics simulation of shale oils. *Chem. Eng. J.* **2021**, *420*, 127578. [CrossRef]

15. Liu, L.; Wang, L.J.; Tian, J.J.; Wu, T.W.; Ma, S. Sedimentary characteristics and cola-accumulation pattern of the Xishanyao Formation in the Toksun Coalfield of Xinjiang. *Acta Geosci. Sin.* **2011**, *32*, 549–558.
16. Li, Z.X.; Wei, J.C.; Lan, H.X.; Han, M.L.; Li, S.C. The high-resolution sequence division in Paleogene faulted basin of Huangxian. *Coal Geol. China* **2000**, *12*, 9–12.
17. Martirosyan, A.V.; Ilyushin, Y.V. Modeling of the Natural Objects' Temperature Field Distribution Using a Supercomputer. *Informatics* **2022**, *9*, 62. [CrossRef]
18. Martirosyan, A.V.; Martirosyan, K.V.; Mir-Amal, A.M.; Chernyshev, A.B. Assessment of a Hydrogeological Object's Distributed Control System Stability. In Proceedings of the 2022 Conference of Russian Young Researchers in Electrical and Electronic Engineering (ElConRus), Saint Petersburg, Russia, 25–28 January 2022; pp. 768–771. [CrossRef]
19. Asadulagi, M.M.; Vasilkov, O.S. The Use of Distributed and Lumped Type Controllers for the Hydro-Lithospheric Process Control System of the Kislovodskoye Field. In Proceedings of the 2019 III International Conference on Control in Technical Systems (CTS), Saint Petersburg, Russia, 30 October–1 November 2019; pp. 7–10. [CrossRef]
20. Tang, Y.G.; Yang, X.Z.; Xie, H.W.; Xu, Z.P.; Wei, H.X.; Xie, Y.N. Tight gas reservoir characteristics and exploration potential of Jurassic Ahe Formation in Kuqa Depression, Tarim Basin. *China Pet. Explor.* **2021**, *26*, 113–124.
21. Wang, K.; Yang, H.J.; Li, Y.; Zhang, R.H.; Ma, Y.J.; Wang, B.; Yu, C.; Yang, Z.; Tang, Y. Geological characteristics and exploration potential of the northern tectonic belt of Kuqa depression in Tarim Basin. *Acta Pet. Sin.* **2021**, *42*, 885–905.
22. Chen, J.P.; Zhao, C.Y.; He, Z.H. Criteria for evaluating the hydrocarbon generating potential of organic matter in coal measures. *Pet. Explor. Dev.* **1997**, *24*, 1–5.
23. Calvert, T.F. Anoxia vs. productivity: What controls the formation of organic carbon rich sediments and sedimentary rocks? *AAPG Bull.* **1990**, *4*, 454–466.
24. Parrish, J.T. Paleogeography of Corg-rich rocks and the preservation versus production controversy. In *Paleogeography, Paleoclimate, and Source Rock*; Huc, A.Y., Ed.; Studies in Geology no. 40; American Association of Petroleum Geologists: Tulsa, OK, USA, 1995.
25. Stein, R. Surface-water paleo-productivity as inferred from sediment deposited in oxic and anoxic deep-water environments of the Mesozoic Atlantic Ocean. *Mitt. Geol. Paläont. Inst. Univ. Hambg.* **1986**, *60*, 55–70.
26. Leeuw, J. Sedimentary organic matter: Organic facies and palynofacies. *Org. Geochem.* **1995**, *23*, 995–996.
27. Katz, B.J. *Controlling Factors on Source Rock Development—A Review of Productivity, Preservation, and Sedimentation Rate*; SEPM Special Publication: Broken Arrow, OK, USA, 2005; Volume 82, pp. 7–16. [CrossRef]
28. Calvert, S.E. Lack of evidence for enhanced preservation of sedimentary organic matter in the oxygen minimum for the gulf California. *Geology* **1992**, *20*, 757–760. [CrossRef]
29. Katz, B.J. Controls on distribution of lacustrine source rocks through time. *AAPG Mem.* **1990**, *50*, 132–139.
30. Arfaoui, A.; Montacer, M.; Kamoun, F.; Rigane, A. Comparative study between Rock-Eval pyrolysis and biomarkers parameters: A case study of Ypresian source rocks in central-northern Tunisia. *Mar. Pet. Geol.* **2007**, *24*, 566–578. [CrossRef]
31. Huang, W.Y.; Meinschein, W.G. Sterols as ecological indicators. *Geochim. Cosmochim. Acta* **1979**, *43*, 739–745. [CrossRef]
32. Moldowan, J.M.; Seifert, W.K.; Gallegos, E.J. Identification of an extended series of tricyclic terpanes in petroleum. *Geochim. Cosmochim. Acta* **1983**, *47*, 1531–1534. [CrossRef]
33. Peters, K.E.; Walters, C.C.; Moldowan, J.M. *The Biomarker Guide: Biomarkers and Isotopes in Petroleum Systems and Earth History*; Cambridge University Press: Cambridge, UK, 2005.
34. Zumberge, J.E. Prediction of source rock characteristics based on terpane biomarkers in crude oils: A multivariate statistical approach. *Geochim. Cosmochim. Acta* **1987**, *51*, 1625–1637. [CrossRef]
35. Tao, S.; Wang, C.; Du, J.; Liu, L.; Chen, Z. Geochemical application of tricyclic and tetracyclic terpanes biomarkers in crude oils of NW China. *Mar. Pet. Geol.* **2015**, *67*, 460–467. [CrossRef]
36. Chen, Z.L.; Liu, G.D.; Wei, Y.Z.; Gao, G.; Ren, J.L.; Yang, F.; Ma, W.Y. Distribution pattern of tricyclic terpanes and its influencing factors in thePermian source rocks from Mahu Depression in the Junggar Basin. *Oil Gas Geol.* **2017**, *38*, 311–322.
37. Jin, Q.; Zha, M.; Zhao, L. Identification of Effective Source Rocks in the Tertiary Evaporate Facies in the Western Qaidam Basin. *Acta Sedimentol. Sin.* **2001**, *19*, 125–129.
38. Lu, L.F.; Cai, J.G.; Liu, W.H.; Teng, G.E.; Wang, J. Occurrence and thermostability of absorbed organic matter on clay minerals in mudstones and muddy sediments. *Oil Gas Geol.* **2013**, *34*, 16–26.
39. Storm, K.M. Land-locked waters and the deposition of black muds, in Trask, Recent marine sediments. *AAPG Bull.* **1969**, *45*, 356–370.
40. Adriano, M.; Michael, K.I.; Anders, N. Complex plumbing systems in the near subsurface: Geometries of authigenic carbonates from Dolgovskoy Mound (Black Sea) constrained by analogue experiments. *Mar. Pet. Geol.* **2008**, *25*, 457–472.
41. Jiao, F.Z.; Zou, C.N.; Yang, Z. Geological theory and exploration & development practice of hydrocarbon accumulation inside continental source kitchens. *Pet. Explor. Dev.* **2020**, *47*, 5–16.
42. Zhu, G.Y.; Jin, Q.; Zhang, S.W.; Zhang, L.Y.; Guo, C.C. Salt lake-saline lake sedimentary combination and petroleum accumulation in the Bonan Sag. *Acta Mineral. Sin.* **2004**, *24*, 25–30.

43. Wu, X.L.; Liu, H.L.; Li, R.X.; Li, D.L.; Zhao, B.S.; Cheng, J.H.; Wei, J.L.; Zhu, Q.P. Progress in researches of development rule and the hydrocarbon generation and explusion characteristics of hydrocarbon source rocks in Terrestrial Evaporite Basins of China. *Geol. Sci. Technol. Inf.* **2017**, *36*, 183–192.
44. Cai, J.G.; Bao, Y.J.; Yang, S.Y.; Wang, X.X.; Fan, D.D.; Xu, J.L.; Wang, A. Occurrence form and enrichment mechanism of organic matter in argillaceous sediments and mudstones. *China Sci. (Earth Sci.)* **2007**, *50*, 92–101.
45. Peters, K.E.; Fraser, T.H.; Amris, W.; Rustanto, B. Geochemistry of Crude Oils from Eastern Indonesia. *AAPG Bull.* **1999**, *83*, 1927–1942.
46. Didyk, B.M.; Simoneit, B.R.T.; Brassell, S.C. Organic geochemical indicators of palaeoenvironmental conditions of sedimentation. *Nature* **1978**, *272*, 216–222. [CrossRef]
47. Sinninghe, J.S.; Kenig, F.; Koopmans, M.P.; Koster, J.; Schouten, S.; Hayes, J.M.; de Leeuw, J.W. Evidence for gammacerane as an indicator of water column stratification. Geochim. *Cosmochim. Acta* **1995**, *59*, 1895–1900. [CrossRef]
48. Zhu, W.L. *Paleolimnology of Offshore Oil-Bearing Basins in China*; Tongji University: Shanghai, China, 2002.
49. Stow, D.A.V.; Huc, A.Y.; Bertrand, P. Depositional processes of black shales in deep water. *Mar. Pet. Geol.* **2001**, *18*, 491–498. [CrossRef]
50. Ibach, L.E. Relationship between sedimentation rate and total organic carbon content in ancient marine sediments. *AAPG Bull.* **1983**, *66*, 170–188.
51. Loutit, T.S. *Condensed Section: The Key to Age Determination and Correlation of Continental Margin Sequences, Sea-Level Changes-an Integrated Approach*; SEPM Special Publication: Broken Arrow, OK, USA, 1988; pp. 183–213.
52. Wignall, P.B. Model for transgressive black shales. *Geology* **1991**, *19*, 167–170. [CrossRef]
53. Zhang, S.C.; Zhang, B.M.; Bian, L.Z.; Jin, Z.J.; Wang, D.R.; Zhang, X.Y.; Gao, Z.; Chen, J. Development constraints of marine source rocks in China. *Earth Sci. Front.* **2005**, *12*, 41–50.
54. Tyson, R.V. Sedimentation rate, dilution, preservation and total organic carbon: Some results of a modelling study. *Org. Geochem.* **2001**, *32*, 333–339. [CrossRef]
55. Ding, X.J.; Liu, G.D.; Zha, M.; Huang, Z.L.; Gao, C.H.; Qu, J.X.; Lu, X.J.; Chen, Z.L.; Guo, J.G. Relationship between sedimentation rate and organic matter abundance of source rocks: A case study of Erlian Basin. *Nat. Gas Geosci.* **2015**, *26*, 1076–1085.
56. Ding, X.J.; Liu, G.D.; Zha, M.; Huang, Z.L.; Gao, C.H.; Lu, X.J.; Sun, M.; Chen, Z.; Liuzhuang, X. Relationship between total organic carbon content and sedimentation rate in ancient lacustrine sediments, a case study of Erlian basin, northern China. *J. Geochem. Explor.* **2015**, *149*, 22–29. [CrossRef]
57. Hao, F.; Zhou, X.; Zhu, Y.; Yang, Y. Lacustrine source rock deposition in response to coevolution of environments and organisms controlled by tectonic subsidence and climate, Bohai Bay basin, China. *Org. Geochem.* **2011**, *42*, 323–339. [CrossRef]
58. Gao, P.; Li, S.; Lash, G.G.; Yan, D.; Zhou, Q.; Xiao, X. Stratigraphic framework, redox history, and organic matter accumulation of an Early Cambrian intraplatfrom basin on the Yangtze Platform, South China. *Mar. Pet. Geol.* **2021**, *130*, 105095. [CrossRef]
59. Gao, P.; He, Z.; Li, S.; Lash, G.G.; Li, B.; Huang, B.; Yan, D. Volcanic and hydrothermal activities recorded in phosphate nodules from the Lower Cambrian Niutitang Formation black shales in South China. *Palaeogeogr. Palaeoclimatol. Palaeoecol.* **2018**, *505*, 381–397. [CrossRef]
60. Liu, S.Y.; Chen, H.Y.; Li, D.Y.; Sun, W.Y.; Li, C.Y. Sedimentary characteristics and source rock development model of the Oligocene Lingshui Formation in Lingshui Sag, Qiongdongnan Basin. *Mar. Orig. Pet. Geol.* **2019**, *24*, 65–72.
61. Bradley, W.H. Green River oil shale-concept of origin extended. *GSA Bull.* **1970**, *81*, 985–1000. [CrossRef]
62. Eugster, H.P.; Surdam, R.C. Depositional environment of the Green River formation: A preliminary report. *GSA Bull.* **1973**, *86*, 319–334. [CrossRef]
63. Fan, P.; Luo, B.J.; Huang, R.C.; Shen, P.; Hui, R.Y.; Shao, H.S.; Wang, Y.X.; Rong, G. Formation and migration of continental oil and gas in China. *Scienitia Sin.* **1980**, *23*, 1286–1295.
64. Ryder, R.T. Lacustrine sedimentation and hydrocarbon occurrences with emphasis on Uinta Basin models. In *AAPG, Fall Education Conference*; AAPG: Houston, TX, USA, 1980; p. 103.
65. Kirkland, D.W.; Evans, R. Source-rock potential of evaporitic environment. *AAPG Bull.* **1981**, *65*, 181–190.
66. Kelts, K. Environments of deposition of lacustrine petroleum source rocks: An introduction. In *Lacustrine Petroleum Source Rocks*; Fleet, A.J., Kelts, K., Talbot, M.R., Eds.; Geological Society Special Publication: London, UK, 1988; Volume 40, pp. 3–26.
67. Cao, J.; Lei, D.W.; Li, Y.W.; Tang, Y.; Imin, A.; Chang, Q.S.; Wang, T. Ancient high-quality alkaline lacustrine source rocks discovered in the Lower Permian Fengcheng Formation, Junggar Basin. *Acta Pet. Sin.* **2015**, *36*, 781–790.
68. Cao, J.; Xia, L.W.; Wang, T.T.; Zhi, D.M.; Tang, Y.; Li, W.W. An alkaline lake in the Late Paleozoic Ice Age (LPIA): A review and new insights into paleoenvironment and petroleum geology. *Earth-Sci. Rev.* **2020**, *202*, 103091. [CrossRef]
69. Ding, X.J.; Liu, G.D.; Zhao, L.M.; Gao, D.K.; Zhang, K.; Kuang, D.Q. Organic Matter Enrichment and Hydrocarbon Source Rock Forming Mechanism in Small-Scale Faulted Lacustrine Basins: A Case from the First Member of Lower Cretaceous Tenger Formation in Erlian Basin. *Xinjiang Pet. Geol.* **2017**, *38*, 650–657.
70. Ding, X.J.; Liu, G.D.; Zha, M.; Huang, Z.L.; Gao, C.H.; Wang, P.G.; Qu, J.; Lu, X.; Chen, Z. Characteristics and origin of lacustrine source rocks in the Lower Cretaceous, Erlian Basin, northern China. *Mar. Pet. Geol.* **2015**, *66*, 939–955. [CrossRef]

71. Barry, J.K. Controlling factors on source rock development-a review of productivity, preservation, and sedimentation rate. In *The Deposition of Organic Carbon Rich Sediments: Models, Mechanisms, and Consequences. Society of Sedimentary Geology*; Harris, N.B., Ed.; SEPM Special Publication: Broken Arrow, OK, USA, 2005; pp. 7–16.
72. Qu, T.; Gao, G.; Xu, X.D.; Liu, F.Y. Control factors of terrestrial organic matter gistribution in delta-shallow sea sedimentary system. *Acta Sedimentol. Sin.* **2020**, *38*, 648–660.

processes

Article

Gas Distribution and Its Geological Factors in the No.5 Coal Seam of the Weibei Field, Southeastern Ordos Basin, North China

Yongkai Qiu [1,2,*], Hui Wang [1,2,*], Guangshan Guo [3], Rujun Mo [4], Zhejun Pan [5], Yidong Cai [1,2] and Abulaitijiang Abuduerxiti [6]

1 School of Energy Resource, China University of Geosciences, Beijing 100083, China; yidong.cai@cugb.edu.cn
2 Coal Reservoir Laboratory of National Engineering Research Center of CBM Development & Utilization, China University of Geosciences, Beijing 100083, China
3 Unconventional Research Center, CNOOC Research Institute, Beijing 100028, China; guogsh2@cnooc.com.cn
4 Shaanxi Yan'an Oil & Gas Co., Ltd., Xi'an 710000, China; mrj_cq@petrochina.com.cn
5 Key Laboratory of Continental Shale Hydrocarbon Accumulation and Efficient Development, Ministry of Education, Northeast Petroleum University, Daqing 163318, China; zhejun.pan@nepu.edu.cn
6 Fengcheng Oilfield Operation District, Xinjiang Oilfield Company, CNPC, Karamay 834000, China; fcabultj@petrochina.com.cn
* Correspondence: qyk@cugb.edu.cn (Y.Q.); 3006200034@cugb.edu.cn (H.W.)

Abstract: The distribution of gas contents in the No.5 coal seams of the Weibei Field of southeastern Ordos Basin, North China, is highly variable and its mechanism remains unclear. In this study, systematic evaluation of the gas content and its geological control factors are conducted based on the field investigation together with 16 coalbed methane (CBM) wells in the Weibei Field, southeastern Ordos Basin, North China. The gas content variability is determined from the perspectives of gas generation, migration, and preservation. The results indicate that the gas generation is largely relevant with the subsidence and fluctuation of the coal seam during the middle-late Yanshanian orogeny, which controls the gas content variability of the Weibei Field. Besides, gas migration and preservation determine the gas content on a regional scale, significantly related to fault types, roof lithology, burial depth, and hydrodynamic conditions, but scarcely affected by the roof thickness. In the Weibei Field, the geological models controlling gas content are identified as (1) hydrodynamic trapping of gas in the deep burial depth and thrust faults, and (2) gas loss by groundwater flushing and normal faults. Basically, the first mechanism causes the high gas content of the east zone, whereas the other one is responsible for the low gas content in the west zone of the study area.

Keywords: coalbed methane; Weibei Field; gas content; geological control factors; Ordos Basin

Citation: Qiu, Y.; Wang, H.; Guo, G.; Mo, R.; Pan, Z.; Cai, Y.; Abuduerxiti, A. Gas Distribution and Its Geological Factors in the No.5 Coal Seam of the Weibei Field, Southeastern Ordos Basin, North China. *Processes* **2022**, *10*, 659. https://doi.org/10.3390/pr10040659

Academic Editor: Aneta Magdziarz

Received: 30 December 2021
Accepted: 24 March 2022
Published: 28 March 2022

1. Introduction

The exploitation of coalbed methane (CBM) not only benefits mining safety but also has great significance for reducing carbon-dioxide emissions [1]. With the increased demand and gas price in recent years, the increasing interest is exhibited in CBM resources, which requires the accurate estimation of the recoverable reserves of CBM resources [2,3]. Basically, CBM resources are abundant in the Ordos Basin, located in one of the most important fossil-fuel energy areas in China [4–11]. The CBM resources are about 9×10^{12} m^3 in the eastern Ordos Basin, while 10.72×10^{12} m^3 in the entire Ordos Basin, which accounts for ~1/3 to 1/4 of the known total CBM resources in China [4].

The Weibei Field, covering an area of 1530 km^2, lies on the southeastern edge of the Ordos Basin, where thousands of CBM production wells have been drilled, and a CBM production field had been established [4]. Therein, the gas production of each CBM well differs because of the different characteristics of reservoirs. Although the key factors controlling CBM production are various in different geological structures, the gas content is widely accepted to be one of the crucial indicators for CBM reservoir productivity [3,12–14]. However, the difficulty remains in the accurate evaluation on the gas content co-determined by multiple geological factors like hydrodynamic factors and coal heterogeneities [3,15,16]. By discussing the geological and hydrodynamic control factors of gas content in the southeastern Ordos Basin (SOB), Yao et al. [4]. found that there is an intimate relationship between hydrodynamics and gas content in the Basin. Furthermore, an investigation of the gas content of the No.2 coal seam in the Yanchuannan Field of the southern SOB revealed that the high gas content is the integrated result of gas generation, migration, and preservation [3]. The previous achievement also indicated that the geological structure types, burial depth, and the groundwater level in the coal seam are the main factors controlling gas content [3]. Yan et al. [17] conducted a simulation aiming to improve knowledge about the deposition-burial, geothermal, and organic maturation evolution and their influence on gas content in the Weibei field. However, the distribution and geological controls of the gas content in the No.5 coal seam are not yet fully understood for the Weibei Field. Therefore, more investigations are needed.

This study conducts a comprehensive investigation on the gas content in the Weibei Field of the SOB. Based on diverse experimental and statistic data, the distribution of gas content and its geological control factors are discussed, including tectonic and depositional evolution, roof/floor lithology and thickness, burial depth of coal seam, and hydrodynamic conditions. Moreover, two typical geological models controlling gas content are established. This work should be helpful in enhancing the knowledge on the geological control mechanism related to the gas content in the No.5 coal seam of Weibei Field and in guiding the potential CBM engineering operations in this area.

2. Geological Setting

2.1. Tectonics of the Basin

The Ordos Basin, with an area of 2.4×10^5 km^2, lies in the western part of North China (Figure 1A) [18]. This is a large asymmetric Basin including a broad, gently dipping eastern limb, and a narrow, steeply dipping western limb, forming an axis in the Tianhuan Sag [19]. The Ordos Basin is surrounded by the Yin Mountain in the north, Lvliang Mountain in the east, Qinling Mountain in the south, and the Liupan and Helan Mountains in the west (Figure 1B). There is six tectonic domains in Ordos Basin, i.e., the Yimeng and Weibei uplift, the Jinxi fold belt, the Yishan slope, the Tianhuan sag, and the western edge thrust belt (Figure 1B). The studied Weibei Field is situated in the northeastern part of the Weibei uplift (Figure 1B), which contains the Hancheng area in the northeast (east zone) and the Heyang area in the southwest (west zone) (Figure 1C). In the study area, there are three dominant thrust faults, including Xuefeng-Bei (F1), Xuefeng-Nan (F2), and Qingao faults (F3) (Figure 1C), formed during the Late Cretaceous Yanshanian Orogency [20].

2.2. Coal-Bearing Strata in the Weibei Field

The Weibei Field preserved the strata of Cambrian, Ordovician, Carboniferous (Pennsylvanian), Permian, Triassic, and Quaternary-aged sequences [17]. In this field, there are two main coal-bearing strata, including the Carboniferous (Pennsylvanian) Taiyuan Formation and Permian Shanxi Formation, where the No.5 coal seam, one of the most important gas-bearing beds, belongs to the Pennsylvanian Taiyuan Formation (Figure 2). Previous studies have reported that the No.5 coal seam dips gently towards the northwest with an angle of 5–15° [21] and has a thickness of 1.2–8.4 m (avg. 5.5 m) and burial depth of 460–1200 m with a mean of 950 m [4,15]. In addition, the maximum vitrinite reflectance is in the range of 1.6–2.5% (avg. 2.0%), and gas content ranges from 2.69–16.15 m^3/t [4].

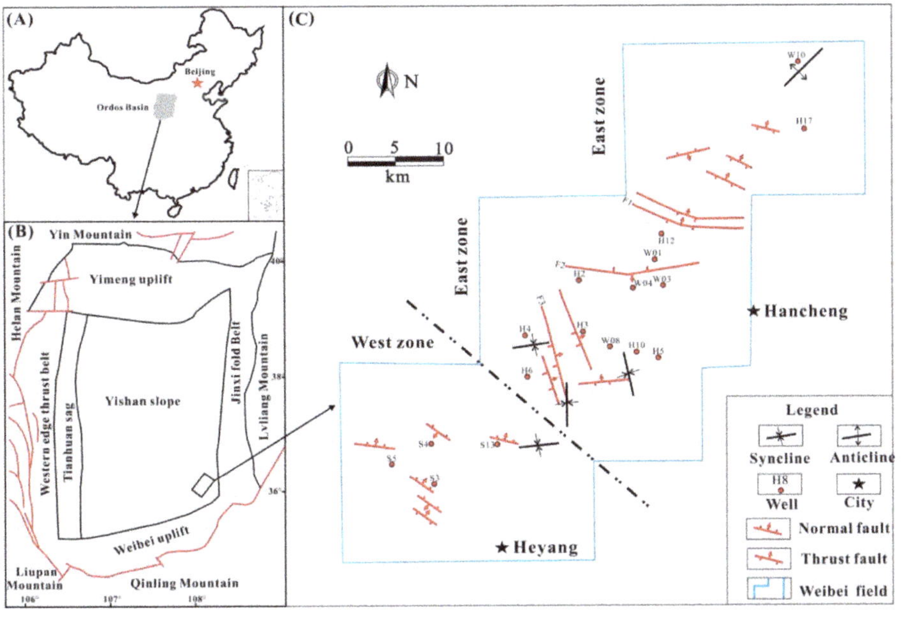

Figure 1. Location of the Ordos Basin in North China (**A**); Location of the Weibei Field (**B**); Geological structures and locations of the exploration wells in the Weibei Field (**C**).

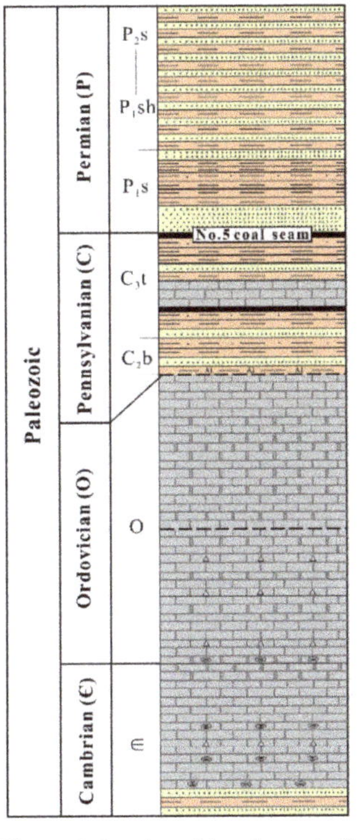

Figure 2. Stratigraphic column in the Weibei Field (Yan et al., 2015 [17]).

3. Methodology

In this study, a total of twenty-two coal samples collected from working coal seams in 18 coal mines were used. The porosity and permeability were measured by the routine core methods following the Chinese Oil and Gas Industry Standard SY/T 5336-1996. Besides, the maceral and vitrinite reflectance were measured by the polished slabs by LABORLUX 12 POL Fluorescence Microscope and Microscope Photometer (MPV-3) following the GB/T 6948-1998 and ICCP system 1994 methods. Isothermal adsorption was carried out on the collected samples following the standard GB/T 19560-2008. Furthermore, the gas content, coal thickness, burial depth, and roof/floor lithology of No.5 coal seam were collected from field test data of 16 exploration wells (Figure 1C). Herein, the measurements of gas content followed the Chinese National Standard GB/T 19559-2004 (2004), and the tectonic characteristics of the No.5 coal seam were interpreted by conventional two-dimensional seismic reflection data. Details regarding the analytical methods are exhibited in our previous study [4]. The burial history, thermal evolution history, and hydrocarbon generation processes were simulated with data of typical wells by BasinMod modeling software. In addition, the microfractures, pore characteristics, permeability, and pressure of the No.5 coal seam are referenced from previous literature [4,21].

4. Results and Discussion

4.1. Gas Content in No.5 Coal Seam

4.1.1. Tectonic and Depositional Evolution Related Gas Generation

The generation of CBM occurred during the coalification process, controlled by various factors, including burial history, tectonic evolution, and paleo-geothermal heating regarding the coal seam [2,22,23]. As for the Weibei Field, it experienced the following four evolution stages: accelerated subsidence (I), fluctuant (II), uplift (III), and terminate (IV) stages (Figure 3).

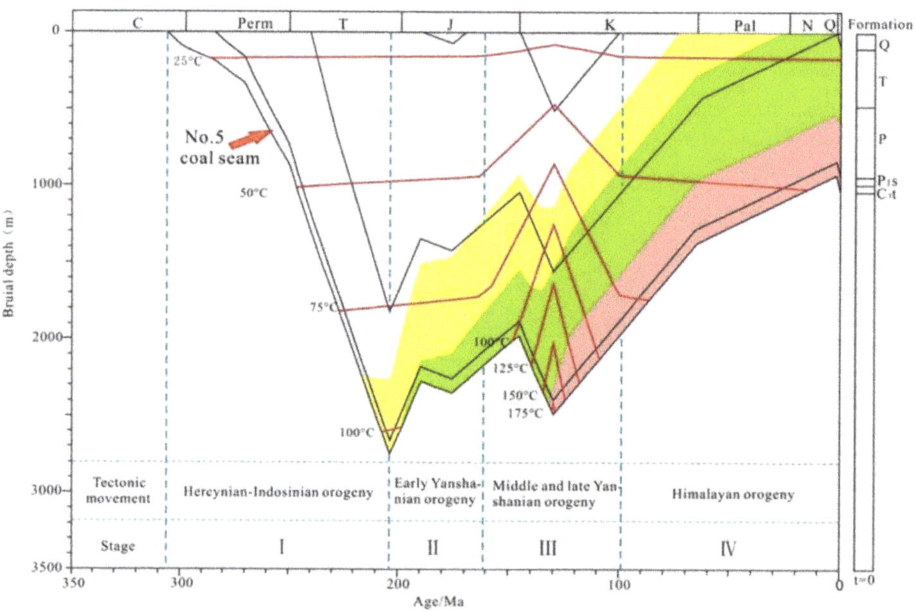

Figure 3. Schematic diagram of tectonic and depositional evolution in the Weibei Field.

The stage I relates to Hercynian-Indosinian orogeny with a gradually increased subsidence speed (Figure 3). In the Weibei Field, the subsidence of coal-bearing strata is approximately 1.5–9 m/Ma and 3–11 m/Ma during the Permian and Triassic, respectively. At this stage, the No.5 coal seam is buried to a depth of ~900 m by the end of Permian and ~2700 m in the Late Triassic (Figure 3). Generally, the temperature of coal reservoir gradually increases with increasing burial depth [2]. In addition, the geothermal gradient in the Ordos Basin was normally ~2.8–4.0 °C/100 m, enabling the maximum temperature of No.5 coal seam to be ~25–36 °C by the end of Permian and to be ~76–108 °C in the late Triassic. Therein, the shallow burial depth of the Permian coal-bearing strata limits the maturation of organic matter, and thus the generated CBM, at this stage, mainly belonged to biogenic gas, which was not well-preserved because of the shallow burial depth [8]. Afterward, the maximum vitrinite reflectance ($R_{o,max}$) of the No.5 coal seam was <0.75% by the end of the Triassic. Due to the relatively strong orogeny during the late Triassic, the CBM migrated at the first adjustment: most of the generated CBM was sealed in place because of the Ordos Basin compressed by the NS trending stress [3].

Stage II relates to the early Yanshanian orogeny with fluctuations in the burial history (Figure 3), where the Weibei Field was compressed by an NS strike stress. In this stage, the uplift and subsidence commonly occurred in the field and resulted in the formation of the EW and SE-NW striking folds and faults [4]. Meanwhile, the temperature of the No.5 coal seam was about 130 °C, and accordingly, the $R_{o,max}$ was up to about 1.0% by the end of Jurassic. Although the coalification was kept more or less stable, rare generated CBM could be preserved in this stage because of thermal fluctuations [3].

Stage III relates to the Middle and Late Yanshanian orogeny. In this stage, the burial depth of the No.5 coal seam reduced rapidly (Figure 3). A magma-heat event resulted in the formation of an abnormal geothermal field as high as 5.5 °C/100 m and some parts were up to 8 °C/100 m. Coalification was significantly affected by the relatively high heat flow from the magmatic intrusions [2]. By the end of the Cretaceous, the $R_{o,max}$ of the No.5 coal seam was about 2.2%, resulting in more gas being generated in this stage.

Stage IV relates to the Himalayan orogeny with the continuously decreasing burial depth of the No.5 coal seam (Figure 3). The No.5 coal seam was uplifted and eroded in the Weibei Field. In addition, the geothermal field reverted to a lower level, and thus the temperature of the No.5 coal seam was continuously decreasing. By the end of the Himalayan orogeny, the gas expulsion of organic matter eventually stopped.

4.1.2. Gas Migration and Preservation

Geological Structures and Related Gas Distribution

The Weibei Field was subdivided into the east and west zones through tectonic complexity, as shown in Figure 1, where mainly thrust faults but few folds existed in the east zone and the faults planes commonly dip south- and eastward. Compared with the east zone, the west zone is extended more intensely with few folds and normal fault development. These faulting and folding activities induce coal deformation, which in turn affects the gas content of the No.5 coal seam. As summarized in Table 1, the gas content of the No.5 coal seam is in the range of 2.69 to 16.15 m^3 /t (avg. 9.95 m^3/t) in the Weibei Field. The gas content is higher (avg. 11.42 m^3/t) in the east zone but lower (avg. 5.56 m^3/t) in the west zone (Figure 4). The compressional deformation is intensified in the east zone and therein the thrust faults are dominated. The compressed stress probably led to the macro- and micro-fractures being close in the coal [21] as well as its roof and floor, creating the seal conditions for gas preservation. On the contrary, extensional deformation is intensified in the west zone, where normal faults and folds are the primary types. The extensional stress caused fractures to develop in the coal as well as its roof and floor, provide open conditions for gas migration. As a result, the CBM content in the east zone is higher than in the west zone in the study area.

Table 1. Coal thickness, burial depth, reservoir pressure, and Langmuir volumes (V_L) data of 16 coalbed methane (CBM) wells in the Weibei Field.

Zone	Well	Burial Depth (m)	Thickness (m)	Reservoir Pressure (MPa)	R_o (%)	Gas Content (m^3/t)	Langmuir Volumes (m^3/t)
East	H3	1013.6	5.6	9.35	1.98	8.99	22.91
	H4	1105.0	4.6	8.51	2.15	12.71	26.12
	H5	587.0	3.2	4.52	2.32	8.91	23.72
	H6	1066.4	5.4	8.22	2.10	11.74	27.77
	H10	715.7	5.0	5.51	2.23	8.59	19.42
	H12	634.1	3.5	4.12	2.32	8.12	23.08
	H17	466.1	2.5	3.72	2.05	11.58	20.63
	W01	638.6	3.1	4.39	2.09	7.42	31.5
	W03	684.3	3.5	6.05	2.15	16.01	28.04
	W04	917.7	4.4	7.06	2.07	16.15	25.83
	W08	883.8	9.0	6.52	2.21	12.38	20.36
	W10	531.6	6.3	4.63	2.23	14.46	24.12
West	S3	607.8	3.8	4.70	2.05	3.22	13.58
	S4	752.2	4.1	6.79	1.73	10.59	22.44
	S5	625.3	2.9	7.43	1.68	2.69	12.43
	S13	854.5	5.3	8.14	1.76	5.72	16.31

Lithology and Thickness Distribution of Roof and Floor

The lithology and thickness of the roof and floor are listed in Table 2, where the floor of the No.5 coal seam is mainly mudstone, while the lithology of the roof varies throughout the entire Weibei Field. In the east zone, the roof of the No.5 coal seam is mainly characterized as mudstone and carbonaceous mudstone, as well as sandstone in some local areas (Table 2 and Figure 5). While in the west zone, the roof is primarily mudstone and argillaceous sandstone, accompanied with silt in some separated areas (Table 2 and Figure 5). Previous studies had reported that the mudstone commonly has a good potential of sealing, which is conducive to CBM storage [3,24], indicating that the floor of the No.5 coal seam has good sealing potential, but variable sealing potential in the roof in the Weibei Field. As a whole, the sealing potential of the roof for generated gas in the east zone is better than that in the west zone.

In addition to lithology, the thickness of the roof and floor directly influences the CBM preservation, where the more thick, the better sealing performance the roof and floor will have [3,24]. In the Weibei Field, it was a shallow marine environment during the Late Pennsylvanian [25], making the thickness of the roof and floor regarding the No.5 coal seam to be stable and within the range of 2.3 to 3.6 m (Table 2), where the mudstone thickness of the floor regarding the No.5 coal seam ranges from 2.3 to 3.5 m (Table 2), representing that there is no significant difference for the floor properties influencing gas content. In addition, the mudstone thickness of the roof for the No.5 coal seam varies from 1.3 to 3.0 m (Table 2). A roof with a mudstone thickness of over 1 m can effectively seal the gas and retain a gas content of over 10 m^3/t [3]. Therefore, there is no obvious relationship between gas content and the roof thickness for mudstone in the Weibei Field (Figure 6). In general, the roof and floor sealing capability of the No.5 coal seam are generally high, which is beneficial to CBM preservation.

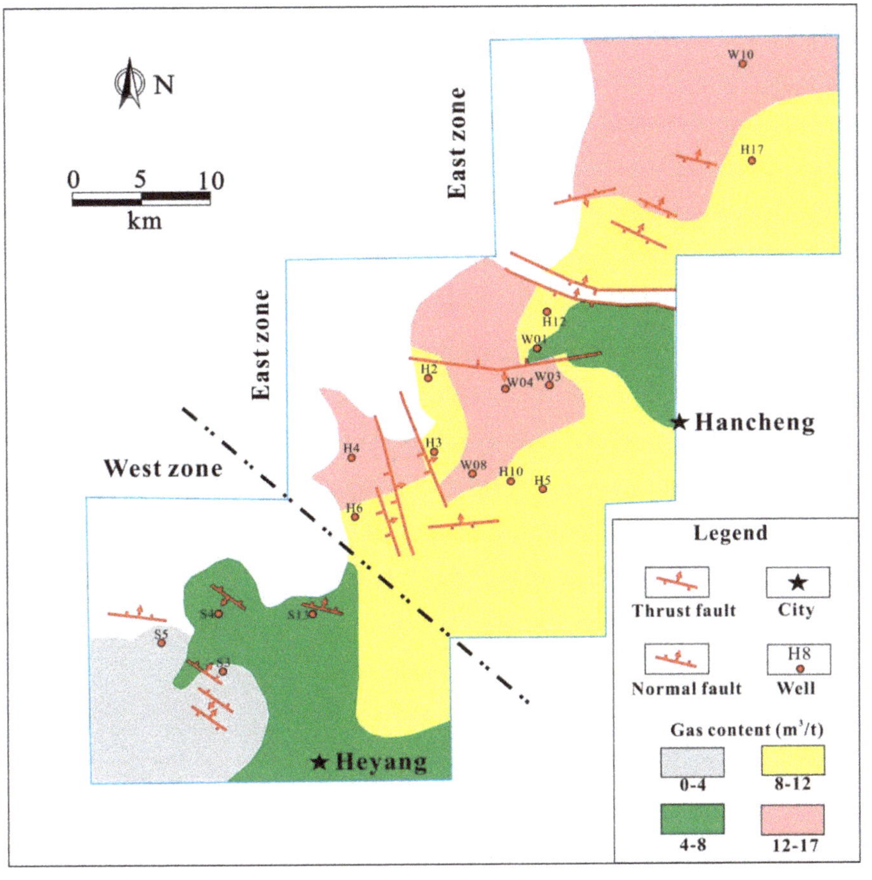

Figure 4. The distribution of gas content of the No.5 coal seam in the Weibei Field.

Table 2. The thickness, roof lithology, and floor lithology of the No.5 coal seam of 16 CBM wells in the Weibei Field. Note: M = Mudstone, S = Sandstone, CM = Carbonaceous mudstone, SM = Silty Mudstone, AS = Argillaceous sandstone.

Zone	Well	Roof			Floor		
		Dominate Lithology	Thickness (m)	M Thickness (m)	Dominate Lithology	Thickness (m)	M Thickness (m)
East	H3	CM	2.7	2.2	M	2.7	2.5
	H4	M	3.2	3.0	M	3.2	3.0
	H5	S	3.6	2.0	M	3.4	3.0
	H6	CM	2.9	2.3	M	2.6	2.2
	H10	CM	3.5	2.8	M	3.2	3.1
	H12	M	3.2	2.9	CM	3.6	3.0
	H17	CM	3.4	2.7	M	2.9	2.8
	W01	M	2.6	2.3	M	3.5	3.3

Table 2. *Cont.*

Zone	Well	Roof			Floor		
		Dominate Lithology	Thickness (m)	M Thickness (m)	Dominate Lithology	Thickness (m)	M Thickness (m)
	W03	M	2.2	2.0	CM	2.8	2.2
	W04	CM	2.5	2.0	M	2.7	2.5
	W08	CM	2.6	2.0	M	3.2	3.1
	W10	CM	3.5	2.7	M	2.8	2.6
West	S3	SM	2.8	2.1	M	2.2	2.0
	S4	AS	2.7	1.3	M	2.5	2.2
	S5	M	2.9	2.6	M	2.6	2.2
	S13	AS	3.2	1.8	M	2.3	2.0

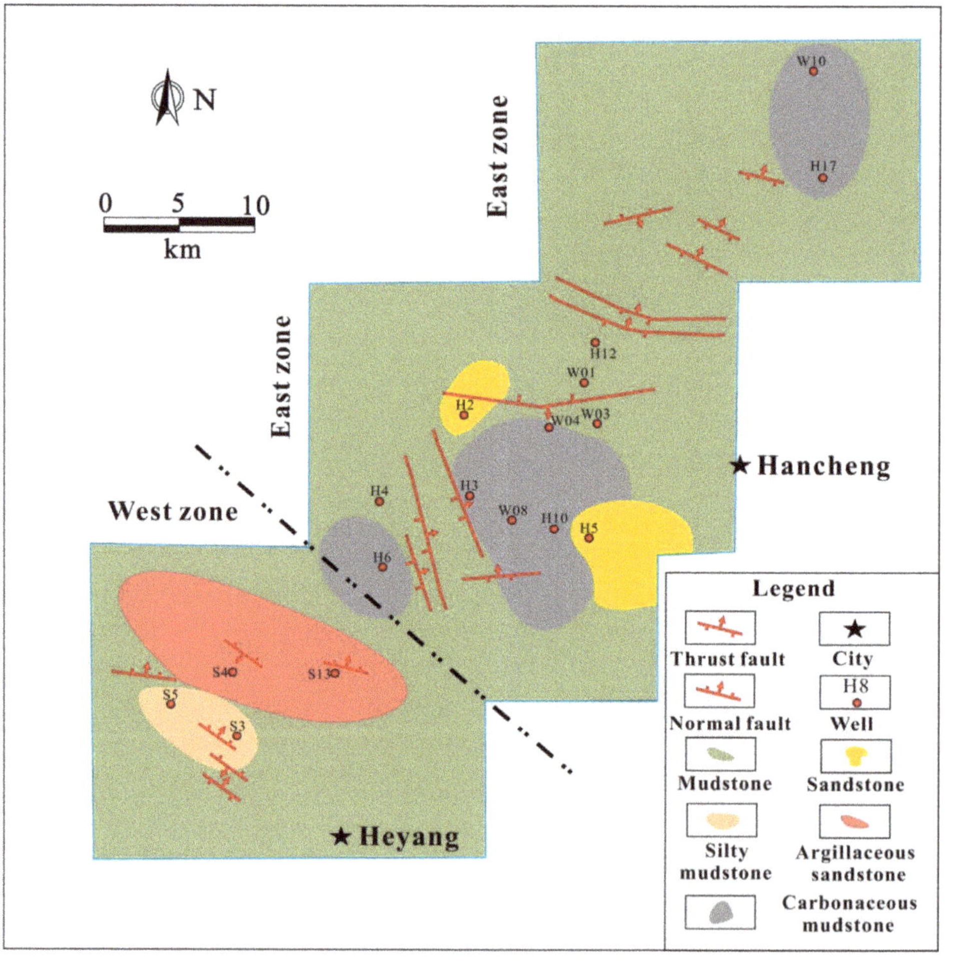

Figure 5. Roof lithology distribution of the No.5 coal seam in the Weibei Field.

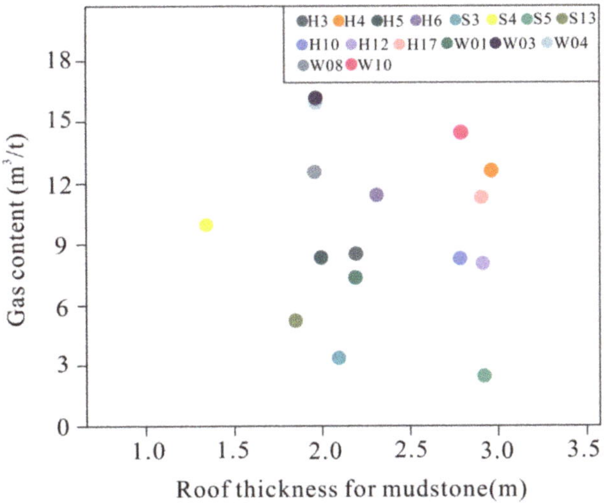

Figure 6. Relationship of roof thickness for mudstone and gas content of the No.5 coal seam in the Weibei Field.

Thickness and Burial Depth

The thickness of the No.5 coal seam ranges from 2.5 to 9.0 m (avg. 4.51 m) and decreases from the center to the margin of the Weibei Field (Figure 7). As listed in Table 1, in the Weibei Field, the No.5 coal seam is generally thicker than 4 m and has a relatively high gas content of over 9 m^3/t. Although coal thickness is not seriously related to current gas content, it can also control gas redistribution during tectonic activities or if magma intrusion occurs [3]. At the margin of the Weibei Field, the thickness of the No.5 coal seam decreases to be less than 4 m, which is a negative factor for gas preservation.

The burial depth of the coal seam has a significant influence on gas content [26,27], because the maximum burial depth during tectonic evolution controls gas generation and coal rank, while the current burial depth affects preservation conditions and gas content [27,28]. To investigate the preservation of the No.5 coal seam, a contour map of current burial depth was illustrated in Figure 8, in which the current burial depth is within 400 to 1200 m in the east zone, and within 500 to 1000 m in the west zone in the Weibei Field. Moreover, the No.5 coal seam burial depth increases from southeast to northwest in the Weibei Field (Figure 8). Besides, it can be observed from Figures 4 and 8 that the gas content of the No.5 coal seam increases with the increase of burial depth. Compared with the west zone, the east zone with a deep burial depth of the No.5 coal seam is favorable for CBM preservation.

Hydrogeological Conditions

Hydrogeological conditions commonly influence the coal properties, gas generation, and gas distribution [4], and are also key controlling factors on the fluid pressure regime and productivity of CBM wells [29,30]. Tectonics, topography, and precipitation are the main factors that affect groundwater migration in the aquifer [4]. According to these conditions, the Weibei Field has three hydrogeological systems, including the Quaternary (Q), Permo-Carboniferous (C–P), and Ordovician (O) aquifers. Previous studies clarified that the Q and O aquifers have no relationship with the No.5 coal seam due to the barrier of one hundred-meter thick strata in the Weibei Field [4]. The groundwater migration in Taiyuan Formation belonging to the C–P aquifer system, however, has an important impact on the No.5 coal seam.

A single C–P aquifer system exists due to a hydrodynamic connection through fractures in the No.5 coal seam, providing abundant produced water for CBM wells in the

Weibei Field [4]. As the thrust fault has a good sealing capability, it can divide a closed local aquifer system in the east zone, for example, thrust faults F2 and F3 in the study area play such a role (Figure 9). In this region, the groundwater with a good hydrodynamic sealing is advantageous for a high gas content in the No.5 coal seam (Figure 4). Compared to the thrust fault, the normal fault commonly has a bad sealing capability. In addition, although the normal fault can interrupt the continuity of the aquifer system, the flow rate of groundwater would significantly accelerate through gravity. These are disadvantages for gas preservation due to the bad hydrodynamic sealing condition as well as some CBM being dissolved in the groundwater and flowing away, for example, well S13 (Figure 9). These phenomena clarify the reasons for the low gas content in the west zone of the Weibei Field.

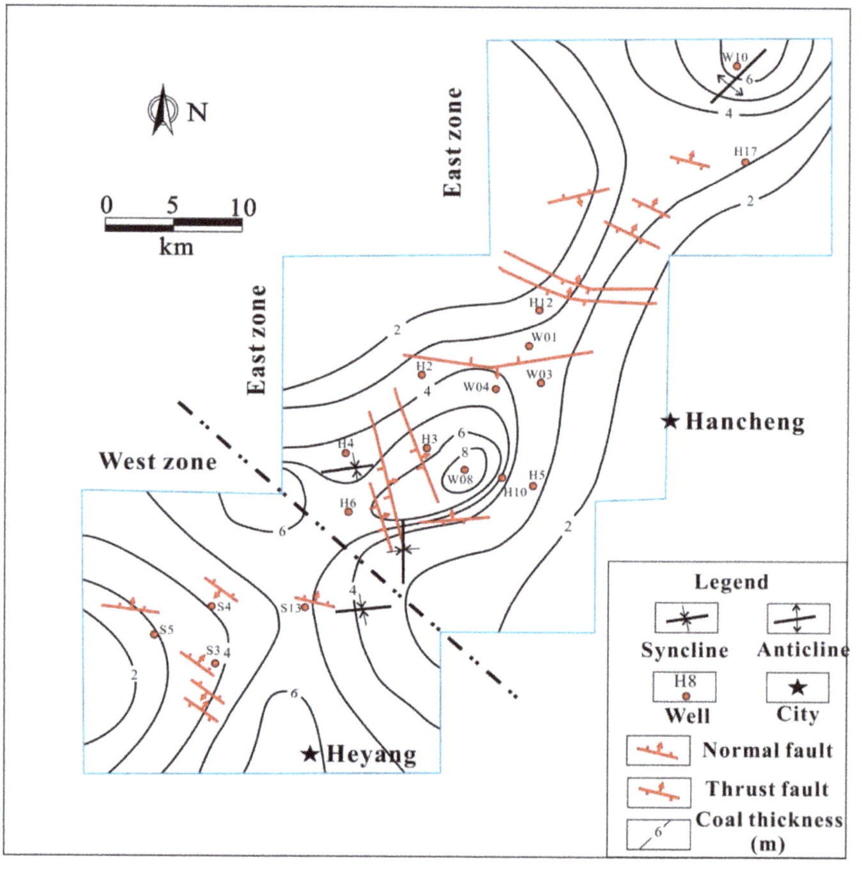

Figure 7. Thickness distribution of the No.5 coal seam in the Weibei Field.

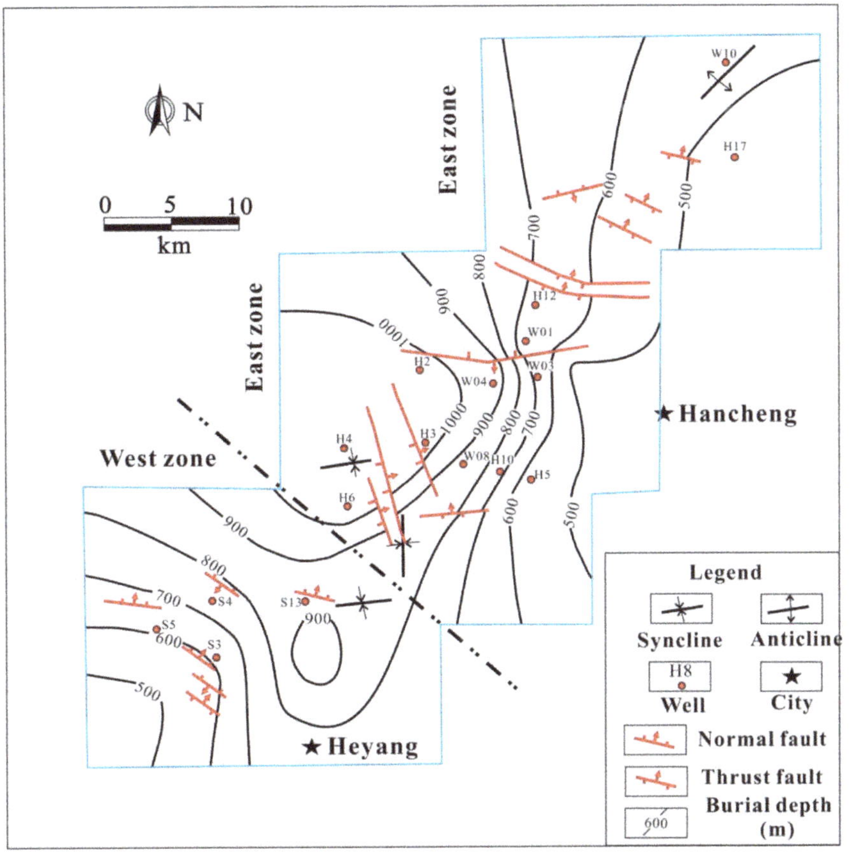

Figure 8. Burial depth distribution of the No.5 coal seam in the Weibei Field.

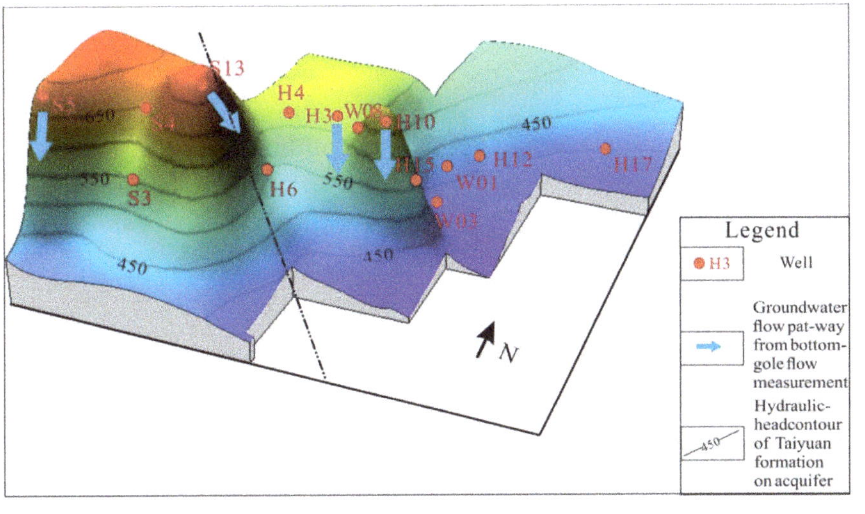

Figure 9. Hydraulic head contour of Taiyuan Formation aquifer.

4.2. Methane Adsorption Capacity of No.5 Coal Seam

CBM is mainly adsorbed on the pore surface of the coal matrix, therefore it is widely accepted that the adsorption capacity has a significant influence on gas content. Table 1 lists the Langmuir volumes (V_L, on an air-dry basis) of the No.5 coal seam in the Weibei Field, suggesting the coal seam has a relatively high methane adsorption capacity revealed by the great V_L varying from 12.43 to 31.50 m^3/t. Moreover, it is lower in the west zone (avg. 16.19 m^3/t) but higher in the east zone (avg. 24.45 m^3/t). These tendencies are consistent with the distribution of gas content in the Weibei Field. Moreover, Figure 10 suggests that the V_L and gas content of the No.5 coal seam exhibits a positive linear relationship in the Weibei Field. Although many factors, such as pressure, temperature, coal composition, ash yield, and pores, affect the methane adsorption capacity of coals, several studies show that the capacity of methane adsorption is related to pore size under the approximate pressure and temperature conditions. Thus, methane is mainly adsorbed on the surface of the pores, which have diameters of <100 nm [3,31].

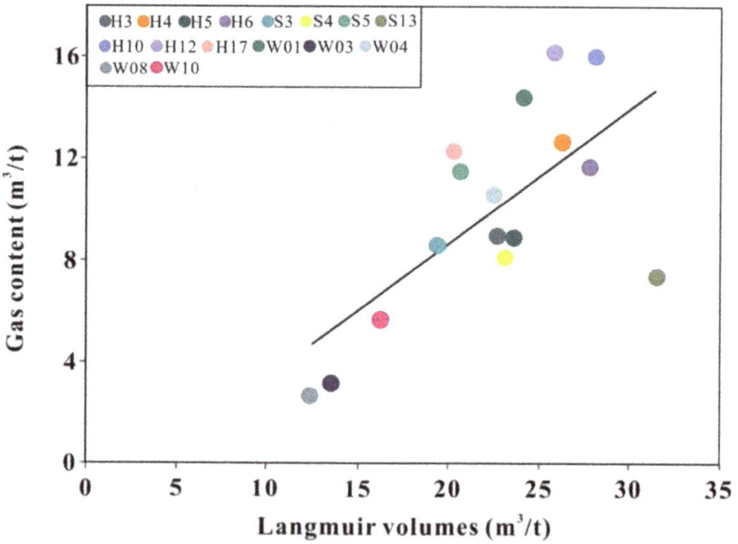

Figure 10. Relationship methane adsorption capacity and gas content of No.5 coal seam in the Weibei Field.

Table 3 lists the pore proportion of the No.5 coal seam in the Weibei Field, showing the proportion of pores with diameters of <100 nm range from 60.31% to 72.04%. Besides, it is higher in the east zone (avg. 67.78%) but lower in the west zone (avg. 62.22%). This situation is consistent with the adsorption capacity distribution characteristics in the Weibei Field. Moreover, the proportion of pores with diameter > 1 μm is lower in the east zone (avg. 5.48%) but higher in the west zone (avg. 24.8%). These phenomena are probably due to the west zone being seriously affected by extensional stress during normal faulting activities, accordingly forming relatively large pores. In contrast, the east zone is seriously affected by compressive stress during thrust faulting activities, which causes the previous large pores to shrink into small ones.

Table 3. The proportion of pores with different diameters of the No.5 coal seam of 16 CBM wells in the Weibei Field.

Zone	Sample No.	Proportion of Pores (%)		
		<100 nm	100 nm to 1 μm	>1 μm
East	E1	72.04	23.61	4.35
	E2	67.06	20.34	8.60
	E3	64.25	32.25	3.50
West	W1	60.31	12.36	27.33
	W2	63.24	13.25	23.51
	W3	63.05	12.40	23.55

4.3. Gas Saturation and Reservoir Pressure

The gas saturation is defined as the percentage of total gas content relative to the maximum capacity of methane adsorption. The gas saturation for a coal sample is determined by reservoir pressure and temperature by comparing the desorption data with an adsorption isotherm [4]. Regarding the calculation of gas saturation, details were exhibited by Pashin [32]. As shown in Figure 11, the gas saturation is in the range of 25% and 94.5% in the Weibei Field. The average gas saturation accounts for ~73% in the east zone, while ~44% in the west zone (Figure 11). This significant difference was affected by secondary processes. For instance, gas migration is related to regional tectonic activities [4]. In the east zone, the compressed stress led to a good sealing capability of the coal seam's roof and floor, resulting in a high gas content; hence, a great gas saturation. In contrast, the extensional stress led to normal faults well development in the west zone, which is negative for gas preservation, resulting in relatively low gas content and gas saturation. In addition, reservoir pressure refers to the fluid pressure acting in the pore-fracture space. The pressure of the No.5 coal seam ranges from 3.72 to 9.35 MPa (Figure 10). A local high-pressure area was formed in the east zone which benefits the occurrence of CBM, while the low-pressure area was formed in the west zone and is a disadvantage for high gas content.

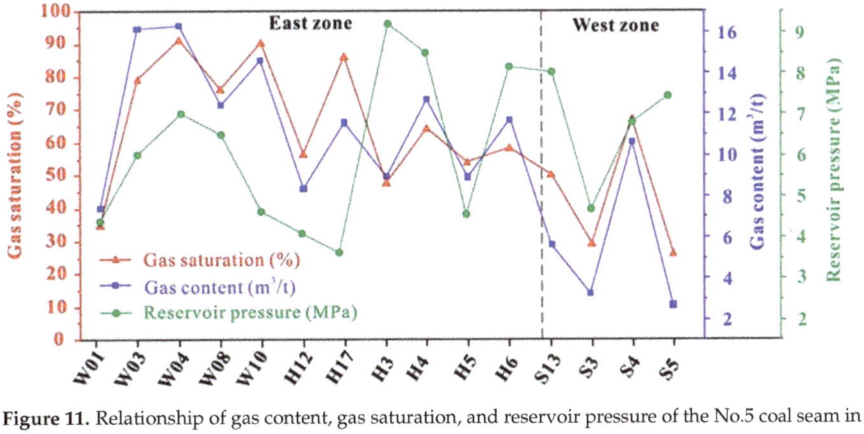

Figure 11. Relationship of gas content, gas saturation, and reservoir pressure of the No.5 coal seam in the Weibei Field.

4.4. Scenarios for Geological Controls on Gas Content

Based on a comprehensive analysis of the effects of geological control factors on gas content in the Weibei Field, the gas content distribution is co-controlled by many factors, including geological structures, roof lithology, burial depth, and hydrodynamic conditions, whereas the coal reservoir properties influence the gas content in a local area. Figure 12 shows two typical geological models, which can give an explanation of the effects of key factors on gas content in the Weibei Field. The first model (Model A) applies to the west zone of the Weibei Field, where the average gas content is <6 m^3/t. Herein, the normal faults are developed, easily allowing the gas to dissipate. Besides, the gas adsorbed in the surface of the coal matrix diffuses into the pore-fracture network and then migrates to the permeable strata by hydrodynamic flushing in the recharge zone.

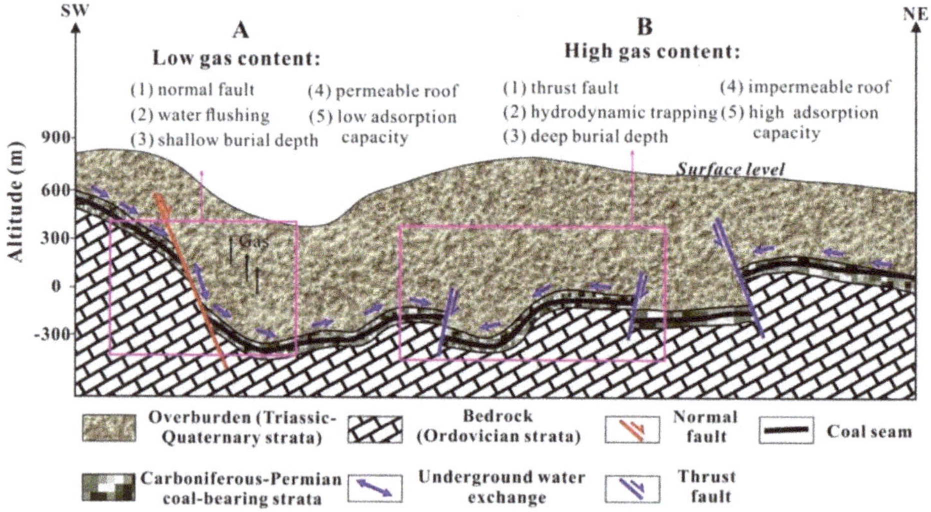

Figure 12. Geological scenarios controlling CBM content of the No.5 coal seam in the Weibei Field. (**A**) shows low gas content model; (**B**) shows high gas content model.

Model B is representative of the east zone in the Weibei Field (Figure 12), where the gas content is high at >11 m^3/t. This model is controlled by the combined effects of hydrodynamic and geological conditions, in which the roof of coal seam developed in the thrust fault zone commonly has an impermeable seal. Moreover, hydrodynamic trapping also increases gas content in the deep burial depth zone, where more gas can be adsorbed in coals with a relatively high reservoir pressure, which is beneficial to CBM preservation.

5. Conclusions

The comprehensive evaluation of gas content and its key controls are conducted based on the working coal seam samples together with 16 CBM wells in the Weibei Field, SOB, north China. The following conclusions are drawn from this study:

(1) The distribution of gas content is directly controlled by the tectonic and depositional evolution, which seriously affects gas generation. In the Weibei Field, the main CBM generation is related to the fluctuation subsidence of the coal seam during the middle and late Yanshanian orogeny.

(2) Highly variable gas contents in the No.5 coal seam of the Weibei Field reflect the combined effects of geological factors, hydrodynamic factors, and the properties of the reservoir. Besides, the fault types, roof lithology, burial depth of coal seam, and hydrodynamic conditions are the key control factors on the distribution of gas content.

(3) In the Weibei Field, the gas production potential for exploration and development is higher in the east zone but lower in the west zone because the recharge and runoff zone of groundwater is located in the west, while the stagnant and weak runoff zone of groundwater is located in the east. According to the analysis of the effects of key control factors on the distribution of gas content, two typical models controlling gas content in the Weibei Field are established.

Author Contributions: Y.Q. and H.W. conceived and designed the experiments; Y.Q. and H.W. performed the experiments and wrote the paper; H.W. and A.A. analyzed the data; Z.P. and Y.C. revised the paper and provided language support; G.G. and R.M. provided technical support. All authors have read and agreed to the published version of the manuscript.

Funding: This research was funded by the National Natural Science Foundation of China (42102227, 42130806, 41922016 and 41830427), and the Fundamental Research Funds for the Central Universities (2652018280).

Institutional Review Board Statement: Not applicable.

Informed Consent Statement: Not applicable.

Data Availability Statement: Restrictions apply to the availability of these data. Data was obtained from China National Petroleum Corporation and are available from the authors with the permission of China National Petroleum Corporation.

Acknowledgments: We would like to thank China National Petroleum Corporation for providing the exploration well data.

Conflicts of Interest: The authors declare no conflict of interest.

References

1. Su, X.B.; Wang, Q.; Lin, H.X.; Song, J.X.; Guo, H.Y. A combined stimulation technology for coalbed methane wells: Part 1. Theory and technology. *Fuel* **2018**, *233*, 592–603. [CrossRef]
2. Cai, Y.D.; Liu, D.M.; Yao, Y.B.; Li, J.Q.; Qiu, Y.K. Geological controls on prediction of coalbed methane of No. 3 coal seam in Southern Qinshui Basin, North China. *Int. J. Coal Geol.* **2011**, *88*, 101–112. [CrossRef]
3. Cai, Y.D.; Liu, D.M.; Zhang, K.M.; Elsworth, D.; Yao, Y.B.; Tang, D.Z. Preliminary evaluation of gas content of the No. 2 coal seam in the Yanchuannan area, southeast Ordos basin, China. *J. Petrol. Sci. Eng.* **2014**, *122*, 675–689. [CrossRef]
4. Yao, Y.B.; Liu, D.M.; Qiu, Y.K. Variable gas content, saturation, and accumulation characteristics of Weibei coalbed methane pilot-production field in the southeastern Ordos Basin, China. *AAPG Bull.* **2013**, *97*, 1371–1393. [CrossRef]
5. Chen, H.; Tian, W.G.; Chen, Z.H.; Zhang, Q.F.; Tao, S. Genesis of Coalbed Methane and Its Storage and Seepage Space in Baode Block, Eastern Ordos Basin. *Energies* **2022**, *15*, 81. [CrossRef]
6. Liu, D.; Li, J.; Liu, J.Q.; Zhang, L. Modeling hydrocarbon accumulation based on gas origin and source rock distribution in Paleozoic strata of the Ordos Basin, China. *Int. J. Coal Geol.* **2020**, *225*, 103486. [CrossRef]
7. Xu, H.; Tang, D.Z.; Liu, D.M.; Tang, S.H.; Yang, F.; Chen, X.Z.; He, W.; Deng, C.M. Study on coalbed methane accumulation characteristics and favorable areas in the Binchang area, southwestern Ordos Basin, China. *Int. J. Coal Geol.* **2012**, *95*, 1–11. [CrossRef]
8. Xu, C.; Yang, G.; Wang, K.; Fu, Q. Uneven stress and permeability variation of mining-disturbed coal seam for targeted CBM drainage: A case study in Baode coal mine, eastern Ordos Basin, China. *Fuel* **2021**, *289*, 119911. [CrossRef]
9. Yao, Y.B.; Liu, D.M.; Tang, D.Z.; Tang, S.H.; Che, Y.; Huang, W.H. Preliminary evaluation of the coalbed methane production potential and its geological controls in the Weibei Coalfield, Southeastern Ordos Basin, China. *Int. J. Coal Geol.* **2009**, *78*, 1–15. [CrossRef]
10. Wei, C.; Qin, Y.; Wang, G.G.X.; Fu, X.; Zhang, Z. Numerical simulation of coalbed methane generation, dissipation and retention in SE edge of Ordos Basin, China. *Int. J. Coal Geol.* **2010**, *82*, 147–159. [CrossRef]
11. Zhao, J.L.; Xu, H.; Tang, D.Z.; Mathews, J.P.; Li, S.; Tao, S. Coal seam porosity and fracture heterogeneity of macrolithotypes in the Hancheng Block, eastern margin, Ordos Basin, China. *Int. J. Coal Geol.* **2016**, *159*, 18–29. [CrossRef]
12. Zhao, J.Z.; Tang, D.Z.; Xu, H.; Lv, Y.M.; Tao, S. High production indexes and the key factors in coalbed methane production: A case in the Hancheng block, southeastern OrdosBasin, China. *J. Petrol. Sci. Eng.* **2015**, *130*, 55–67. [CrossRef]
13. Wang, H.; Yao, Y.B.; Huang, C.C.; Liu, D.M.; Cai, Y.D. Fault Development Characteristics and Their Effects on Current Gas Content and Productivity of No. 3 Coal Seam in the Zhengzhuang Field, Southern Qinshui Basin, North China. *Energy Fuels* **2021**, *35*, 2268–2281. [CrossRef]
14. Wang, H.; Yao, Y.B.; Liu, D.M.; Pan, Z.J.; Yang, Y.H.; Cai, Y.D. Fault-sealing capability and its impact on coalbed methane distribution in the Zhengzhuang field, southern Qinshui Basin, North China. *J. Nat. Gas Sci. Eng.* **2016**, *28*, 613–625. [CrossRef]
15. Chen, S.D.; Tang, D.Z.; Tao, S.; Xu, H.; Li, S.; Zhao, J.L.; Ren, P.F.; Fu, H.J. In-situ stress measurements and stress distribution characteristics of coal reservoirs in major coalfields in China: Implication for coalbed methane (CBM) development. *Int. J. Coal Geol.* **2017**, *182*, 66–84. [CrossRef]
16. Clarkson, C.R.; Rahmanian, M.; Kantzas, A. Relative permeability of CBM reservoirs: Controls on curve shape. *Int. J. Coal Geol.* **2011**, *88*, 204–217. [CrossRef]
17. Yan, T.T.; Yao, Y.B.; Liu, D.M. Critical tectonic events and their geological controls on gas generation, migration, and accumulation in the Weibei coalbed methane field, southeast Ordos basin. *J. Nat. Gas Sci. Eng.* **2015**, *27*, 1367–1380. [CrossRef]

18. Chen, Y.H.; Zhu, Z.W.; Zhang, L. Control actions of sedimentary environments and sedimentation rates on lacustrine oil shale distribution, an example of the oil shale in the Upper Triassic Yanchang Formation, southeastern Ordos Basin (NW China). *Mar. Petrol. Geol.* **2019**, *102*, 508–520. [CrossRef]
19. Qiao, J.Q.; Littke, R.; Zieger, L.; Jiang, Z.X.; Fink, R. Controls on gas storage characteristics of Upper Paleozoic shales from the southeastern Ordos Basin. *Mar. Petrol. Geol.* **2020**, *117*, 104377. [CrossRef]
20. Yang, Q.; Liu, D.M.; Huang, W.H.; Che, Y.; Hu, B.; Yu, Y. *Geology and Resources Evaluation of Coalbed Methane in Northwest China*; Geological Publishing House: Beijing, China, 2005; p. 303, (In Chinese with English Abstract).
21. Cai, Y.; Li, Q.; Liu, D.; Zhou, Y.; Lv, D. Insights into matrix compressibility of coals by mercury intrusion porosimetry and N_2 adsorption. *Int. J. Coal Geol.* **2018**, *200*, 199–212. [CrossRef]
22. Moore, T.A. Coalbed methane: A review. *Int. J. Coal Geol.* **2012**, *101*, 36–81. [CrossRef]
23. Flores, R.M. Coalbed methane: From hazard to resource. *Int. J. Coal Geol.* **1998**, *35*, 3–26. [CrossRef]
24. Yielding, G.; Freeman, B.; Needham, D.T. Quantitative fault seal prediction. *AAPG Bull.* **1997**, *81*, 897–917.
25. Wang, X.H.; Liu, D.M.; Yao, Y.B.; Xie, C. Carboniferous-Permian coal-bearing depositional system and its controlling role of coalbed methane in Hancheng area in Ordos Bain. *Petrol. Geol. Exp.* **2013**, *35*, 646–650. (In Chinese)
26. Hou, S.H.; Wang, X.M.; Wang, X.J.; Yuan, Y.D.; Zhuang, X.G.; Wang, X.M. Geological controls on gas saturation in the Yanchuannan coalbed methane field, southeastern Ordos Basin, China. *Mar. Petrol. Geol.* **2016**, *78*, 254–270. [CrossRef]
27. Li, Z.T.; Liu, D.M.; Ranjith, P.; Cai, Y.D.; Wang, Y. Geological controls on variable gas concentrations: A case study of the northern Gujiao Block, northwestern Qinshui Basin, China. *Mar. Petrol. Geol.* **2018**, *92*, 582–596. [CrossRef]
28. Ruppert, L.F.; Hower, J.C.; Ryder, R.T.; Levine, J.R.; Trippi, M.H.; Grady, W.C. Geologic controls on thermal maturity patterns in Pennsylvanian coal-bearing rocks in the Appalachian basin. *Int. J. Coal Geol.* **2010**, *81*, 169–181. [CrossRef]
29. Ayers, W.B., Jr. Coalbed gas systems, resources, and production and a review of contrasting cases from the San Juan and Powder River Basins. *AAPG Bull.* **2002**, *86*, 1853–1890.
30. Yao, Y.B.; Liu, D.M.; Yan, T.T. Geological and hydrogeological controls on the accumulation of coalbed methane in the Weibei field, southeastern Ordos Basin. *Int. J. Coal Geol.* **2014**, *121*, 148–159. [CrossRef]
31. Liu, Z.S.; Liu, D.M.; Cai, Y.D.; Yao, Y.B.; Pan, Z.J.; Zhou, Y.F. Application of nuclear magnetic resonance (NMR) in coalbed methane and shale reservoirs: A review. *Int. J. Coal Geol.* **2020**, *218*, 103261. [CrossRef]
32. Pashin, J.C. Variable gas saturation in coalbed methane reservoirs of the Black Warrior Basin: Implications for exploration and production. *Int. J. Coal Geol.* **2010**, *82*, 135–146. [CrossRef]

processes MDPI

Article

Organic Matter Source, Fluid Migration, and Geological Significance of Stylolites in Organic-Lean Carbonate Rocks: A Case from the Sichuan Basin

Shengnan Liu [1,2,3], Shiju Liu [1,2,3], Gang Gao [2,3,*] and Rukai Zhu [1]

1 PetroChina Research Institute of Petroleum Exploration & Development, Beijing 100083, China; lsj_19942021@163.com (S.L.); zrk@petrochina.com.cn (R.Z.)
2 School of GeoSciences, China University of Petroleum, 18 Fuxue Road, Beijing 102249, China
3 State Key Laboratory of Petroleum Resource and Prospecting, China University of Petroleum, 18 Fuxue Road, Beijing 102249, China
* Correspondence: gg28192023@126.com

Abstract: Carbonate rocks exhibit significant heterogeneity as both a source rock and reservoir. Stylolite formation plays a crucial role in the enrichment of organic matter and the migration of geofluids within carbonate rocks. In order to study the enrichment mechanism of organic matter and the geofluid migration mode in the stylolites developed in carbonate rocks, stylolite-bearing core samples from the Dachigan structural belt in the eastern Sichuan Basin were collected. The stylolites and matrix were subjected to the total organic carbon (TOC) test and Rock-Eval pyrolysis, thin-section observation under fluorescent light, whole-rock X-ray diffraction, carbon and oxygen isotope analysis, and scanning electron microscopy. The organic matter occurring in the stylolites is mainly in the form of three types: soluble organic matter, pyrobitumen, and bitumen. This suggests that the organic matter within the stylolites mainly consists of secondary migrated organic matter. The stylolites also exhibit well-developed secondary dolomite and pyrite resulting from late-diagenetic recrystallization. These minerals contribute to the preservation of intercrystalline pore spaces and fractures, providing favorable conditions for oil and gas accumulation and migration. The strong cementation observed at the contact between the stylolites and matrix, along with the presence of secondary minerals nearby, may be attributed to the fractionation of light and heavy oil components during the migration of hydrocarbon fluids from the matrix to the stylolites. The thicknesses of the stylolites vary within the bulk, indicating severe diagenesis in thinner areas. Consequently, this leads to significant fractionation effects. The fractionation of crude oil components by stylolites poses challenges for the study of definitive oil–source correlations. To overcome these challenges, future research could investigate biomarker compounds to attempt oil–source correlations. Additionally, future efforts should take into consideration the spatial variation in the crude oil properties. Understanding the role of stylolites in organic matter enrichment and geofluid migration is crucial for optimizing exploration strategies in the Sichuan Basin, a region of growing importance in the energy industry. Moreover, our findings shed light on the complex interactions within stylolite-bearing rocks, which are not limited to this specific basin. These insights offer valuable contributions to the broader field of geology and reservoir characterization, enhancing our ability to predict and interpret similar geological formations globally.

Keywords: carbon and oxygen isotopes; organic matter; fluid migration; stylolites; carbonate rock

Citation: Liu, S.; Liu, S.; Gao, G.; Zhu, R. Organic Matter Source, Fluid Migration, and Geological Significance of Stylolites in Organic-Lean Carbonate Rocks: A Case from the Sichuan Basin. *Processes* **2023**, *11*, 2967. https://doi.org/10.3390/pr11102967

Academic Editors: Ping Gao, Yidong Cai, Yingfang Zhou and Quan Gan

Received: 25 June 2023
Revised: 8 September 2023
Accepted: 25 September 2023
Published: 13 October 2023

1. Introduction

In the realm of geological investigation, the distinctiveness of carbonate rock as both a source rock and reservoir type has garnered considerable attention from scholars [1–8]. However, the intrinsic attributes of carbonate rocks, marked by low organic matter content and intricate heterogeneity [9–13], pose challenges, resulting in reservoirs characterized by their inherent tightness. This intricacy has led to a prolonged discourse encompassing hydrocarbon generation, expulsion, and migration within carbonate formations [14–19].

Of particular intrigue are stylolites, a distinctive feature inherent to carbonate rocks [18]. The inception of their observation by Mylius in 1751 instigated an array of research endeavors aimed at unraveling their origins, giving rise to diverse genetic theories, including those rooted in organics, crystallization, pitch, pressure, gas, denudation, dissolution, and contraction pressure [20–22]. Among these, the pressure-solution theory has gained broad acceptance. In the evolution of stylolites, the enrichment of the varying minerals within these structures has undergone meticulous scrutiny [23–26], as has the presence of organic matter [27–32].

Concurrently, a pivotal discourse questions whether the characteristics of hydrocarbon generation and migration mirror those prevalent in mudstone [4,9,15,33]. Emerging studies suggest that stylolites, distinguished by a heightened presence compared to fractures in carbonate rocks, contain reservoir-related materials, like petroleum and bitumen, indicating a discernible role in hydrocarbon migration [33–37]. Perspectives offered by Lind et al. (1994), Heap et al. (2014), and Rustichelli et al. (2015) [38–40] propose that stylolites indeed serve as migration conduits rather than impediments. Complementing this, micro-CT data by Heap et al. (2018) [41] spotlight the stylolite porosity, further endorsing their potential as pathways for oil and gas migration. Biomarker compound analysis also lends insight into the impact of stylolites on migration, with Liu et al. (2020) [42] juxtaposing the organic fluid migration from the matrix to the stylolites and along these structures, revealing pronounced migration fractionation, particularly along the latter.

In summation, the ubiquity of stylolites within carbonate formations, distinguished by their assorted morphologies and pronounced variations in thickness, length, and composition, emerges from the interplay of stress and dissolution processes. While investigations into stylolites span a spectrum of themes—distribution, typology, composition, and provenance—their implications for reservoir porosity and fluid migration have encountered considerable exploration. Yet, an appreciable void persists in the realm of studies concerning the petrological nuances of stylolites, the origins of their organic matter, and the mechanisms governing hydrocarbon generation and expulsion in contexts characterized by low organic matter content.

In this study, we examined the stylolites and matrix in Carboniferous and Ordovician marine carbonate rocks from the Sichuan Basin. Several analyses were performed, including total organic carbon content (TOC) testing, Rock-Eval pyrolysis, X-ray diffraction (XRD), carbon and oxygen isotope composition analyses, as well as petrological analysis. The objective was to investigate the mechanism behind the accumulation and migration of the organic matter within the stylolites.

2. Geologic Setting and Stratigraphy

The Sichuan Basin is situated in the middle and eastern parts of Sichuan Province, encompassing a significant portion of Chongqing as well. Geologically, it is a morphologically rhomboic basin. The basin is surrounded by fault belts, adjacent to the Longmenshan platform margin depression zone in the east and bounded by the Dabashan and Daloushan platform margin depression zones in the west, closely adjacent to the Daliangshan fault zone, and connected to the Micangshan platform fold belts in the north, totally covering about 260,000 km^2 (Figure 1A) [43]. Structurally, the Sichuan Basin belongs to an important first-level unit in the western part of the Yangtze platform. It is a large petroliferous basin developed on the foundation of the Upper Yangtze Craton, which has undergone multiple tectonic cycles [44]. Throughout its geological history, the Sichuan Basin has experienced various periods of tectonic movement, primarily characterized by subsidence and sedimentation. These stages can be classified into two main phases of sedimentary and tectonic evolution: (1) The Sinian–Middle Triassic, which was the sedimentary–tectonic evolution stage of the marine craton basin with the deposition of carbonate rocks; the basin is dominated by vertical movement; (2) the sedimentary–tectonic evolution stage, which was dominated by continental clastic rocks that had entered the foreland basin since the Late Triassic, forming a "foreland-cratonic" type of superimposed basin [45].

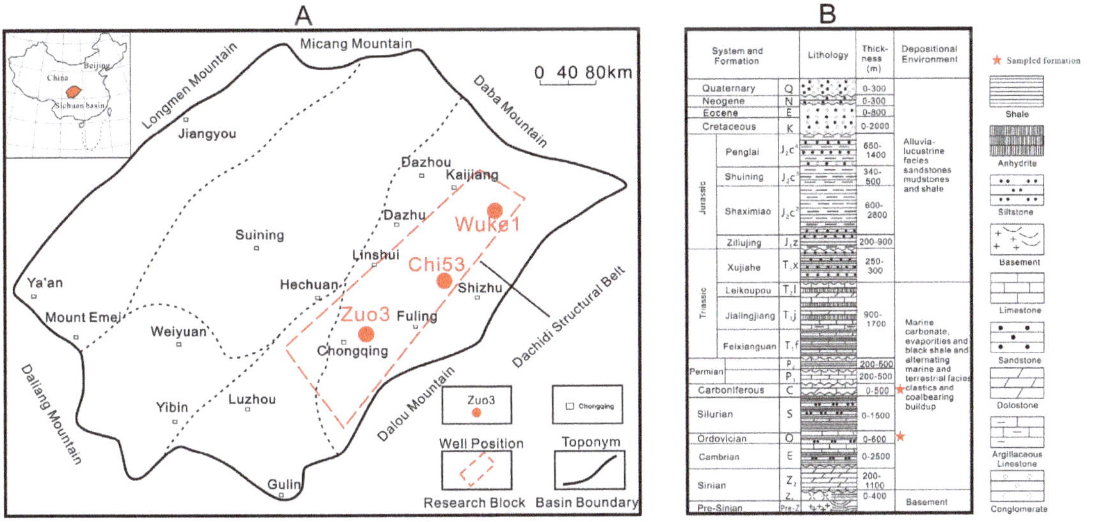

Figure 1. (**A**) Sampling location of the Carboniferous and Ordovician carbonate rocks and (**B**) stratigraphic column in the Sichuan Basin.

The strata in the Sichuan Basin are mainly composed of marine carbonate rocks deposited from the Sinian to the Middle Triassic (Figure 1B). These strata have a total thickness of several thousand meters, and while some layers contain continental sandstone and carbonate rocks, the main reservoirs consist of marine carbonate rock intervals [46]. Evaporite deposits, such as open-platform facies, restricted-platform facies, and intra-platform beaches, are prevalent in the basin. These beach facies anomalies are conducive to the development of early karsts, which is conducive to the occurrence of the pressure solution in carbonate rocks to form stylolites. For this study, the researchers focused on the Dachigan structural belt located in the eastern part of the Sichuan Basin. This area was chosen due to the discovery of thicker stylolites in the carbonate rocks. The carbonate samples from this region exhibit a rich occurrence of stylolites, making them suitable for the purposes of this study.

3. Samples and Methods

3.1. Samples

A total of 7 core samples of carbonate rocks were collected from three drilling wells in the Dachigan structural belt in the eastern Sichuan Basin: Zuo3 (3), Chi53 (2), and Wuke1 (2) (Figure 1, Table 1). In the study, the stylolites and corresponding matrix on both sides were compared and analyzed. Therefore, it was essential to separate the stylolites and matrix in the samples. The collection of pure matrix samples is relatively simple. [47] showed that the porosity and specific area are different close to stylolites but remain unaffected beyond a distance of 1 cm. Hence, breaking the core and grinding it with pure matrix debris near the stylolites (within 1 cm) was sufficient to obtain the matrix sample. However, collecting stylolites proved more challenging due to their irregularities and varying thicknesses in different parts. In the process of collecting, we tried to select the thicker parts of the stylolites, and we first disintegrated them along the stylolites with a dentist drill. Then, special pliers were used to separate the filling in the stylolites and remove the evident matrix with a file for a finer separation. The pure stylolite filling was then ground with a mortar. Subsequently, all rock samples were crushed and sieved using an 80-mesh sieve before various tests were conducted. They were then placed in sample bags, appropriately numbered, and prepared for further experiments (Table 1). In addition, sections were sliced along the vertical stratification plane of the rock with a thickness of about 0.5 cm, and fluorescent thin slices and scanning electron microscope samples were prepared with this part of the sample.

Table 1. Basic information and numbers of samples.

Well ID	Layer	Depth (m)	No.
Zuo3	O_2^2	4316.30 m	1-Ssr 1-Msr
		4318.84 m	2-Ssr 2-Msr
		4317.60 m	3-Ssr 3-Msr
Chi53	C_2	3030.13 m	4-Ssr 4-Msr
		3053.88 m	5-Ssr 5-Msr
WuKe1	C_2	4254.28 m	6-Ssr 6-Msr
		4251.67 m	7-Ssr 7-Msr

3.2. Methods

3.2.1. TOC and Rock-Eval Experiments

Using a balance sensitive to one ten-thousandth of a gram, weigh the samples (0.1 g for suture samples, 0.25 g for matrix samples) and place them into crucibles. Then, add 10% diluted hydrochloric acid to decompose the carbonate in the samples until no gas bubbles are produced. Subsequently, rinse the samples with distilled water, taking care to prevent any sample overflow from the crucibles. After drying the samples in the crucibles, use the Leco CS-230 instrument to determine the organic carbon content [32,48,49]. Rock-Eval analyses were performed on an OGE-II instrument, which was developed by the Experimental Center of Petroleum Geology of the Research Institute of China Petroleum Exploration and Development, and it can mainly obtain the data of the S_1 (mg HC/g rock) (<300 °C), S_2 (mg HC/g rock) (300 °C < t < 600 °C), and Tmax (°C) (the maximum temperature at which S_2 pyrolyzate can generate) [50,51]. The standard sample used for the TOC was LOT NO.0602 (CARBON% = 0.691 ± 0.006), and the standard samples used for the Rock–Eval pyrolysis were IFP 160000 and Chinese standard reference materials.

3.2.2. Petrographic Analyses

In the study, a total of 7 carbonate rock samples containing stylolites were used to prepare thin sections, and these samples were observed with the transmitted light, reflection, and fluorescence of the Leica microscope. Blue fluorescence was used to observe the organic matter of different samples, with an excitation wavelength ranging from 420 nm to 485 nm and an emission wavelength of 515 nm [52]. Based on the observation of thin sections, the fresh surfaces and argon ion polishing of three samples were selected and observed under the scanning electron microscope. The scanning electron microscope model was a Hitachi SU8010 field-emission SEM-EDS. The experiment was carried out in the State Key Laboratory of the China University of Petroleum, Beijing.

3.2.3. XRD Analysis

The X-ray diffraction (XRD) was analyzed via a Bruker D2 PHASER diffractometer system with operating conditions of 30 kV, 10 mA, a scanning rate of 2° 2θ/min, and with Cu Kα radiation and whole-rock random-powder patterns recorded from 4.5° to 50° 2θ, which were conducted at the State Key Laboratory of Petroleum Resource and Prospecting, China University of Petroleum, Beijing and State Key Laboratory, and the dosage of the experimental sample was about 100 mg [53].

3.2.4. Isotope Composition Analyses

The carbon and oxygen isotope analyses were conducted at the State Key Laboratory of Petroleum Resource and Prospecting, China University of Petroleum (Beijing), and performed using a MAT 253 gas mass spectrometer equipped with a Kiel IV carbonate device [54]. The results, based on replicate analyses of GBW 04405, are given using conventional $\delta^{13}C$ and $\delta^{18}O$ notations with respect to the Vienna Pee Dee Belemnite (VPDB) standard, with the precision and reproducibility better than ±0.030‰ and ±0.080‰, respectively [55].

4. Results

4.1. TOC and Rock-Eval Pyrolysis

The data on the TOC content and pyrolysis parameters for the stylolites and matrix samples of carbonate rocks are illustrated in Table 2. The TOC values of the stylolites varied between 0.63 wt.% and 0.88 wt.%, with an average of 0.77 wt.%. The TOC values of the matrix varied between 0.03 wt.% and 0.24 wt.%, with an average of 0.12 wt.% (Table 3). The Rock-Eval pyrolysis S_1 (free hydrocarbon), S_2 (pyrolyzed hydrocarbon), and $S_1 + S_2$ of the stylolites, respectively, were in the ranges of 0.09~1.05 mg HC/g rock, 1.92~6.50 mg HC/g rock, and 2.29~6.64 mg HC/g rock, with corresponding averages of 0.39 mg HC/g rock, 4.27 mg HC/g rock, and 4.66 mg HC/g rock, respectively. The Rock-Eval pyrolysis S_1, S_2, and $S_1 + S_2$ of the matrix were, respectively, in 0.05~0.53 mg HC/g rock, 0.16~0.82 mg HC/g rock, and 0.21~1.17 mg HC/g rock, with corresponding averages of 0.23 mg HC/g rock, 0.35 mg HC/g rock, and 0.57 mg HC/g rock, respectively (Table 3). The ratios of the HCI (S_1/TOC) values of the stylolites were 10.2~148.5 mg HC/g TOC, with an average of 566.5 mg HC/g TOC (Table 3). The ratios of the HCI of the matrix were 120.5~307.2 mg HC/g TOC, with an average of 185 mg HC/g TOC (Table 3). The PI ($S_1/(S_1 + S_2)$) values of the stylolites were 0.02~0.35, with an average of 0.1, and those of the matrix were 0.21~0.75, with an average of 0.39 (Table 3). The Tmax values of the stylolites were 373 °C~500 °C (averaging 420 °C), and the Tmax values of the matrix varied from 420 °C to 600 °C (averaging 481 °C) (Table 3).

Table 2. TOC and Rock-Eval pyrolysis of stylolites and matrix in carbonate rock samples.

Sample No.	TOC (wt.%)	Tmax (°C)	S_1 (mg HC/g Rock)	S_2 (mg HC/g Rock)	$S_1 + S_2$ (mg HC/g Rock)	PI	HI (mg HC/g TOC)	HCI (mg HC/g TOC)	ΔTmax	ΔTOC
1-Ssr	0.71	373	1.05	1.92	2.97	0.35	271.5	148.5	47	0.53
1-Msr	0.17	420	0.53	0.18	0.71	0.75	104.3	307.2		
2-Ssr	0.88	424	0.09	2.2	2.29	0.04	1221.5	50.0	116	0.64
2-Msr	0.24	540	0.35	0.82	1.17	0.30	343.5	146.6		
3-Ssr	0.81	391	0.45	4.34	4.79	0.09	535.8	55.6	62	0.70
3-Msr	0.10	453	0.2	0.29	0.49	0.41	280.5	193.4		
4-Ssr	0.73	431	0.14	6.50	6.64	0.02	893.6	19.2	11	0.67
4-Msr	0.05	442	0.09	0.23	0.32	0.28	443.2	173.4		
5-Ssr	0.81	500	0.19	5.91	6.1	0.03	733.4	23.6	100	0.77
5-Msr	0.03	600	0.05	0.16	0.21	0.24	462.4	144.5		
6-Ssr	0.63	440	0.14	4.94	5.08	0.03	786.9	22.3	42	0.50
6-Msr	0.12	482	0.15	0.56	0.71	0.21	449.8	120.5		
7-Ssr	0.81	380	0.64	4.05	4.69	0.14	494.4	78.1	53	0.71
7-Msr	0.10	433	0.21	0.19	0.4	0.53	189.2	209.2		

Note: PI = $S_1/(S_1 + S_2)$; HI = S_2/TOC; HCI = S_1/TOC.

Table 3. Statistical data on TOC and Rock-Eval parameters for the stylolites and matrix in carbonate rock samples.

Index	Stylolites				Matrix			
	Min	Max	Mean	No.	Min	Max	Mean	No.
TOC (wt.%)	0.63	0.88	0.77	7	0.03	0.24	0.12	7
S_1 (mg HC/g rock)	0.09	1.05	0.39	7	0.05	0.53	0.23	7
S_2 (mg HC/g rock)	1.92	6.5	4.27	7	0.16	0.82	0.35	7
$S_1 + S_2$ (mg HC/g rock)	2.29	6.64	4.66	7	0.21	1.17	0.57	7
PI	0.02	0.35	0.1	7	0.21	0.75	0.39	7
HI (mg HC/g TOC)	271.5	893.6	566.5	7	104.3	462.4	324.7	7
HCI (mg HC/g TOC)	10.2	148.5	51.2	7	120.5	307.2	185	7
Tmax (°C)	373	500	420	7	420	600	481	7

4.2. Petrographic Analyses

The petrological characteristics record the heterogeneity of the sedimentation and diagenesis, which can be used to initially evaluate the preservation of the original sedimentary environment [56,57]. From the handpicked core samples, the thicknesses of the stylolites are not uniform, showing toruloid distribution (Figure 2B). Peeling off the cores along the stylolite lines, it could be observed that the surfaces of the stylolites are uneven (Figure 2A). Under microscopic observation, it could be seen that the particles in the matrix are densely cemented; however, the stylolites were dark-colored and showed toruloid distribution (Figure 2C), indicating that they were formed via pressure dissolution. Under reflective light, the distribution of secondary pyrite was observed (Figure 2D,E) in massive occurrence, which is quite different from the framboidal occurrence indicating the formation of primary pyrite. Under the condition of blue fluorescence excitation, the stylolites are characterized by extremely strong fluorescence luminescence with a green color, indicating that they are rich in light oil (Figure 2F,G). Through the SEM observation of the fresh surfaces of the samples, a large amount of secondary organic matter (oil, bitumen, and pyrobitumen) was observed in the stylolites, but no primary organic matter was observed (Figure 3), which is consistent with the phenomenon observed in the thin sections. In addition, a large number of intercrystalline pores and fractures could also be observed with the scanning electron microscope (Figure 3).

4.3. XRD and Isotopic Analyses

The data on the XRD of the stylolites and matrix for the same carbonate rock samples are listed in Table 4. The matrix contained more calcite than stylolites; however, the dolomite, quartz, pyrite, and clay minerals were more abundant in the stylolites (Table 4). The carbon and oxygen isotopic compositions of the stylolites and matrix for the same carbonate rock samples are shown in Table 5. The obtained $\delta^{13}C_{V\text{-}PDB}$ of the stylolite and matrix had characteristic values of −4~3‰ and −2.7~3.2‰ (Table 5). In different carbonate rock samples, the $\delta^{18}O_{V\text{-}PDB}$ of the stylolites was evidently lower than that of the matrix (Table 5). The $\delta^{18}O_{V\text{-}PDB}$ of the stylolites and matrix were, respectively, −12.6~−1.3‰ and −13.2~−2‰ (Table 5).

Figure 2. (**A**) Photograph of core showing the uneven surfaces of stylolites in Well Z3, 4317.60 m, O_2^2; (**B**) photograph of core showing stylolites are beaded in Well Wk1, 4251.68 m, C_2; (**C**) photomicrograph showing carbonate rock samples containing stylolites under plane-polarized light (the red color is to highlight the stylolites), Wk1, 4251.68 m, C_2; (**D**) photomicrograph of a closed zone of stylolite-rich laminae in (**C**) under plane-polarized light; (**E**) same field as (**D**) under reflective light; (**F**) photomicrograph of a closed zone of stylolite-rich laminae in (**C**) under plane-polarized light; (**G**) same field as (**F**) under fluorescent light.

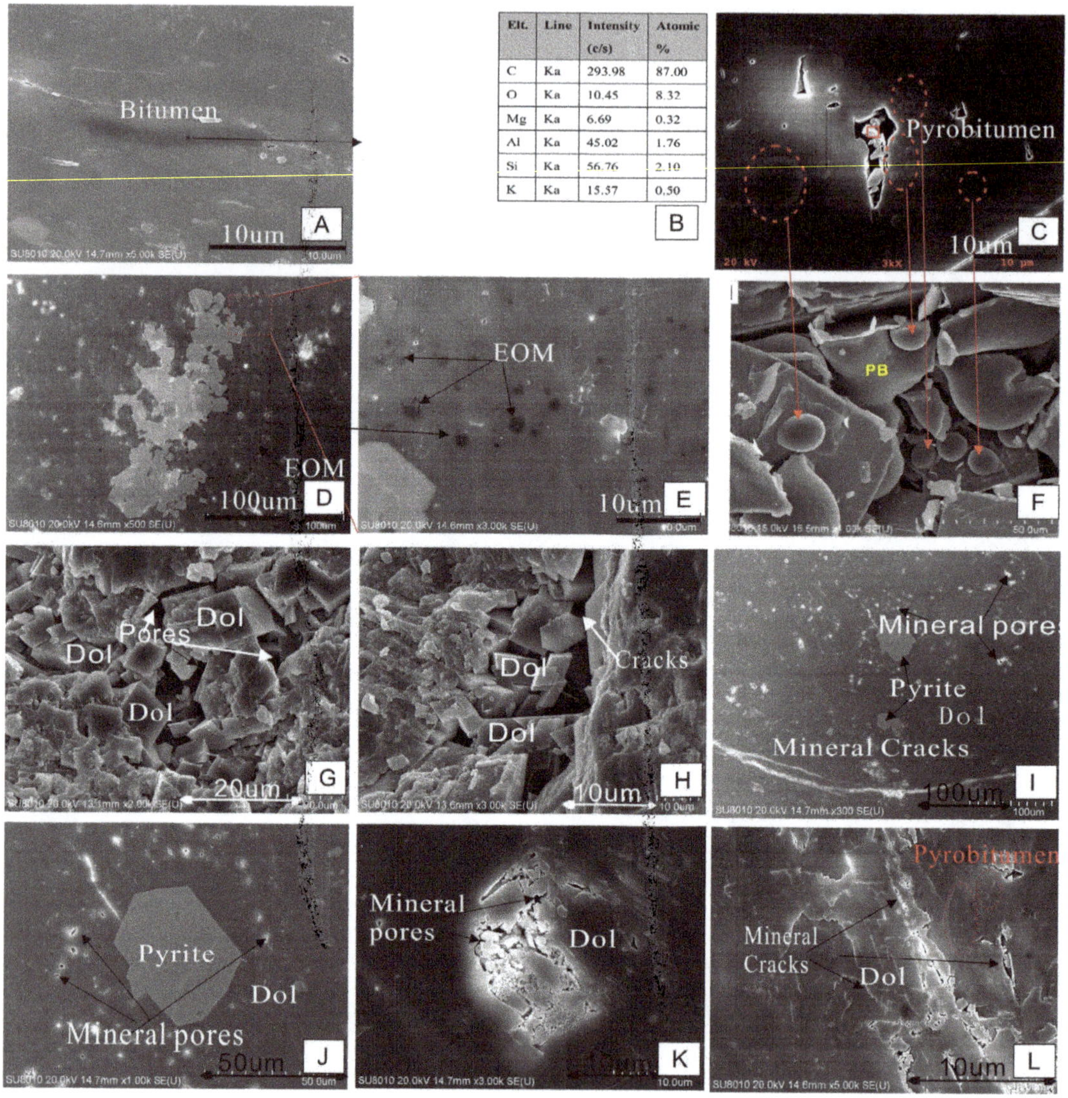

Figure 3. Microphotographs showing (**A–F**) the occurrence of organic matter and (**G–L**) the pores and cracks in stylolites: (**A**) bitumen with smooth surface, WK1, 4251.68 m, C_2; (**B**) the energy spectrum of the OM of A; (**C**) solid globular pyrobitumen pieces with a rough surface, pasted on the crystal face of dolomite, WK1, 4251.68 m, C_2; (**D**) dispersed EOM; when the energy spectrum is excited, the black spot disappears and bulges appear, which are combined with the fluorescent flakes and are considered to be soluble organic matter and secondary pyrite, WK1, 4251.68 m, C_2; (**E**) photomicrograph of a closed zone of dispersed soluble organic matter in (**D**); (**F**) photograph of pyrobitumen pieces with a rough surface, Adapted with permission from Ref. [58]. 2018, Chengyu Yang; (**G**) intergranular pores of secondary dolomite and with euhedral morphology, Z3, 4317.60 m, O_2^2; (**H**) the crack formed by secondary minerals, WK1, 4251.68 m, C_2; (**I–L**) intergranular pores and cracks of secondary dolomite and secondary pyrite in stylolites (the sample with argon ion polishing).

Table 4. The data from the XRD for the stylolites and matrix in carbonate rock samples.

Sample No.	Type and Content of Minerals (%)						Clay Minerals (%)
	Quartz	Plagioclase	Calcite	Dolomite	Pyrite	Anhydrite	
1-Ssr	5.1	2.6	11.8	47.6	14.1	0	17.5
1-Msr	2.4	2.6	56.1	27.2	3.4	0	5.3
2-Ssr	5.7	1.8	8.3	48.1	12.5	1.7	19.2
2-Msr	2.4	2.5	66.1	17.3	2.6	0	7.1
3-Ssr	5.3	2.1	12.4	45.5	10.6	0	22.5
3-Msr	2.6	1.5	58.7	26.4	5.7	0	3.1
4-Ssr	6.9	1.9	11.8	43.4	11.5	0	22.7
4-Msr	3.4	2.4	68.4	16.8	3.2	1.1	4.1
5-Ssr	6.6	2.7	11.5	41.3	9.3	0	26.3
5-Msr	2.4	1.4	59.3	27.9	2.2	0	2.8
6-Ssr	6.2	2.9	12.2	48.6	8.7	1.5	18.5
6-Msr	1.6	2.1	66.7	15.2	1.8	0	8.2
7-Ssr	6.8	2.8	9.7	44.8	13.1	0	18.1
7-Msr	2.2	2.7	63.4	18.1	2.7	0	6.9

Table 5. The δ^{13}C and δ^{18}O for the stylolites and matrix in carbonate rock samples.

Sample No.	$\delta^{13}C_{V\text{-}PDB}$ (‰)	$\delta^{18}O_{V\text{-}PDB}$ (‰)	$\Delta\delta^{13}$C (‰)	$\Delta\delta^{18}$O (‰)
1-Ssr	1.4	−10.8	0.4	−0.3
1-Msr	1.8	−11.1		
2-Ssr	1.6	−10.6	0.6	−0.4
2-Msr	2.2	−11		
3-Ssr	3	−11.1	0.2	0.6
3-Msr	3.2	−10.5		
4-Ssr	1	−12.6	0	−0.6
4-Msr	1	−13.2		
5-Ssr	−2.3	−8.1	−0.4	−0.4
5-Msr	−2.7	−8.5		
6-Ssr	0.3	−1.3	0.3	−0.7
6-Msr	0.6	−2		
7-Ssr	−4	−6.4	1.7	2.9
7-Msr	−2.3	−3.5		

5. Discussion

5.1. Source of Organic Matter in Stylolites

5.1.1. Comparison of Organic Geochemical Characteristics between Stylolites and Matrix

Organic matter is fundamental to hydrocarbon generation from source rocks, whether in conventional or unconventional petroleum systems, and including in emerging hydrocarbon resources, such as gas hydrates [1,59–61]. Generally, the parameters used to assess the organic matter abundance of source rocks include the total organic carbon (TOC), Rock-Eval pyrolysis $S_1 + S_2$, and organic solvent extract content [1,62–64]. According to data obtained from the samples analyzed (Tables 2 and 3), it is evident that the TOC, S_2, and $S_1 + S_2$ values of the stylolites were higher than those of the corresponding matrix in the same carbonate rock (Figure 4a,b,d). The values of the Rock-Eval pyrolysis S_1 between the stylolites and matrix for the same sample were not consistent; this may be attributed to the influence of migrated hydrocarbons in the matrix or the poor hydrocarbon expulsion efficiency in the matrix [42,65].

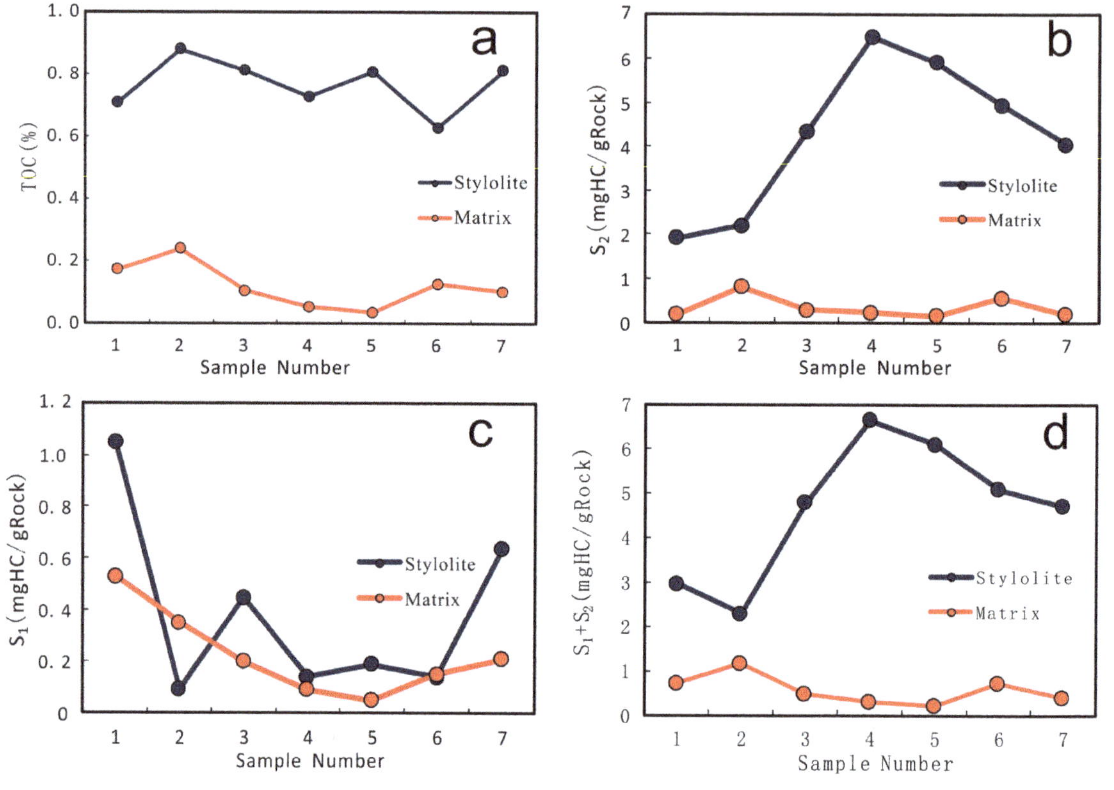

Figure 4. Comparison of geochemical parameters between different samples: (**a**) TOC (wt.%); (**b**) S_2 (mg HC/g rock); (**c**) S_1 (mg HC/g rock); (**d**) $S_1 + S_2$ (mg HC/g rock).

Parameters such as the Rock-Eval $S_1 + S_2$ (hydrocarbon generation potential) and HI (hydrogen index) were used as auxiliary indices to assess the hydrocarbon generation potential of the source rocks [1,59,66]. The residual hydrocarbon content of source rocks, a crucial parameter characterizing the generation and expulsion of hydrocarbons, can be reflected by the Rock-Eval pyrolysis S_1. When no hydrocarbon expulsion occurs, it can represent the total hydrocarbon amount generated in the source rock and reflect the hydrocarbon-generating characteristics of organic matter. Therefore, for source rocks with the same organic matter type and maturity, if no hydrocarbon expulsion occurs, then the S_1 values increase with the increasing TOC content, whereas there is no variation and slight variations in the ratios of the HCI (S_1/TOC). When hydrocarbon expulsion occurs, the S_1 values, solvent extract contents, and ratios of the HCI decrease [32]. In Figure 5A,B, some samples show extremely low TOC values for the matrix, indicating higher HI and hydrocarbon generation potential ($S_1 + S_2$), suggesting that the matrix near the stylolites was influenced by migrated hydrocarbons. Figure 5C reveals that the HCI (S_1/TOC) of the stylolites was lower than that of the matrix in the same sample, even though the corresponding TOC content was higher than that of the matrix. This indicates that the hydrocarbon expulsion efficiency of the stylolites is much higher than that of the matrix. In carbonate source rocks, stylolites serve as the primary channel for hydrocarbon discharge; therefore, the hydrocarbons formed in the matrix aggregated near the stylolites and caused a higher hydrocarbon index of the matrix adjacent to the stylolites [42].

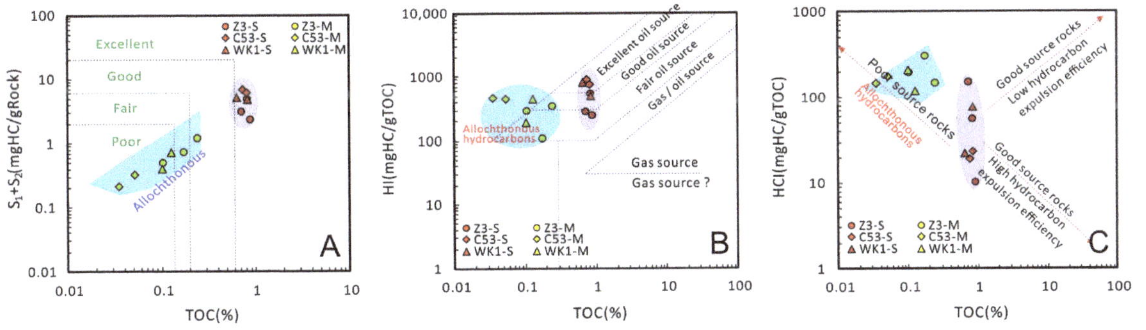

Figure 5. Diagrams of (**A**) TOC vs. $S_1 + S_2$, (**B**) TOC vs. HI, and (**C**) TOC vs. HCI of stylolites and matrix in the same carbonate rock samples.

5.1.2. Occurrence and Source of Organic Matter in Stylolites

Organic matter can be indirectly observed through fluorescence microscopes and scanning electron microscopes. Under the excitation of blue fluorescence, different types of organic matter will show different fluorescence characteristics. The fluorescence color of the hydrocarbon-generating parent material often varies with the evolution of the source rock. The fluorescence luminescence of organic matter tends to vary from yellow to brown to black as the thermal maturity of the organic matter evolves from low to high maturity [67–69]. Light oils tend to be green when excited by blue fluorescence, while heavier oils tend to be reddish brown. In scanning electron microscopy observation, organic matter can be directly identified through the analysis of energy spectrum data. From Figure 2F,G, it can be seen that the fluorescence intensity in the stylolite was extremely strong with green luminescence, indicating an enrichment in the light oil in the stylolite. The observation of the same sample under a scanning electron microscope revealed that the organic matter in the stylolite mainly migrated from the carbonate rock matrix, also known as migrated organic matter. For example, the organic matter observed in Figure 3A shows strong homogeneity and a high carbon content (Figure 3B), characteristic of typical solid bitumen. Figure 3C shows a spherical morphology, consistent with the intermediate state of bitumen, as previously published (Figure 3F) [58]. Figure 3D,E represents a relatively common form observed under scanning electron microscopy, characterized by small black dots. When excited by the energy spectrum, the dots disappear and bulges appear. Combined with the observation of thin sections under fluorescent luminescence, these dots are likely to be soluble organic matter adsorbed onto mineral surfaces. Ref. [42] compared the biomarkers of a stylolite and the matrix, and their findings suggested that the organic matter in the stylolite was inherited from the matrix, where it formed liquid hydrocarbons that migrated to the stylolite. The Rock-Eval pyrolysis Tmax is used as an auxiliary index for figuring out the thermal-maturity levels [1,59,66]. However, a high content of soluble organic matter can lead to a decrease in the Tmax [70]. Figure 6A shows that the Tmax values of the stylolites were lower than those of the matrix, and the ΔTmax displays a positive correlation with the ΔTOC, which is consistent with the understanding that the organic matter of stylolites is the result of the migration and accumulation of liquid hydrocarbon from the matrix to the stylolite.

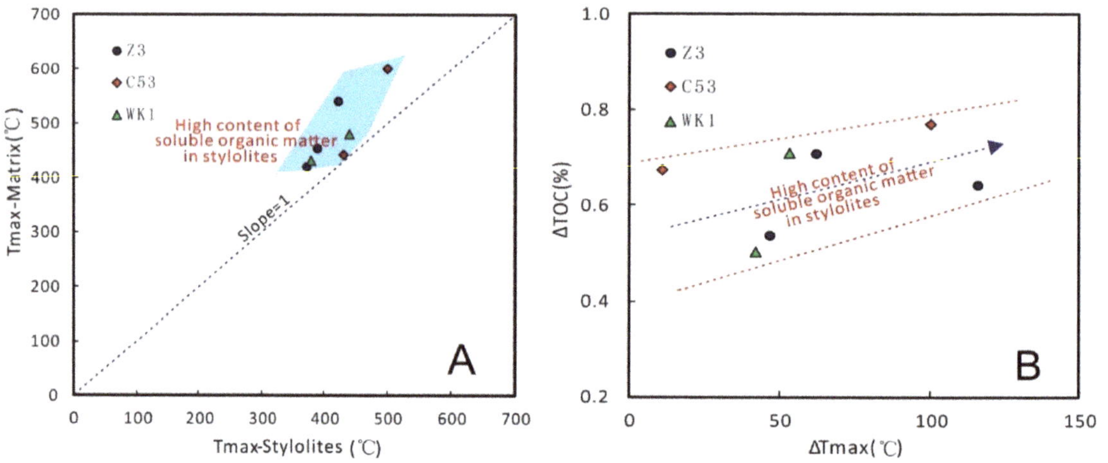

Figure 6. Diagrams of (**A**) Tmax between stylolites and matrix and (**B**) ΔTmax vs. ΔTOC for the same carbonate rock sample (ΔTmax = (Tmax-M) − (Tmax-S), ΔTOC = (TOC-S) − (TOC-M)).

5.2. Influence Mechanism of Stylolites on Migration and Accumulation of Geofluid

5.2.1. Mineral Composition Characteristics of Stylolites and Matrix

Through the study of the characteristics and genesis of carbonate rock stylolites, it is universally realized that the formation of stylolites is a process of the continuous dissolution of carbonate rocks and the gradual enrichment of insoluble residues [71]. Most of the stylolites in carbonate rocks are distributed in a beaded shape, with highly uneven thicknesses [39] (Figure 2). Thicker stylolites tend to have zones of stress relief, dissolution, and mineral precipitation, while thinner stylolites often exhibit stress concentration zones with strong diagenesis [39]. The dissolution and precipitation zones in stylolites provide favorable conditions for the formation of secondary minerals during the later stage of diagenesis. As stylolites form and carbonate rocks continuously dissolve, the salinity within the stylolites increases, creating a conducive environment for the formation of dolomite [72]. Numerous studies have explored the material compositions of stylolites, revealing that they are mainly composed of insoluble clay minerals, solid organic matter, asphalt, secondary pyrite, and dolomite [73]. In carbonate rocks, the precipitation of dolomite requires specific spatial conditions for the growth of soluble minerals and a high-salinity environment [74–76]. By comparing the mineral compositions of the stylolite and matrix in the same sample, it becomes evident that the matrix has a higher content of calcite, while the stylolite contains higher levels of insoluble quartz, clay minerals, and secondary dolomite and pyrite (Figure 7). The presence of dolomite and pyrite, with well-defined crystalline morphologies observed under scanning electron microscopy, shows the secondary genesis of minerals (Figure 3). Also, the clay minerals that are more abundant in stylolites can catalyze oil cracking [77].

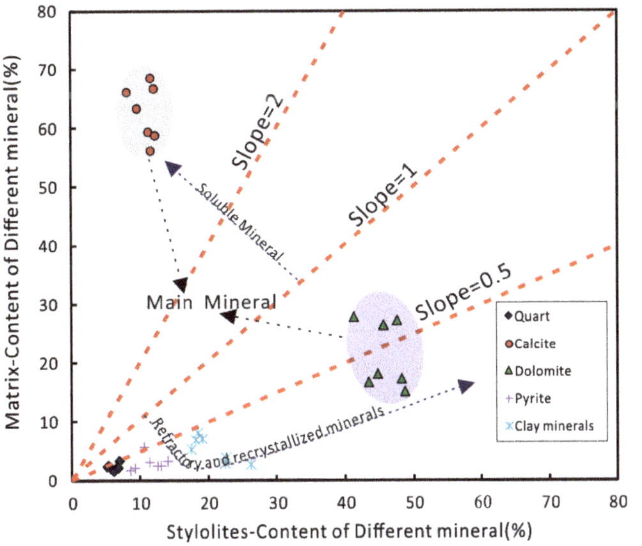

Figure 7. Diagrams of the contents of different minerals between stylolites and matrix for the same carbonate rock sample.

5.2.2. $\delta^{13}C$ and $\delta^{18}O$ Characteristics of Stylolites and Matrix

The analysis of stable isotopes, such as $\delta^{13}C$ and $\delta^{18}O$, is valuable for reconstructing the fluid properties during dolomite precipitation, especially the $\delta^{13}C$ value, which is influenced by the diagenesis after deposition [78–81]. The $\delta^{18}O$ value alongside the associated crossplot of $\delta^{13}C$ and $\delta^{18}O$, as the oxygen isotope exchanges between seawater/marine carbonate and meteoric/burial water, has tended to occur easily relative to other stable isotopes during diagenetic alteration [79,80,82–86]. By comparing the carbon isotope characteristics of the stylolites and matrix, it is found that the carbon isotope characteristics of the stylolites are lighter than that of the matrix (Figure 8A). During the formation of stylolites, the dissolution of carbonate minerals in the matrix leads to the release of lighter $\delta^{12}C$, resulting in the enrichment of lighter stable carbon isotopes in the stylolites. The formation of secondary dolomite in the stylolites indicates the high salinity of the geofluid, which can be reflected by the salinity of the diagenetic fluid [87]. The high content of secondary dolomite in the stylolites suggests higher $\delta^{18}O$ values than those of the matrix, but there were some outsiders for the variation law of the $\delta^{18}O$ values in the stylolites and matrix (Figure 8B). This is possibly due to the precipitation of extremely light oxygen isotopes in the matrix, resulting in greater $\delta^{18}O$ values. Additionally, the diagenesis is not as strong as the isotopic fractionation in certain cases. The $\Delta\delta^{13}C$ and $\Delta\delta^{18}O$ values of the stylolites and matrix show a certain positive correlation (Figure 8C), further supporting the differences in the compositions of the carbon and oxygen isotopes of the studied samples.

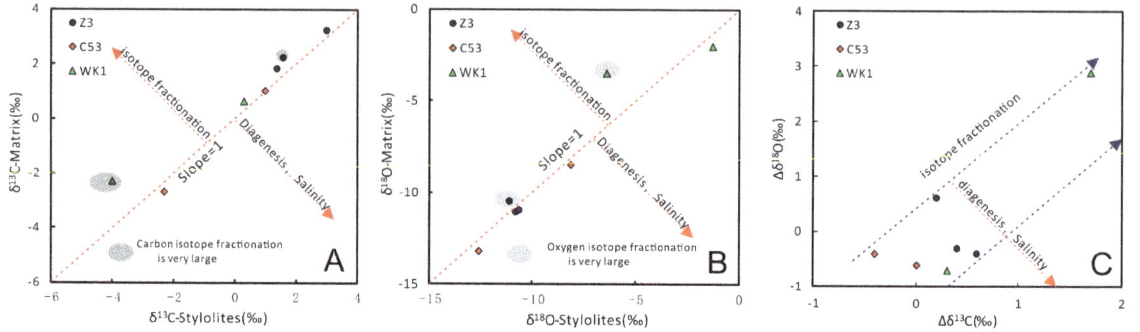

Figure 8. Diagrams of (**A**) $\delta^{13}C$ and (**B**) $\delta^{18}O$ between stylolites and matrix and (**C**) $\Delta\delta^{13}C$ versus $\Delta\delta^{18}O$ for the same carbonate rock samples ($\Delta\delta^{13}C = (\delta^{13}C - S) - (\delta^{13}C - M)$, $\Delta\delta^{18}O = (\delta^{18}O\text{-}S) - (\delta^{18}O\text{-}M)$).

5.2.3. Microscopic Characteristics and Migration Effect of Stylolites

A stylolite is a three-dimensional structure with a certain thickness, and its ability to serve as an enrichment zone and migration channel of organic fluid can be deciphered by the observation of the microscopic characteristics of its pore structure. Scanning electron microscopy is an effective method for studying the microscopic pore structure [88,89]. From Figure 3, it is evident that the stylolites show well-developed secondary dolomites with intact crystalline morphologies (Figure 3G,H), which can retain good intercrystalline pores. The brittleness of dolomite allows for the formation of many intergranular cracks due to tectonic stress (Figures 2G and 3H,I,L). These secondary pores and cracks serve as important storage spaces and migration channels for geofluid.

Liu et al. (2020) [32] compared the molecular markers of a matrix and stylolites and found that the light-to-weight ratio parameters in saturated hydrocarbons (C_{21}^-/C_{22}^+, tricyclic/pentacyclic terpane, and ($C_{21} + C_{22}$) pregnane/regular sterane index) revealed that small molecular compounds are more likely to enter stylolites than macromolecular compounds, and it has been found that the migration differentiation is more pronounced when hydrocarbons migrate along stylolites. During the formation of stylolites, fluids carrying dissolved carbonate move along the stylolites or to the matrix adjacent to the stylolites. When conditions change, the dissolved mineral ions will recrystallize and act as cementation, leading to a decrease in the porosity of the matrix around the stylolites [90,91]. By observing the thin sections under reflected light, the precipitation of tight minerals can be observed at the contact between the stylolites and matrix (Figure 9A,B), accompanied by the formation of secondary pyrite (Figure 9B). The strong cementation at the contact interface and the formation of secondary minerals nearby may be the cause of the fractionation effect when the hydrocarbon fluid migrates from the matrix to the stylolites. Considering an individual stylolite as a whole, its thickness is not uniform. Through the observation of the stylolite under fluorescence light, it can be found that the wider part of the stylolite exhibits a stronger fluorescence intensity, and the central part of the stylolite is evidently stronger than the contact between the stylolite and matrix (Figure 9C,D). Therefore, the regions where the stylolite is thinner will cause an evident fractionation effect (Figure 10).

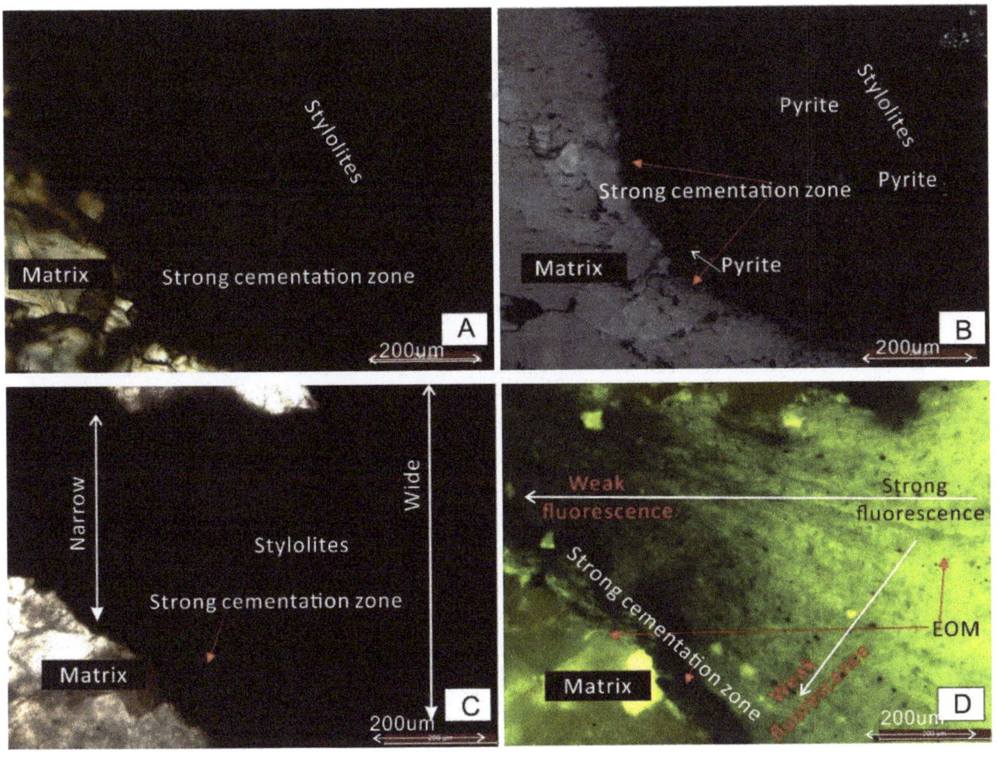

Figure 9. (**A**,**B**) Thin-section photomicrograph showing the structure of stylolites under plane-polarized light (**A**) and reflected light, Well Z3, 4317.60 m, O_2^2; (**C**) photomicrograph showing the structure of stylolites under plane-polarized light, Well Wk1, 4251.68 m, C_2; (**D**) same field as (**C**) under fluorescent light.

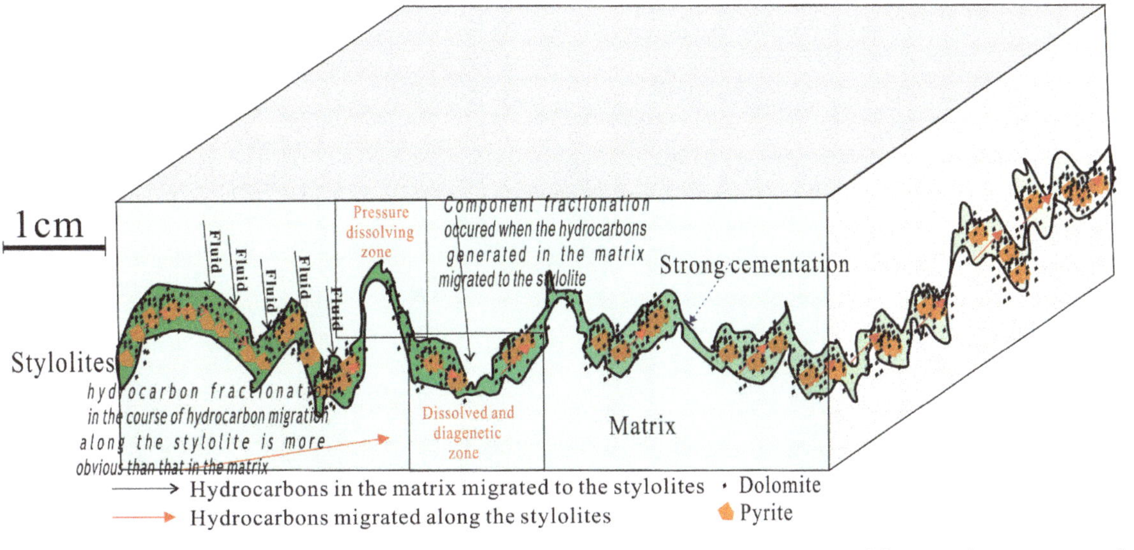

Figure 10. Model of organic matter source and fluid migration of stylolites in carbonate source rock.

5.3. Geological Significance of Stylolites in Carbonate Rocks for Oil and Gas–Source Correlation

Stylolites are relatively developed in carbonate rocks. The abovementioned research shows that stylolites are beneficial to the enrichment and migration of soluble organic matter. In carbonate rock oil and gas reservoirs, when stylolites are well developed, the thermal cracking of the enriched soluble organic matter can lead to the generation of significant amounts of natural gas. The source of natural gas in the gas reservoir is affected by the natural gas formed by the thermal cracking of the organic matter in the stylolites, resulting in the enrichment of the $\delta^{13}C$ in the gas reservoir, which exerts a certain influence on the identification of natural gas sources [92,93].

Stylolites are essentially important for the migration of crude oil, and they can form a network channel for fluid migration by combining with fracture. However, crude oil undergoes a strong fractionation effect during the migration process in stylolites. As the migration distance increases, the properties and biomarkers of crude oil change to a certain extent, and the contents of the light components and light molecules change, making it challenging to determine the sources of crude oils accurately. In such cases, it is safer to comprehensively explain the spatial variation in the crude oil properties.

6. Conclusions

Stylolites are a common geological phenomenon found in carbonate rocks. The analyses of the total organic carbon, Rock-Eval pyrolysis, and petrography show that the organic matter in the stylolites is primarily composed of EOM, pyrobitumen, and bitumen, indicating that the organic matter in the stylolites is mainly secondary organic matter.

Comparisons of the mineral composition, $\delta^{13}C$ and $\delta^{18}O$ characteristics, and microscopic characteristics between the stylolites and the matrix reveal that the secondary dolomite and pyrite minerals in the stylolites are more developed, and the secondary intercrystalline pores and intercrystalline fractures in the stylolites provide storage space and migration channels for oil and gas. The contact interface between the stylolites and matrix has a strong cementation, which impedes the migration of oil and gas from the matrix to the stylolites. This causes the migration fractionation effect.

For the oil and gas in carbonate rock reservoirs, the influence of stylolites on the oil and gas properties should be considered. The enrichment of soluble organic matter in stylolites can lead to the continuous production of thermally cracked gas at high maturity levels, impacting the group composition and isotope characteristics of the natural gas in the gas reservoir. When stylolites serve as effective migration channels, the fractionation effect on the crude oil during the migration process can cause the loss of the biomarkers of crude oil, making it challenging to determine the oil–source relationship.

Author Contributions: Writing—original draft, S.L. (Shengnan Liu); Writing—review & editing, S.L. (Shiju Liu) and G.G.; Supervision, R.Z. All authors have read and agreed to the published version of the manuscript.

Funding: This research was funded by Natural Science Foundation of China (No. 411372142 and No. U22B6004) and research project of PetroChina Research Institute of Exploration and Development (2022yjcq03). G.G. (Gang Gao) is the founder of Natural Science Foundation of China (No. 411372142); W.Z. (Wenzhi Zhao) is the founder of Natural Science Foundation of China (No. U22B6004); R.Z. (Rukai Zhu) is the founder of research project of PetroChina Research Institute of Exploration and Development (2022yjcq03).

Data Availability Statement: The data presented in this study are available on request from the corresponding author. The data are not publicly available due to ethical restrictions.

Acknowledgments: Thanks to the funds provided by the Natural Science Foundation of China (No. 411372142 and No. U22B6004) and the research project of the PetroChina Research Institute of Exploration and Development (2022yjcq03), and to the guidance provided by the China University of Petroleum (Beijing), as well as the experiment and testing support sponsored by the State Key Laboratory of Petroleum Resources and Prospecting, Beijing.

Conflicts of Interest: The authors declare no conflict of interest.

References

1. Tissot, B.P.; Welte, D.H. *Petroleum Formation and Occurrence*, 2nd ed.; Springer: New York, NY, USA, 1984; ISBN 978-3-642-96448-0.
2. Hunt, J.M. *Petroleum Geochemistry and Geology*; Freeman and Company: San Francisco, CA, USA, 1979; pp. 5–24. ISBN 0716724413.
3. Hunt, J.M.; Whelan, J.K.; Eglinton, L.B.; Cathles, L.M., III. Gas generation—A major cause of deep Gulf Coast overpressures. *Oil Gas J.* **1994**, *92*. [CrossRef]
4. Palacas, J.G. Characteristics of carbonate source rocks of petroleum. In *Petroleum Systems of the United States*; Magoon, L.B., Ed.; Geology Survey Bulletin; Department of the Interior: Washington, DC, USA, 1989; Volume 1870, pp. 20–25.
5. Hao, S.; Wang, F.; Gao, G.; Gang, W. Characteristics and evaluation of high over mature hydrocarbon source rocks in the Lower Paleozoic. *Explor. Oil Gas* **1996**, *1*, 25.
6. Lu, S.F.; Zhong, N.N.; Xue, H.T.; Pan, C.; Li, J.; Li, H. Chemical kinetics study of hydrocarbon regeneration from organic matter in carbonate source rocks and its significance. *Sci. China* **2007**, *50*, 536–543. [CrossRef]
7. Liu, Q.Y.; Jin, Z.J.; Liu, W.H.; Lu, L.F.; Meng, Q.X.; Tao, Y.; Han, P. Presence of carboxylate salts in marine carbonate strata of the Ordos Basin and their impact on hydrocarbon generation evaluation of low TOC, high maturity source rocks. *Sci. China Earth Sci.* **2013**, *56*, 2141–2149. [CrossRef]
8. Chen, F.; Lu, S.; Ding, X.; He, X.; Xing, X. The splicing of backscattered scanning electron microscopy method used on evaluation of microscopic pore characteristics in shale sample and compared with results from other methods. *J. Pet. Sci. Eng.* **2018**, *160*, 207–218. [CrossRef]
9. Jones, R.W. *Comparison of Carbonate and Shale Source Rocks: Abstract*; Bulletin; AAPG: Tulsa, OK, USA, 1984. [CrossRef]
10. Sassen, R.; Moore, C.H.; Meendsen, F.C. Distribution of hydrocarbon source potential in the Jurassic Smackover Formation. *Org. Geochem.* **1987**, *11*, 379–383. [CrossRef]
11. Sassen, R. Geochemistry of carbonate source rocks and crude oils in Jurassic salt basins of the gulf coast. In *Classic Petroleum Provinces*; Brooks, J., Ed.; Speciel Publication; Geological Society: London, UK, 1990; Volume 50, pp. 265–277. [CrossRef]
12. Radwan, A.E.; Trippetta, F.; Kassem, A.A.; Kania, M. Multi-scale characterization of unconventional tight carbonate reservoir: Insights from October oil filed, Gulf of Suez rift basin, Egypt. *J. Pet. Sci. Eng.* **2021**, *197*, 107968. [CrossRef]
13. Balaky, S.M.; Al-Dabagh, M.M.; Asaad, I.S.; Tamar-Agha, M.; Ali, M.S.; Radwan, A.E. Sedimentological and petrophysical heterogeneities controls on reservoir characterization of the Upper Triassic shallow marine carbonate Kurra Chine Formation, Northern Iraq: Integration of outcrop and subsurface data. *Mar. Pet. Geol.* **2023**, *149*, 106085. [CrossRef]
14. Dunnington, H.V. Aspects of diagenesis and shape change in stylolitic limestone reservoirs. In *World Petroleum Congress*; WPC: Beijing, China, 1967; p. WPC-12129.
15. Leythaeuser, D.; Borromeo, O.; Mosca, F.; di Primio, R.; Radke, M.; Schaefer, R.G. Pressure solution in carbonate source rocks and its control on petroleum generation and migration. *Mar. Pet. Geol.* **1995**, *12*, 711–733. [CrossRef]
16. Aharonov, E.; Katsman, R. Interaction between pressure solution and clays in stylolite development: Insights from modeling. *Am. J. Sci.* **2009**, *309*, 607–632. [CrossRef]
17. Farooq, U.; Caetano, H.; Radwan, E.S. The Impact of Dolomitization on Reservoir Quality Evolution of the Fractured Carbonate Reservoir Upper Cretaceous, Onshore Abu Dhabi Oilfield, UAE. In *SPE Reservoir Characterisation and Simulation Conference and Exhibition?* SPE: Richardson, TX, USA, 2017; p. D031S013R001.
18. Toussaint, R.; Aharonov, E.; Koehn, D.; Gratier, J.P.; Ebner, M.; Baud, P.; Rolland, A.; Renard, F. Stylolites: A review. *J. Struct. Geol.* **2018**, *114*, 163–195. [CrossRef]
19. Magoon, L.B.; Dow, W.G. The petroleum system. In *The Petroleum System—From Source to Trap*; Magoon, L.B., Dow, W.G., Eds.; American Association of Petroleum Geologists Memoir 60; AAPG: Tulsa, OK, USA, 1994; pp. 3–24.
20. Gordon, C.H. On the nature and origin of the stylolitic structure in tennessee marble. *J. Geol.* **1918**, *26*, 561–568. [CrossRef]
21. Jose, R.; Andrade, R. Stylolites measurement of rock loss. *Rev. Bras. Geociênc.* **2000**, *30*, 432–435.
22. Amstutz, G.C.; Park, W.C. Stylolites of diagenetic age and their role in the interpretation of the southern Illinois fluorspar deposits. *Miner. Depos.* **1967**, *2*, 44–53. [CrossRef]
23. Bathurst, R.G.C. *Carbonate Sediments and Their Diagenesis*; Robin, G.C., Ed.; Developments in Sedimentology 12; Elsevier: New York, NY, USA, 1975; ISBN 9780080869230.
24. Ricken, W. *Diagenetic Bedding*; Lecture Notes Earth Science 6; Springer: Berlin, Germany, 1986; ISBN 978-3-540-16494-4.
25. Scholle, P.A.; Halley, R.B. Burial diagenesis: Out of sight, out of mind. *Carbonate Sedimentol. Petrol. Short Course Geol.* **1989**, *4*, 135–160. [CrossRef]
26. Moore, C.H. *Carbonate Diagenesis and Porosity*; Moore, C.H., Ed.; Development in Sedimental 48; Elsevier: New York, NY, USA, 1997; ISBN 9780080869605.
27. Sassen, R.; Chinn, E.W. Effects of elemental sulfur during programmed pyrolysis of kerogen. *Org. Geochem.* **1989**, *14*, 475–477. [CrossRef]
28. Sassen, R.; Moore, C.H. Framework of hydrocarbon generation and destruction in eastern Smackover Trend. *AAPG Bull.* **1988**, *72*, 649–663. [CrossRef]
29. Sassen, R. Geochemical and carbon isotopic studies of crude oil destruction, bitumen precipitation, and sulfate reduction in the deep Smackover Formation. *Org. Geochem.* **1988**, *12*, 351–361. [CrossRef]

30. Gao, G. Effects of Stylolites on the Generation, Discharge, Migration and Accumulation of Hydrocarbon in Carbonate Rock. *J. Xi'an Shiyou Univ. (Nat. Sci. Ed.)* **2000**, *15*, 32–36, (In Chinese with English abstract).
31. Gao, G. Research Status and Oil & Gas Geology Significance of Carbonate Rock Stylolite. *Nat. Gas Geosci.* **2013**, *24*, 218–226.
32. Gao, G.; Xuan, W.; Zhang, W.W.; Dong, Y.; Wang, C.C.; Li, J.Y. Significance of stylolite on organic matter enrichment and fluid migration: A case study on stylolite in the Ordovician carbonate rocks of Ordos Basin. *Pet. Sci. Bull.* **2017**, *2*, 1–11. [CrossRef]
33. Primio, R.; Leythaeuser, D. Quantification of the effect of carbonate redistribution by pressure solution in organic-rich carbonates. *Mar. Pet. Geol.* **1995**, *12*, 735–739. [CrossRef]
34. Ramsdom, R.M. Stylolites and oil migration. *AAPG Bull.* **1952**, *36*, 2185–2186.
35. Котяхов, ф.И. *Трещиноватые породы и из коллекторские свойства* (Fractured Rocks and Their Collector Properties); Государственное научно-техническое издательство нефтяной и горно-топливной литературы (State Scientific and Technical Publishing House of Oil and Mining Fuel Literature): Moscow, Russia, 1958.
36. He, B. Oil seedlings and asphalt in Guangxi. *Earth Sci.* **1959**, *5*, 8–13.
37. Zhou, D.; Li, G.; Luo, P.; Zhang, D.; Tang, H. Study on Diageneis of Silurian Shiniulan Formation, Southeastern Sichuan Basin. *Nat. Gas Technol. Econ.* **2009**, *3*, 6–11.
38. Lind, I.; Nykjaer, O.; Priisholm, S.; Springer, N. Permeability of Stylolite-Bearing Chalk. *J. Pet. Technol.* **1994**, *46*, 986–993. [CrossRef]
39. Heap, M.J.; Baud, P.; Reuschlé, T.; Meredith, P.G. Stylolites in limestones: Barriers to flow? *Geology* **2014**, *42*, 51–54. [CrossRef]
40. Rustichelli, A.; Tondi, E.; Korneva, I.; Baud, P.; Vinciguerra, S.; Agosta, F.; Reuschlé, T.; Janiseck, J. Bedding parallel stylolites in shallow-water limestone successions of the Apulian Carbonate Platform (central-southern Italy). *Ital. J. Geosci.* **2015**, *134*, 513–534. [CrossRef]
41. Heap, M.; Reuschlé, T.; Baud, P.; Renard, F.; Iezzi, G. The permeability of stylolite-bearing limestone. *J. Struct. Geol.* **2018**, *116*, 81–93. [CrossRef]
42. Liu, S.; Gao, G.; Qu, T.; Dang, W.; Zhang, W.; Yang, S.; Zhu, K. Implications of organic matter source and fluid migration from geochemical characteristics of stylolites and matrix in carbonate rocks: A case study from the Carboniferous and the Ordovician in the Sichuan Basin, SW China. *J. Pet. Sci. Eng.* **2020**, *186*, 106606. [CrossRef]
43. Rong, Y.; Hu, M.; Tan, Y. Carboniferous tectonic evolution and hydrocarbon accumulation relationship analysis of Kaijiang Area in eastern Sichuan basin. *J. Chongqing Univ. Sci. Technol. (Nat. Sci. Ed.)* **2010**, *12*, 24–29.
44. Wu, T.; Xie, S.; Zhang, D.; Li, P.; Jiao, C. Geochemical characteristics and fluid origin of the Dengying Formation dolomites in southern Sichuan Basin. *Oil Gas Geol.* **2016**, *37*, 721–730. [CrossRef]
45. Liu, J.; Zhu, X.; Wang, S.; Wei, J. Geologic-tectonic evolutional characteristics and prospecting potential for ISL-amenable sandstone-type uranium deposits in Sichuan basin. *Uranium Geol.* **2005**, *21*, 321–330. [CrossRef]
46. Guo, T. Charateristics and exploration potential of Ordovician reservoirs in Sichuan Basin. *Oil Gas Geol.* **2014**, *35*, 372–378. [CrossRef]
47. Baud, P.; Rolland, A.; Heap, M.; Xu, T.; Nicolé, M.; Ferrand, T.; Reuschlé, T.; Toussaint, R.; Conil, N. Impact of stylolites on the mechanical strength of limestone. *Tectonophysics* **2016**, *690*, 4–20. [CrossRef]
48. Zhan, H.; Dong, W.; Chen, S.; Hu, D.; Zhou, H.; Luo, J. Improved test method for convection heat transfer characteristics of carbonate fractures after acidizing etching. *Adv. Geo-Energy Res.* **2021**, *5*, 376–385. [CrossRef]
49. Liao, Q.; Xue, L.; Wang, B.; Lei, G. A new upscaling method for microscopic fluid flow based on digital rocks. *Adv. Geo-Energy Res.* **2022**, *6*, 357–358. [CrossRef]
50. Liu, S.; David, M.; Gang, W.; Li, J.; Jin, J.; Duan, Y.; Xiang, B.; Gao, G.; Zhang, Y.; Wang, M.; et al. Evaluation of the tight oil "sweet spot" in the Middle Permian Lucaogou Formation (Jimusaer Sag, Junggar Basin, NW China): Insights from organic petrology and geochemistry. *Org. Geochem.* **2023**, *177*, 104570. [CrossRef]
51. Liu, C.; Zhang, L.; Li, Y.; Liu, F.; Martyushev, D.A.; Yang, Y. Effects of microfractures on permeability in carbonate rocks based on digital core technology. *Adv. Geo-Energy Res.* **2022**, *6*, 86–90. [CrossRef]
52. Shiju, L.; Misch, D.; Gao, G.; Jin, J.; Gang, W.; Yanjuan, D.; Wu, X.; Xiang, B.; Wang, M.; Luo, Q. Physical properties of lacustrine shale oil: A case study on the lower member of the Lucaogou Formation (Jimusaer Sag, Junggar Basin, NW China). *Mar. Pet. Geol.* **2022**, *145*, 105888. [CrossRef]
53. Ying, R.; Zhong, D.; Gao, C.; Liang, T.; Sun, H.; Wu, D.; Zheng, X. High-resolution carbon isotope records and correlations of the lower Cambrian Longwangmiao formation (stage 4, Toyonian) in Chongqing, South China. *Palaeogeogr. Palaeoclimatol. Palaeoecol.* **2017**, *485*, 572–592. [CrossRef]
54. Dong, S.; Chen, D.; Qing, H.; Jiang, M.; Zhou, X. In situ stable isotopic constraints on dolomitizing fluids for the hydrothermally-originated saddle dolomites at Keping, Tarim Basin. *Sci. Bull. Engl. Version* **2013**, *58*, 2877–2882. [CrossRef]
55. Ying, R.; Zhong, D.; Gao, C.; Li, B.; Cao, X.; Wang, A.; Dong, Y.; Yan, T. The paleoenvironmental evolution of the Cambrian Longwangmiao Formation (Stage 4, Toyonian) on the Yangtze Platform, South China: Petrographic and geochemical constrains. *Mar. Pet. Geol.* **2019**, *100*, 391–411. [CrossRef]
56. Frimmel, H.E. An evaporitic facies in Neoproterozoic post-glacial carbonates: The Gifberg Group, South Africa. *Gondwana Res.* **2008**, *13*, 453–468. [CrossRef]
57. Zhang, P.; Liu, W.; Hua, H. Isotopic and REE evidence for the paleoenvironmental evolution of the late Ediacaran Dengying Section, Ningqiang of Shaanxi Province, China. *Precambrian Res.* **2014**, *242*, 96–111. [CrossRef]

58. Yang, C.; Ni, Z.; Li, M.; Wang, T.; Chen, Z.; Hong, H.; Tian, X. Pyrobitumen in South China: Organic petrology, chemical composition and geological significance. *Int. J. Coal Geol.* **2018**, *188*, 51–63. [CrossRef]
59. Peters, K.E.; Walters, C.C.; Moldowan, J.M. *The Biomarker Guide*, 2nd ed.; Biomarkers and Isotopes in Petroleum Systems and Earth History (II); Cambridge University Press: Cambridge, UK, 2005; ISBN 9780521837620.
60. Burton, Z.F.M. Sediment organic contents required for gas hydrate formation: A survey of published basin and hydrocarbon system models. *Fuels* **2022**, *3*, 580–587. [CrossRef]
61. Burton, Z.F.; Dafov, L.N. Testing the sediment organic contents required for biogenic gas hydrate formation: Insights from synthetic 3-D basin and hydrocarbon system modelling. *Fuels* **2022**, *3*, 555–562. [CrossRef]
62. Huang, D.F.; Li, J.C.; Gu, X.Z. *Evolution and Hydrocarbon-Generating Mechanism of Terrigenous Organic Matter*; Petroleum Industry Press: Beijing, China, 1984.
63. Baban, D.H. Sedimentary organic matter and source rock potential of the Paleocene Aaliji Formation in Qumar Oil Field, NE Iraq. *Arab. J. Geosci.* **2013**, *7*, 4733–4744. [CrossRef]
64. El Diasty, W.S. Khatatba Formation as an active source rock for hydrocarbons in the northeast Abu Gharadig Basin, north Western Desert, Egypt. *Arab. J. Geosci.* **2015**, *8*, 1903–1920. [CrossRef]
65. Gao, G.; Hao, S.; Wang, H. Characteristics of hydrocarbon generation and expulsion in matrix and stylolite of carbonate rocks. *Sci. China Ser. D Earth Sci.* **1999**, *42*, 202–206. [CrossRef]
66. Cheng, K.M. *Oil and Gas Generation of Tuha Basin*; Petroleum Industry Press: Beijing, China, 1994; ISBN 7502121498.
67. Teichmuller, M.; Wolf, M. Application of fluorescence microscopy in coal petrology and oil exploration. *J. Microsc.* **1977**, *109*, 49–73. [CrossRef]
68. Xiao, X. Alteration patterns of fluorescence intensity of mineral-bituminous groundmass in oil source rocks and its relationship with maturation level. *Prog. Nat. Sci.* **1991**, *1*, 240–246.
69. Xiao, X. Organic petrology and its applocation to the evaluation of oil and gas generation. *Adv. Earth Sci.* **1992**, *7*, 39.
70. Zhang, Z.; Wu, L.; Shu, N. Cause analysis of abnormal Tmax values on Rock-Eval pyrolysis. *Pet. Explor. Dev.* **2006**, *33*, 72. [CrossRef]
71. Burgess, C.J.; Peter, C.K. Formation, Distribution, and Prediction of Stylolites as Permeability Barriers in the Thamama Group, Abu Dhabi. In *SPE Middle East Oil and Gas Show and Conference*; SPE: Richardson, TX, USA, 1985. [CrossRef]
72. Zhang, B.; Guan, Q.; Pan, L. Compaction of carbonate rocks in ordos basin. *Earth Sci.* **1995**, *20*, 299–305.
73. Liu, H. The formation of suture and its significance in petroleum geology. *Geol. Rev.* **1959**, *19*, 367–379.
74. Wei, X.; Zhu, Y.J.; Xu, H.; Zhao, G. Discussion on Neogene dolostone forming condition in Xisha Islands: Evidences from isotope C and O and fluid inclosures. *Acta Petrol. Sin.* **2006**, *22*, 2394–2404. [CrossRef]
75. Huang, Q.; Zhang, S.; Ye, N.; Li, Y. Petrologic, geochemical characteristics and origin of the Lower Ordovician dolomite in Yubei area. *Oil Gas Geol.* **2014**, *35*, 391–400. [CrossRef]
76. Jiang, W.; Hou, M.; Xing, F.; Xu, S.; Ling, L. Diagenetic characteristics of dolomites in the Cambrian Loushanguan Group in southestern Sichuan Basin. *Pet. Geol. Exp.* **2016**, *38*, 311–319.
77. Qi, Y.; Cai, C.; Sun, P.; Wang, D.; Zhu, H. Crude oil cracking in deep reservoirs: A review of the controlling factors and estimation methods. *Pet. Sci.* **2023**, *20*, 1978–1997. [CrossRef]
78. Keith, M.L.; Weber, J.N. Carbon and oxygen isotopic composition of selected limestones and fossils. *Geochim. Cosmochim. Acta* **1964**, *28*, 1787–1816. [CrossRef]
79. Kaufman, A.J.; Hayes, J.M.; Knoll, A.H.; Gerard J., B. Isotopic compositions of carbonates and organic carbon from upper Proterozoic successions in Namibia: Stratigraphic variation and the effects of diagenesis and metamorphism. *Precambrian Res.* **1991**, *49*, 301–327. [CrossRef] [PubMed]
80. Veizer, J.; Ala, D.; Azmy, K.; Bruckschen, P.; Buhl, D.; Bruhn, F.; Carden, G.A.; Diener, A.; Ebneth, S.; Goddéris, Y.; et al. 87Sr/86Sr, δ13C and δ18O evolution of Phanerozoic seawater. *Chem. Geol.* **1999**, *161*, 59–88. [CrossRef]
81. Knauth, L.P.; Kennedy, M.J. *Adelaide Research and Scholarship: The Late Precambrian Greening of the Earth*; Nature Publishing Group: Berlin, Germany, 2009.
82. Brand, U.; Veizer, J. Chemical diagenesis of a multicomponent carbonate system-2, stable isotopes. *J. Sediment. Pet.* **1981**, *51*, 987–997. [CrossRef]
83. Burdett, J.W.; Grotzinger, J.P.; Arthur, M.A. Did major changes in the stable-isotope composition of Proterozoic seawater occur? *Geology* **1990**, *18*, 227–230. [CrossRef]
84. Frank, T.D.; Lyons, T.W.; Lohmann, K.C. Isotopic evidence for the paleoenvironmental evolution of the Mesoproterozoic Helena Formation, Belt Supergroup, Montana, USA. *Geochim. Cosmochim. Acta* **1997**, *61*, 5023–5041. [CrossRef]
85. Kumar, B.C.; Sharma, S.; Sreenivas, B.; Rao, M.S. Carbon, oxygen and strontium isotope geochemistry of Proterozoic carbonate rocks of the Vindhyan Basin, central India. *Precambrian Res.* **2003**, *121*, 289–291. [CrossRef]
86. Frank, T.D.; Lyons, T.W. *The Integrity of δ18O Records in Precambrian Carbonates: A Mesoproterozoic Case Study*; SEPM (Society for Sedimentary Geology): Tulsa, OK, USA, 2000. [CrossRef]
87. Rostek, F.; Ruhlandt, G.; Bassinot, F.C.; Muller, P.J.; Labeyrie, L.D.; Yves, L.; Bard, E. Reconstructing sea surface temperature and salinity using sigma 18O and alkenone records. *Nature* **1993**, *364*, 319. [CrossRef]
88. Lim, S.; Lee, H.S.; Kawashima, S. Pore structure refinement of cement paste incorporating nanosilica: Study with dual beam scanning electron microscopy/focused ion beam (SEM/FIB). *Mater. Charact.* **2018**, *145*, 323–328. [CrossRef]

89. Ma, Y.; Ardakani, O.H.; Zhong, N.; Liu, H.; Zhang, C. Possible pore structure deformation effects on the shale gas enrichment: An example from the lower cambrian shales of the eastern upper yangtze platform, south china. *Int. J. Coal Geol.* **2020**, *217*, 103349. [CrossRef]
90. Elizabeth, A.F.; Bruce, H.W. Stylolitization as source of cement in Mississippian Salem limestone, west-central Indiana. *AAPG Bull.* **1990**, *74*, 174–186. [CrossRef]
91. Cai, J.X. Characteristics of sutures and their forming mechanism. *Acta Petrol. Sin.* **1990**, *5*, 51–60.
92. Hu, T.; Pang, X.; Jiang, F.; Zhang, C.; Wu, G.; Hu, M.; Jiang, L.; Wang, Q.; Xu, T.; Hu, Y.; et al. Dynamic continuous hydrocarbon accumulation (DCHA): Existing theories and a new unified accumulation model. *Earth-Sci. Rev.* **2022**, *232*, 104109. [CrossRef]
93. He, T.-H.; Li, W.-H.; Lu, S.-F.; Yang, E.-Q.; Jing, T.-T.; Ying, J.-F.; Zhu, P.-F.; Wang, X.-Z.; Pan, W.-Q.; Zhang, B.-S.; et al. Quantitatively unmixing method for complex mixed oil based on its fractions carbon isotopes: A case from the Tarim Basin, NW China. *Pet. Sci.* **2023**, *20*, 102–113. [CrossRef]

processes

MDPI

Article

A Method for Defining Sedimentary Characteristics and Distributions and Its Application in Qinnan Depression, Bohai Bay Basin

Zehua Zhang [1], Chunqiang Xu [2], Chenjie Wang [2], Hong Li [2], Wensen Zhu [2], Hongliang Wang [1,*], Kaixuan Liang [1] and Yong Su [1]

[1] School of Energy Resources, China University of Geosciences (Beijing), Beijing 100083, China; 3006200026@cugb.edu.cn (Z.Z.); 2106200039@email.cugb.edu.cn (K.L.); 2006200017@email.cugb.edu.cn (Y.S.)

[2] Tianjin Branch of China National Offshore Oil Company Ltd., Tianjin 300452, China; xuchq2@cnooc.com.cn (C.X.); wangchj34@cnooc.com.cn (C.W.); lihong31@cnooc.com.cn (H.L.); zhuws@cnooc.com.cn (W.Z.)

* Correspondence: whliang@cugb.edu.cn; Tel.: +86-010-82321865

Abstract: A new method incorporating geophysical analysis and geological analysis is proposed to define the sedimentary characteristics and distributions in basins with few drilling wells to promote the exploration of reservoirs. This method is applied to a study, through which its principles, closed-loop workflow and technologies are introduced in detail and the sedimentary characteristics and distributions of the study area are accurately defined. During the application process of the method, a compatible geological model is established, based on which the seismic data are interpreted and the results derived from the interpretation are further verified via seismic forward modeling. The study results exhibit a successive sand-rich deposition from the retrogradational gully-filling gravity flow deposition including near-shore fans, slope fans and basin-floor fans delimited by different slope break belts in transgressive sequences to the progradational delta deposition in a retrogressive sequence including braided river deltas with a long extension distance and fan deltas developed along a steep slope belt. And the potential reservoirs are located at the point-out sites of sand bodies with lower average *P*-wave velocities than those of muddy sediments. The proposition and application of this method are of great significance for oil and gas exploration.

Keywords: geological model; seismic facies; paleogeomorphology; stratigraphic pattern; seismic forward modeling; Qinnan depression

Citation: Zhang, Z.; Xu, C.; Wang, C.; Li, H.; Zhu, W.; Wang, H.; Liang, K.; Su, Y. A Method for Defining Sedimentary Characteristics and Distributions and Its Application in Qinnan Depression, Bohai Bay Basin. *Processes* **2023**, *11*, 2539. https://doi.org/10.3390/pr11092539

Academic Editor: Qingbang Meng

Received: 17 July 2023
Revised: 21 August 2023
Accepted: 23 August 2023
Published: 24 August 2023

1. Introduction

Accurately defining the sedimentary distributions and characteristics in a basin is fundamental for the exploration of oil and gas reservoirs [1–5]. Over the past years, lithofacies and logging and seismic data have been used jointly to recognize the distributions and characteristics of potential reservoirs [6–10], of which the procedures usually include (1) becoming acquainted with the geological setting of the study area and identifying various seismic facies (e.g., external geometry and internal configuration shown in seismic sections) and logging facies mainly indicated by logging curves and (2) converting geophysical facies to sedimentary facies via borehole calibration to define the sedimentary characteristics and distributions. The above process was proven to be a valid method for defining the sedimentary characteristics and distributions, but has its limitations because of the difficulty and high cost of obtaining all types of data in actual exploration. In order to successfully discover reservoirs with few wells in the early exploratory stage, sedimentologists and geophysicists have effectively analyzed seismic data from the aspects of geological recognition and data digitization and visualization [11]. Based on sedimentology, sequence stratigraphy and geomorphology, sedimentologists usually recognize the sedimentary distributions and characteristics through a comprehensive analysis of the paleogeomorphology (restored via multiple methods such as the residual thickness method, the impression method and the sequence stratigraphic method) [12–14] as well as external geometries (e.g., wedge, lens and fan), internal configurations (e.g., chaotic, parallel and progradational) and reflection termination patterns (i.e., onlap, toplap, downlap and truncation) displayed in seismic sections [15–22]. However, the analysis results of sedimentology primarily obtained via visual observation present uncertainty and non-uniqueness. Geophysicists prefer to focus on seismic data processing, such as phase shifting, the stacking of multiple attributes and the auto-classification of seismic facies [10,23–29]. Many seismic parameters including amplitude, frequency, interval velocity and waveform have been extracted or recalculated to reflect the characteristics and distributions of sedimentary bodies [10,24,28,30–33]. And seismic forward modeling has been used to speculate the seismic reflection features of sedimentary bodies for facilitating the recognition of sedimentary distributions and characteristics [34–40]. Although the reprocessing or modeling of seismic data is conducive to improving data readability and diminishing subjective effects, calculation error and the absence of geological theory can lead to inaccurate analysis results and distort depositional models. Consequently, a method to accurately define the characteristics and distributions of sedimentary bodies with the inadequate drilling data obtained is still required.

A new method integrating geological analysis and geophysical analysis is proposed to resolve the difficulty in accurately recognizing the sedimentary characteristics and distributions in basins with few drilling wells. This study introduces the principle and the workflow of the method in detail by applying the method to a case study on the northern Qinnan depression at the edge of the Bohai Bay Basin to define its sedimentary characteristics and distributions, and sheds light on oil and gas exploration in similar basins with few wells.

2. Geological Setting

The Qinnan depression, the study area, is located at the edge of the Bohai Sea in China with a gross area of approximately 2300 km^2, and is surrounded by the Qinnan uplift, Liushouying uplift, Liaoxi uplift and Shijiutuo uplift. The study area can be divided into the eastern, western and southeastern sags (Figure 1) [40,41]. Among them, the western sag is characterized by a steep slope induced by fault activities, and the eastern sag features a gentle slope. Controlled by the Qinnan I growth fault (F1) and the Qinnan II growth fault (F2), the Qinnan depression has gone through the syn-rift stage (65.0–24.6 Ma) and the post-rift thermal subsidence stage (24.6 Ma–present) [42,43]. The syn-rift stage is further divided into four substages: the strong rift subsidence (65.0–50.5 Ma) during the Paleocene Kongdian formation (Ek), the strong rift subsidence with a weak strike-slip component (50.5–38.0 Ma) from the fourth member of the Eocene Shahejie formation (Es_4) to the third member of the Eocene Shahejie formation (Es_3), the rift subsidence with a strike-slip component (38.0–30.0 Ma) from the second member of the Eocene Shahejie (Es_2) formation to the third member of the Oligocene Dongying formation (Ed_3), and the rift subsidence with a strong strike-slip component (30.0–24.6 Ma) (Figure 1d) [44,45].

The lacustrine sediments mainly consist of sandstones and mudstones deposited during the Eocene Shahejie formation (Es) (Figure 1e) [46–48]. The deposits in the Es covered by fluvial deposits constitute important source rocks and reservoirs of the Qinnan depression. Despite the three wells exhibiting low oil development values in the northern Qinnan depression, the amount of oil and gas that was developed from the southeastern sag indicates that the northern Qinnan depression still has potential oil and gas resources today [49]. Therefore, it is of great significance to understand the sedimentary characteristics and distributions of the northern Qinnan depression with little wells.

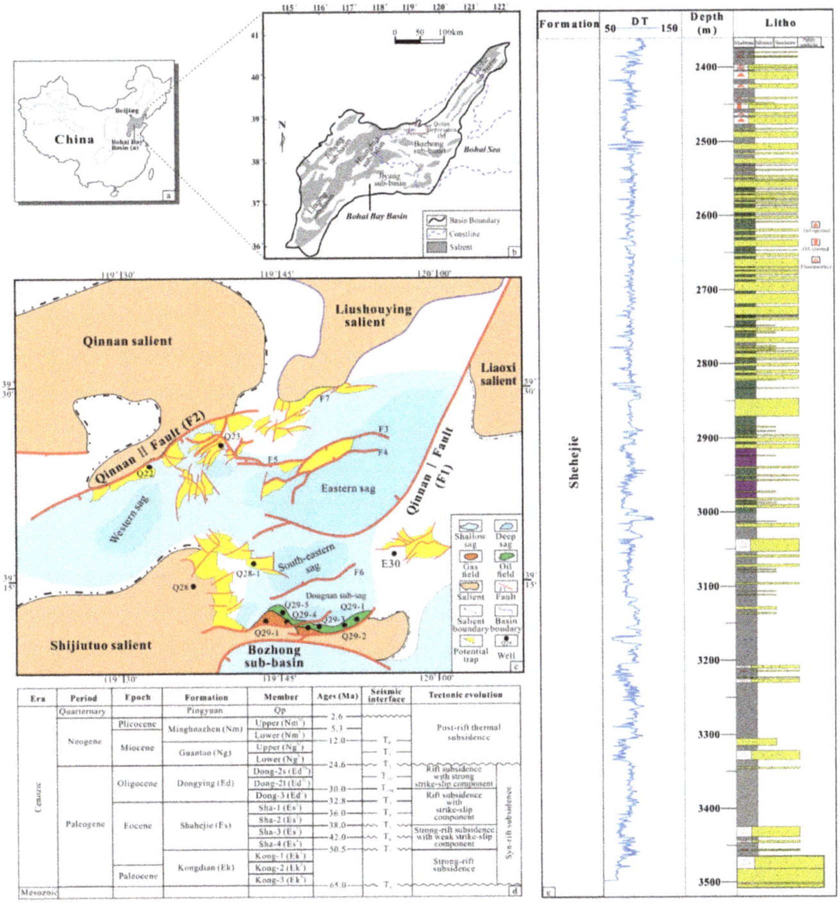

Figure 1. (**a**) Location of Bohai Bay Basin, (**b**) location of Qinnan depression, (**c**) geological framework of the Qinnan depression, (**d**) generalized stratigraphic column and tectonic evolution of the Qinnan depression, and (**e**) lithology and acoustic logging of the Eocene Shahejie formation in Qinnan depression.

Considering that the Es_3 is the main hydrocarbon-bearing strata in the Bohai Bay Basin and the Es_4 is absent in the eastern sag, the Es_3 was chosen for the application of the new method proposed in this study.

3. Data and Method

3.1. Data

A combination of seismic data and logging data of at least one well or a combination of seismic data and acoustic velocities of outcrop is necessary for the utilization of the new method proposed in this paper. Provided by the China National Offshore Oil Corporation Limited (CNOOC), Tianjin Branch, three-dimensional (3D) seismic data, the logging data of one well and the time–depth relation are leveraged in this study. The dominant frequency of 3D seismic data processed by the pre-stack time migration is 35 Hz. Three kinds of software including Landmark, ResForm and Tesseral are used.

3.2. Method

The new method put forward in this paper relies upon the construction of paleogeomorphology, the analysis of stratigraphic pattern, the interpretation of seismic data and the verification via seismic forward modeling (henceforth known as the CAIM method). During the application of the CAIM method, all analysis processes follow the principles of sedimentology, sequence stratigraphy and seismic geomorphology. And the interpretation of seismic data revolves around well-recognized criteria which have been summarized in some existing publications [50,51]. The method is to apply the constructed model based on the analysis results of paleogeomorphology and sequence stratigraphy to conduct seismic interpretation, then verify the results of seismic interpretation and, finally, accurately define the sedimentary characteristics and distributions according to the verified seismic interpretation results (Figure 2). The specific steps are as follows:

1. Based on the geological background, choose a suitable method with less workload (e.g., residual thickness) to construct paleogeomorphology for understanding the features and distributions of low-lying areas (gullies), flowpaths and slope break belts.
2. Based on the principle of sequence stratigraphy, establish the stratigraphic framework of the study area to define the stacking pattern of sedimentary bodies in different system tracts.
3. Based on the paleogeomorphic features and the sequence stratigraphic framework of the target interval, modify some existing geological models of basins presenting similar geological settings with the study area to construct a compatible geological model matching the geological characteristics of the study area.
4. According to the seismic interpretation criteria summarized by Veeken et al. (2013) and Xu et al. (2020) [50,51] as well as the compatible geological model, interpret the seismic data using multiple parameters including but not limited to seismic facies.
5. Extract a two-dimensional (2D) sedimentary model from a typical seismic section. And acquire the P-wave velocities of different sedimentary bodies from acoustic logging data. Use the constructed velocity model to carry out seismic forward modeling. When the section obtained from seismic forward modeling is consistent with the seismic section, the interpretation of seismic data can be regarded as valid.
6. Based on the interpretation results of seismic data in combination with the plane distribution of sedimentary bodies' peculiar seismic attributes, the sedimentary characteristics and distributions in the target interval of the study area are accurately defined.

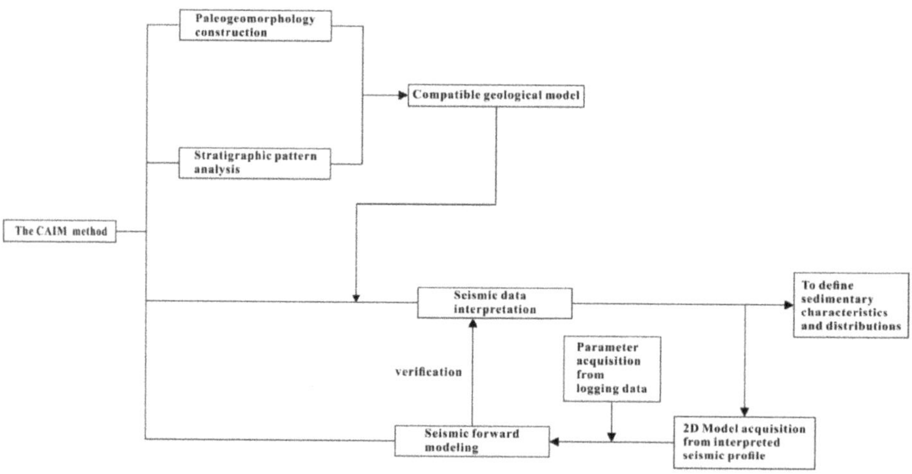

Figure 2. Flowchart of the CAIM method for defining the sedimentary characteristics and distributions.

The application of the CAIM method guarantees the validity of seismic data interpretation and the precision of the descriptions of the characteristics and distributions of potential reservoirs in the study area. Being applied to the study on the Es_3 in the northern Qinnan depression, the working details of this method are introduced, meanwhile, the sedimentary characteristics and distributions of potential reservoirs in the study area are defined.

4. Application of the Method

4.1. Geological Model

The paleogeomorphic feature and stratigraphic pattern jointly control the sedimentary distribution [52]. Establishing a compatible geological model according to the analysis of the paleogeomorphology and stratigraphic pattern is a precondition of interpreting the seismic data.

4.1.1. Paleogeomorphology Construction

To select the construction method of paleogeomorphology in a study area with few wells, the geological setting should be taken into account. For a depression with a single provenance, the application of sequence stratigraphy with the restoration of denudation is ideal for identifying the flowpaths and depocenters in detail. However, when it comes to multi-provenance depression, it is difficult to restore denudation. In order to identify the flowpaths and local depositional centers swiftly, the application of residual thickness is considered to be effective and convenient.

Restored by using residual thickness, the paleogeomorphology of Qinnan depression before the Es_3 exhibits five high-lying regions, four low-lying regions and multiple slopes (Figure 3a). The sediments in the western low-lying region corresponding with the western sag source from the Qinnan and Shijiutuo uplifts and the sediments in the middle low-lying region near the Q23 well and in the eastern low-lying region corresponding with the eastern sag are primarily sourced from the Liushouying uplift (Figures 1 and 3a). Seven drainages are distinguished by six watersheds. The drainage originating from the Qinnan uplift has two trends: a northeast–southwest direction and a near north–south direction. The three northeast–southwest trending flowpaths divided by four watersheds imply that there are three sedimentary bodies at least in the middle low-lying region. Different from the middle low-lying region, the north of the eastern low-lying region is divided into three drainages by two watersheds. Among them, the altitude of the western drainage is higher than the others, which suggests that the sediment deposition in the western drainage formed after the other two drainages were filled up. Additionally, the trends of six gullies in three drainages of the eastern low-lying region indicate four near north–south trending and northeast–southwest trending flowpaths (Figure 3a), which represents the axial direction of sedimentary bodies.

The faults and topographic flexures jointly result in the formation of multiple slope break belts, which control the accommodation change and affect the location of the local sedimentary centers. In the Qinnan depression, step-fault slopes are the main type of slopes, including multi-level step-fault gentle slopes in the eastern sag and step-fault steep slopes in the western sag (Figure 3). Three distinct slope break belts result in three local low-lying regions (Figure 3b). Among them, the first-order slope break belt near the northern source areas is caused by the fault and topographic flexures. The drop of the western first-order slope break belt is higher than that of the eastern first-order slop break belt. The second-order slope break belt is mainly brought about by faults, of which the fault amplitude in the middle low-lying region is much higher than that in the eastern sag. The third-order slope break belt with a high drop corresponding to the F3 fault and the steep slope fault belt caused by the F1 fault jointly result in the formation of the subsiding center of the eastern sag (Figure 3b).

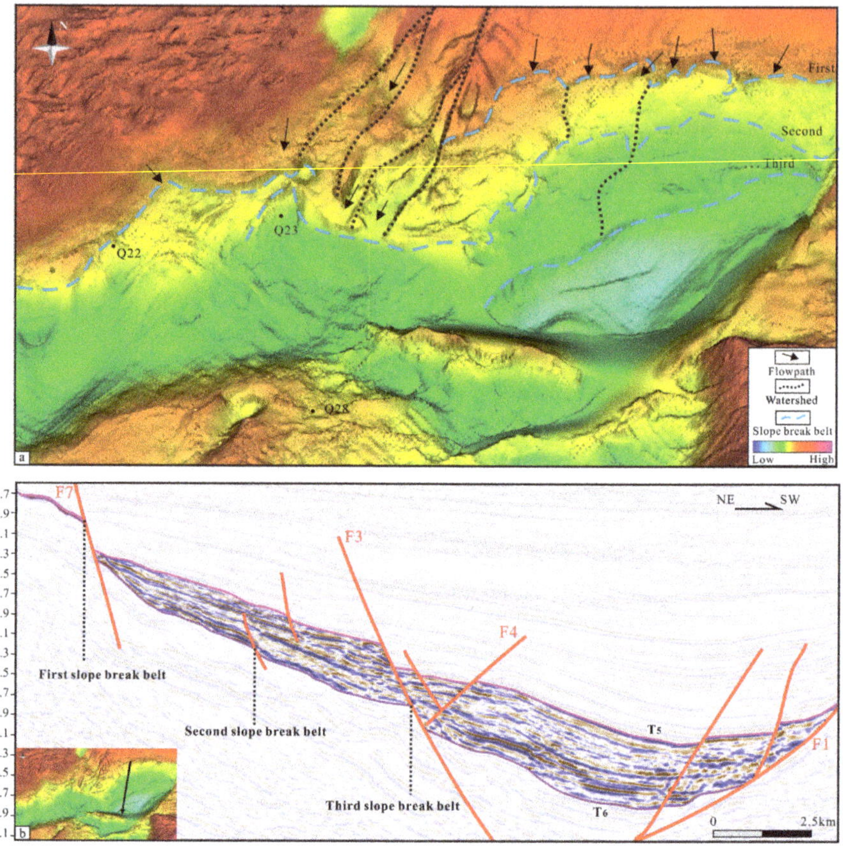

Figure 3. (**a**) The paleogeomorphic map before the third member of Eocene Shahejie formation (Es_3) in Qinnan depression. (**b**) A seismic section showing topographic features of the bottom of the Es_3 and three slope break belts in the Es_3.

In general, the flowpaths indicated by the trends of the gullies and local low-lying regions influenced by multi-level slope break belts suggest that sedimentary bodies fill up the local low-lying regions in the early stage of the Es_3 and have north–south and northeast–southwest trends in the Qinnan depression.

4.1.2. Stratigraphic Pattern

The stratigraphic pattern consists of a sequence stratigraphic framework and its internal depositional configuration [15]. The establishment of the sequence stratigraphic framework is the key for predicting the characteristics and distributions of sedimentary bodies. The Es_3 is divided into the upper sequence (Es_{3U}) and the lower sequence (Es_{3L}) by identifying three sequence boundaries and one maximum flooding surface. The Es_{3U} is thicker than the Es_{3L}. The T_6 and T_5 sequence boundaries correspond to the bottom surface and top surface of the target interval, respectively, which are identified on the seismic reflections via truncation and the toplap below them as well as the onlap above them; the T_{6M} sequence boundary is the bottom boundary of the Es_{3U}, and is characterized by the onlap above it; and the maximum flooding surface (MFS) of both the Es_3 and Es_{3U} is a downlap surface, which divides the Es_{3U} into two system tracts including the lowstand to the transgressive systems tract (LST and TST) and the highstand systems tract (HST) (Figure 4). The T_{6M} and MFS surfaces both terminate on T_6 between the first-order slope break belt and the second-order slope break belt (Figure 4). Therefore, the Es_{3L} and the LST and TST of the Es_{3U} distribute in the south of the first-order slope break belt.

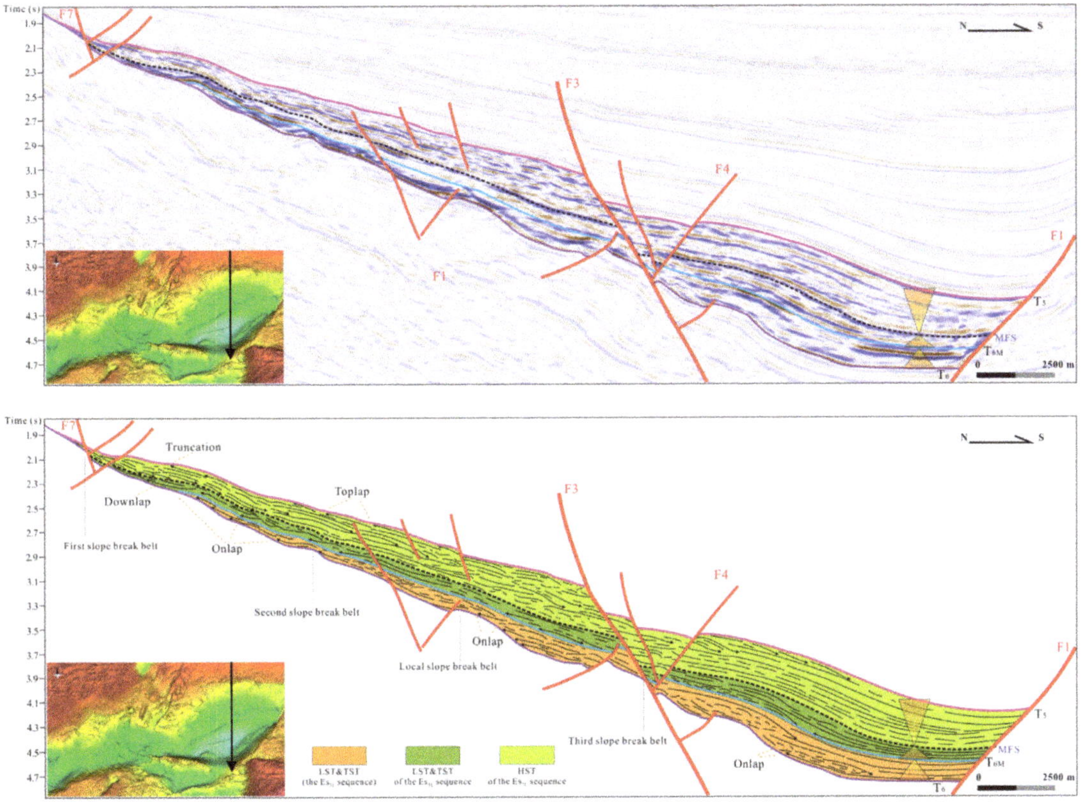

Figure 4. The sequence stratigraphic framework of the Es_3 in Qinnan depression.

The absence of progradation reflection in the upper Es_{3L} implies that the HST is not developed. The retrogradation reflection in the Es_{3L} indicates the rise in the base level (Figure 4). Three lentoid sedimentary bodies with chaotic reflections or chaotic/hummocky reflections fill up the low-lying lands (relatively high accommodation) induced by different slope break belts in dip sections (Figure 4), which signifies subaqueous deposits in different locations of the slope. The onlap spot of the seismic reflection axes gradually moves to the top of the slope, indicating retrogradation and transgression in the Es_{3L}. A local high accommodation indicated by the thickening of sequences is always accompanied by the development of faults crossing the Es_{3L} (Figure 4), suggesting that the development of faults occurs before the formation of the Es_{3L}.

The LST and TST are thinner than the HST in the Es_{3U} sequence. Inheriting from the Es_{3L}, three slope break belts lead to the formation of areas with a relatively high accommodation in the LST and TST of the Es_{3U} (Figure 4). Hummocky reflections with a weak to medium amplitude are present between the first-order slope break belt and the second-order slope break belt as well as between the third-order slope break belt and the F1 fault; clean blocky reflections with a weak amplitude are shown between the second-order slope break belt and the F3 fault (Figure 4). Similar to the Es_{3L}, faults crossing the LST and TST of the Es_{3U} induce the increase in local accommodation, indicating that the faults develop before the formation of the LST and TST of the Es_{3U}. The bidirectional onlap reflection is observed in local low-lying regions occasionally. The characteristics of the seismic reflections in the LST and TST the Es_{3U} are similar to those in the Es_{3L}, indicating the filling of the gullies, the retrogradation of the sedimentary bodies and transgression. Convergent progradational seismic reflection axes with a medium amplitude are shown in the HST, which suggests the progradation of sedimentary bodies and a decreasing accommodation (Figure 4). The sedimentary pattern in the HST is less influenced by slope break belts. Different from the LST and TST of the Es_{3U}, faults crossing the HST do not lead to the thickening of sequences (expect for the F7, F3 and F1 faults), which indicates that the faults develop after the formation of the HST and have little influence on the sedimentary characteristics of the LST and TST of the Es_{3U} (Figure 4). The development of the F7, F3 and F1 faults induce the formation of slope break belts and increase the thickness of the sequences in the Es3. Therefore, the F7, F3 and F1 faults develop before the formation of the Es_3 and control deposition in the Qinnan depression.

Based on the sequence stratigraphic framework, the vertical sedimentary pattern of the Es_3 includes a successive stacking of retrogradational sedimentary bodies indicated by filled gullies from the Es_{3L} to the LST and TST of the Es_{3U} and overlying progradational sedimentary bodies in the HST of the Es_{3U}. The faults control the change in the topography and the sequence thickness. The growth of the F1 fault causes the formation of the Qinnan depression and an eastern gentle slope [44,45]. Thus, it is inferred that the F1 fault, as a synsedimentary fault, controls the change in the accommodation and topography and, in turn, controls the distribution, amount and characteristics of sequence stratigraphy.

4.1.3. Compatible Model

The geological model's high applicability tends to lead to the seismic interpretation's high accuracy. Basins with similar geological settings usually have similar geological models. Therefore, the existing geological models of basins with a geological setting similar to that of the study area can be drawn upon after certain modification to produce the geological model of the study area. Drawing upon and modifying existing geological models of sedimentary basins with similar geological settings to establish a compatible geological model by leveraging the sequence stratigraphic framework and paleogeomorphic features of the study area is a key procedure of the CAIM method proposed in this paper, which will facilitate seismic data interpretation and further improve the validity of the sedimentary characteristics and distributions.

Up to now, a lot of geological models of the sags or depressions of rift basins have been established [51], which present retrogradational clinoforms in the LST and TST and progradational sand bodies in the HST. Based on the paleogeomorphic features and sequence stratigraphy of the Es_3 in the Qinnan depression, a compatible geological model modified from some existing depositional models of other depressions in the Bohai Bay Basin (a rift basin) is established (Figure 5). The flowpaths indicated by the trends of the gullies suggest that the axial direction of the sedimentary bodies is mainly from the northeast to the southwest. Under the MFS, two sets of sedimentary bodies in two transgressive sequences are predicted, which, respectively, incorporates three sedimentary bodies filling up three low-lying lands controlled by three slope break belts. In view of the tectonic evolution in the Es_3 and the gully-filling seismic facies without progradation reflections, the three sedimentary bodies of one set can be regarded as subaqueous gravity flow deposits and are divided into near-shore deposition near source areas, slope deposition between the first-order and second-order slope break belts and basin-floor deposition in the subsiding center. The progradation reflections overlying the MFS converge on the center of three low-lying regions (Figures 3 and 5), indicating the progradational sedimentary bodies with a gradually falling base level.

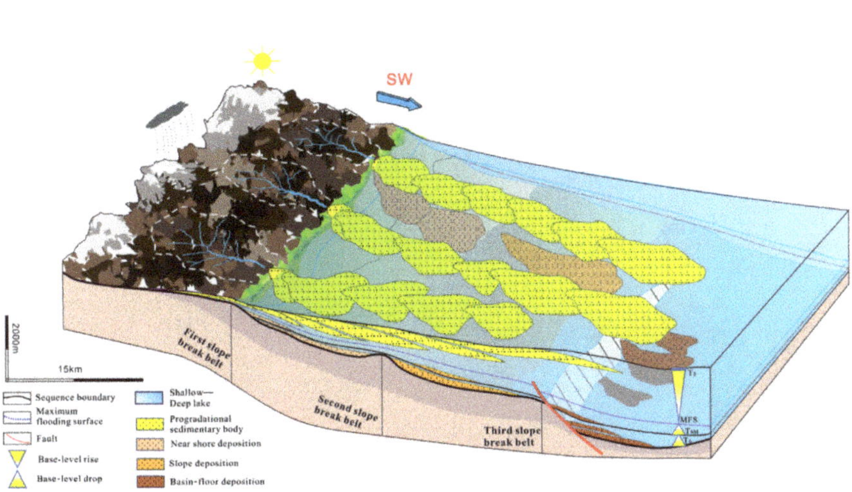

Figure 5. A compatible geological model suggested by the sequence stratigraphic framework and paleogeomorphic features.

4.2. Seismic Interpretation

Using seismic parameters to analyze lithology and sedimentary facies is crucial for defining the sedimentary characteristics and distributions of a study area [11]. With the development of computer science and technology, scholars are more willing to interpret seismic data by reprocessing multiple seismic parameters than by using the view of geology [9,20]. By contrast, the CAIM method highlights the conversion from seismic data to geological messages through the interpretation of seismic data according to a compatible geological model and the extraction of seismic attributes. Based on the geological model established in this study, the seismic facies analysis, as a primary unit of the seismic interpretation section of the CAIM method, is leveraged to define the sedimentary characteristics and distributions of the Es_3 in the Qinnan depression. Xu et al. (2020) [51] provided criteria for the interpretation of seismic facies in the Bohai Bay Basin, which are used in this study to interpret the seismic data of the Qinnan depression in the Bohai Bay Basin.

4.2.1. Facies

Six sedimentary facies have been identified by the positions and system tracts of sedimentary bodies and seismic features such as the amplitude, internal configurations and external geometries exhibited in seismic sections. The six sedimentary facies involve braided river delta, fan delta, near-shore subaqueous fan, slope fan, basin-floor fan and lake (Figure 6). Apart from the lake facies, other sedimentary facies are dominated by sandstones. As the fan delta, braided river delta and subaqueous gravity flow deposition are in the shape of a lens with a chaotic or stratiform reflection in the off-axis seismic sections, the sedimentary facies are distinguished through seismic features shown in the along-axis sections rather than in the off-axis sections.

Being observed in the HST of the Es_{3U}, the wedged progradation reflection in the dip sections indicates delta deposition. In the Qinnan depression, the fan delta and braided river delta are distinguished by different seismic features and development locations. Characterized by a sigmoid progradation reflection with a medium to strong amplitude, the braided river delta develops at the gentle slope of the eastern sag. The stratiform seismic reflection axis in the braided river delta indicates a successive deposition of the interbedded sandstones and mudstones (Figure 6). As opposed to the braided river delta, the fan delta is developed at the steep slopes caused by fault activities in the western sag, with the high accommodation caused by fault activities, providing a favorable geological setting for its development. In the dip seismic sections, the wedged fan delta with an overall progradation reflection is indicated by the weak stratiform or chaotic reflection with a weak amplitude (Figure 6), confirming that the fan delta is characterized by thick sandstones and sandy deposits interbedded with muddy deposits.

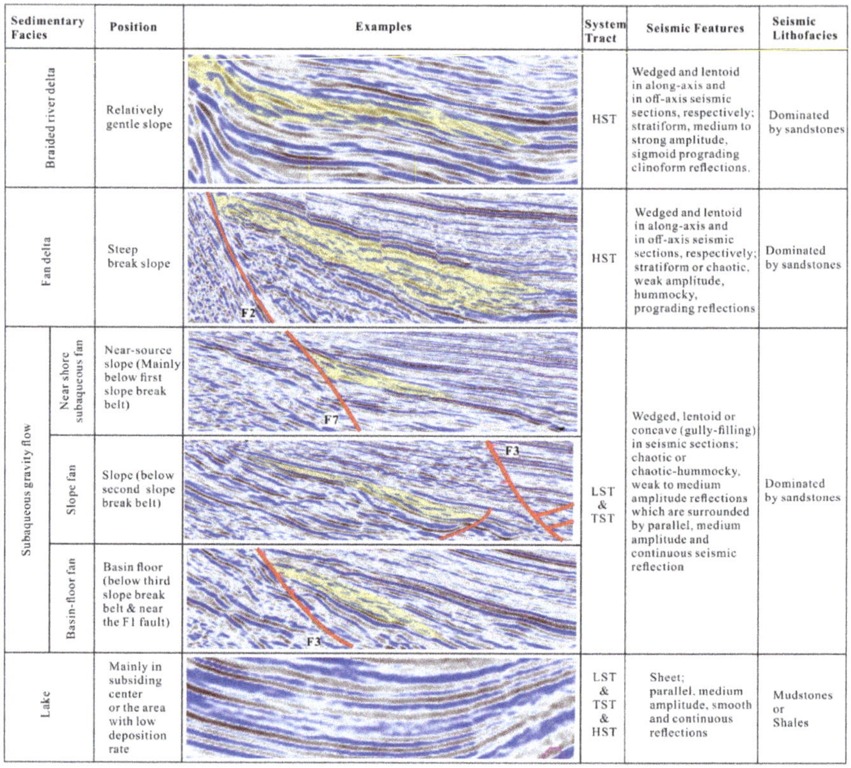

Figure 6. The distinctive seismic features, development positions, dominated system tracts and seismic lithofacies of different sedimentary facies in the Es_3 of Qinnan depression. All examples derive from the northeast–southwest seismic sections.

Wedged, lentoid or gully-filling chaotic reflections with a weak to medium amplitude in along-axis sections indicate inhomogeneous subaqueous gravity flow deposition dominated by sandstones in the Es_{3L} and in the LST and TST of the Es_{3U} in the Qinnan depression (Figures 6 and 7). The weakly continuous retrogradation reflection is present in different types of subaqueous gravity flow deposits. Depending on the location, the subaqueous gravity flow deposits surrounded by continuous reflections are divided into three facies, i.e., near-shore subaqueous fan developing from the top of the slope, the slope fan at the slope controlled by the second-order slope break belt and the basin-floor fan terminating at the bottom of the slope (Figures 3 and 5–7).

The smooth, continuous parallel reflection with a medium amplitude indicates lake facies dominated by mudstones or shales, and represents continuous and stable deposition with a low deposition rate. The parallel reflection indicating muddy deposits in the lake widely exists in all sequences of the Es_3 in the Qinnan depression, which usually overlies the subaqueous gravity flow deposition with a lentoid chaotic reflection and underlies the delta deposition with a progradation reflection.

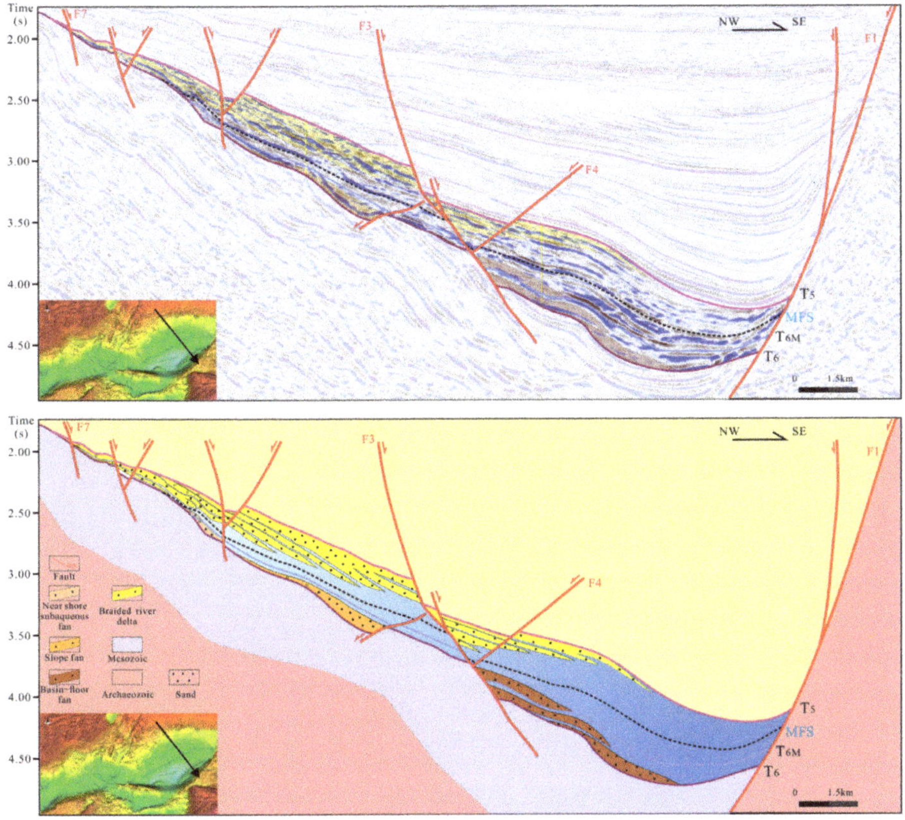

Figure 7. The longitudinal distribution of different sedimentary facies dominated by sandstones in the Es_3 of Qinnan depression.

4.2.2. Facies in Section

Seismic sections reflect the vertical stacking of different sedimentary bodies, of which the analysis is fundamental for defining the sedimentary characteristics and distributions. Referring to the compatible geological model, the seismic sections of the Qinnan depression are interpreted according to the configuration of seismic facies that indicate sedimentary facies.

In seismic sections, a continuous parallel reflection of the lake facies represents a muddy deposition and suggests that lithological changes between sandstones and mudstones generally induce an abrupt increase in the amplitude; thus, a continuous parallel reflection with a relatively strong amplitude in a wedged or lentoid sedimentary body signifies mudstone interlayers, which can distinguish sand-rich portions from mud-rich portions in a sedimentary body. In the interpretation of sedimentary facies, corresponding the sand-rich portion of a sedimentary body to sedimentary facies can improve the accuracy of the exploration (Figure 7). In the Qinnan depression, the retrogradational basin-floor fan and the retrogradational near-shore subaqueous fan, both with chaotic reflections in a transgressive sequence represented by the Es_{3L} and the LST and TST of the Es_{3U}, are developed from the third-order slope break belt to the F1 fault and from the first-order slope break belt to the second-order slope break belt, respectively (Figure 7). Influenced by faults, the wedged reflection of the near-shore subaqueous fan and of the basin-floor fan converges southward to the sequence boundaries (Figure 7). The slope fan with a chaotic/hummocky reflection only develops in the Es_{3L} from the second-order to the third-order slope break belts, of which the wedged reflection converges northward to the sequence boundaries and reflects a gully-filling deposition (Figure 7). Overall, slope break belts result in a relatively high local accommodation to form subaqueous gravity flow deposits. The MFS, indicating a widely spread mudstone layer, covers all the subaqueous gravity flow depositions. The progradational sand-rich delta deposition above the MFS extends to the center of the low-lying regions (the depocenter of sags in Qinnan depression) and converges on the T_5 boundary, of which the average thickness is thicker than that of the subaqueous gravity flow deposition.

Generally speaking, the seismic sections show a successive deposition from the retrogradational gully-filling gravity flow deposition delimited by different slope break belts in a transgression sequence to the progradational delta deposition with a long extension distance (Figure 7). During the Es_{3L} and the LST and TST of the Es_{3U}, the faults control the change in the local accommodation and, in turn, control the distribution of the gravity flow depositions. The F1 fault controls the distribution of the subsidence center with a high accommodation, and thus, determines the sedimentary facies types in the subsidence center and their distribution. During the HST of the Es_{3U}, the weakening F1 fault activities induce the decrease in the relative accommodation and, in turn, promote the development of delta facies and influence their distribution.

4.3. Verification

Seismic forward modeling has been widely used to probe the characteristics of reservoirs, verify seismic inversion algorithms and provide guidelines for seismic interpretation [32,37,38]. Apart from that, seismic forward modeling as a part of the CAIM method can also be used to verify the results of seismic interpretation to ensure the accuracy of the sedimentary characteristics and distributions. The attainment of a velocity model from a representative seismic section showing the sedimentary characteristics and distributions of the study area is the key for the utilization of seismic forward modeling. The verification mainly consists of four steps: (1) extract a sedimentary model in depth domain from a representative seismic section displaying sedimentary characteristics and distributions, and establish a typical velocity model by obtaining the *P*-wave velocity (Vp) of different rocks on buried depth from the logging data; (2) set up the framework of seismic forward modeling incorporating the seismic source parameter, source signal and seismic observation system; (3) process point-shot records via gathering, normal move out (NMO), stacking and migration to attain a synthetic seismic section; and (4) compare the synthetic seismic section with the actual seismic section to verify the correctness of the seismic interpretation. It is worth noting that the framework set in step (2) should be as similar to the framework of the seismic data acquisition as possible, and that the processing of the point-shot records must be in accordance with the processing of the actual seismic data.

The seismic interpretation in the Qinnan depression is verified by seismic forward modeling. A seismic section in the eastern sag is selected in the first place, which exhibits representative sedimentary characteristics and distributions including the retrogradational subaqueous gravity flow deposition with a chaotic/hummocky reflection in transgressive sequences, the progradational delta deposition with progradation reflection in regressive sequences and the muddy deposition indicated by a parallel reflection (Figure 7). This section in time domain is interpreted as a sedimentary model in depth domain by the relationship between depth and time as follows:

$$\text{depth (m)} = -1.5661\ \text{time (ms)} + 706.71 \quad (1)$$

Based on the extracted sedimentary model and the relationship between the *P*-wave velocities obtained from the logging data of well Q23 and the burial depth (Figure 8), the sedimentary velocity model is established for simulation (Figure 9a), during which it is scaled down to half its original size to prevent the distortion of velocity assignment. The spot interval, geophone interval (trace interval) and spatial sampling are set as 30 m, 10 m and 4 ms, respectively. The resolution of the seismic data is usually one-quarter of the dominant wavelength (λ); therefore, the source signal is set as zero-phase Riker with a peak frequency of 70 Hz according to the seismic data's dominant frequency of 35 Hz and the 280 average ratio of the sand-rich sedimentary bodies' Vp to thickness (λ/4). Given that the actual seismic sections are obtained by the processing (i.e., gathering, NMO, stacking and pre-stack time migration) of the original seismic data, the point-shot records attained from seismic forwarding modeling are also processed in the same way to produce the synthetic seismic section corresponding to the sedimentary velocity model (Figure 9b).

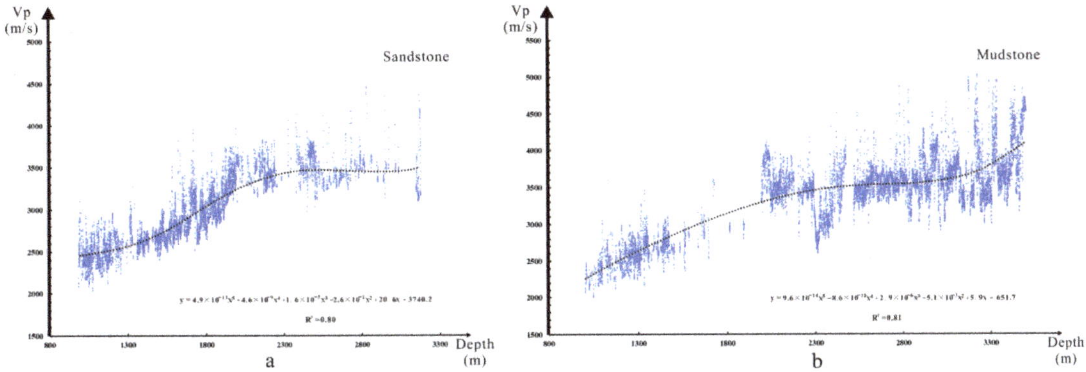

Figure 8. Dependence of rocky *P*-wave velocity on buried depth in northern Qinnan depression. In this figure, y is the rocky *P*-wave velocity (Vp) and x is the depth. (**a**) Dependence of the Vp of sandstone on buried depth. (**b**) Dependence of the Vp of mudstone on buried depth.

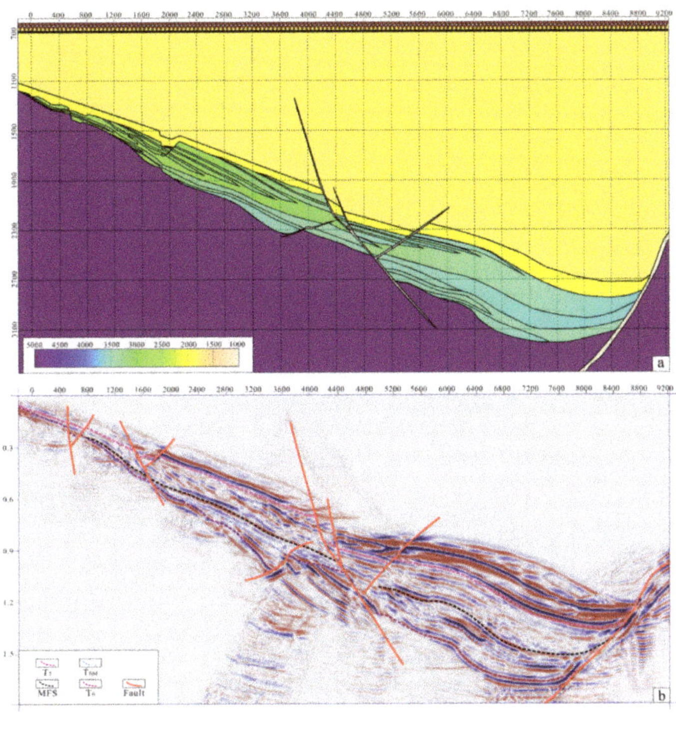

Figure 9. (**a**) A sedimentary velocity model extracted from the typical seismic section shown in Figure 7. (**b**) The synthetic seismic section produced via seismic forwarding modeling.

The synthetic seismic section obtained via seismic forwarding modeling is similar to the actual seismic section (Figures 7 and 9b), suggesting that the results derived via seismic data interpretation are correct. As a result, the understanding of the sedimentary characteristics and distributions in the study area is facilitated, and the exploration accuracy is enhanced.

4.4. Distribution

The CAIM method focuses on recognizing the distributional features based on the correct interpretation of results verified by the simulation and the distribution of extracted attributes that indicate sedimentary bodies so as to accurately define the sedimentary distribution and discover potential reservoirs in a study area.

Representing the lithofacies' transition surfaces, the peak reflection and trough reflection are, respectively, regarded as the interface from low to high impedance and the interface from high to low impedance [53]. Thus, it is most appropriate to leverage the maximum-peak amplitude to distinguish sand-rich deposition from mud-rich deposition, and then to depict the plane distribution of the sand-rich deposition. Based on the average thickness of sedimentary bodies in different sequences, the attribute extractions are compiled by extracting maximum-peak amplitudes from a −20 ms window above the T_6 boundary, a −15 ms window above the T_{6M} boundary and a +50 ms window below the T_5 boundary to understand the distribution of the gravity flow deposition in the Es_{3L}, the LST and TST of the Es_{3U} and the distribution of the delta deposition in the HST of the Es_{3U}. The scopes of the sedimentary bodies are indicated by the areas with value anomalies of maximum peak in the plane maps (Figure 10). In general, the distributional area of the sedimentary bodies decreases firstly and then increases from the Es_{3L} to the HST of the Es_{3U}.

In the HST of the Es_{3U}, multiple progradational braided river deltas with three flowpaths cover almost the entire northern Qinnan depression. Five interconnected braided river deltas with the near north–south flowpath develop in the eastern sag, of which the inflow mounds are closely related to the development of gullies; three interconnected braided river deltas with the northeast–southwest flowpath are present in the middle low-lying region and connect with a fan delta in the north; three progradational braided river deltas with the southwest–northeast flowpath fill in the western sag, which is suggested by the lentoid reflection below the T_5 boundary and the extracted maximum-peak amplitude map; and four fan deltas sourced from the Qinnan uplift are arranged along the Qinnan II fault (F2), of which the tail end connects with the braided river deltas (Figures 3 and 10a,b).

In the LST and TST of the Es_{3U}, four near-shore subaqueous fans are arranged along the first-order slope break belts. Multiple slope fans of the northeast–southwest trend are found in the middle low-lying region, of which the distribution is controlled by the second-order slope break belt and gullies. Five basin-floor fans are deposited in the depocenter of the eastern sag, which are delimited between the third-order slope break belt and the F1 fault (Figure 10c,d).

The subaqueous gravity flow deposition has a larger area in the Es_{3L} than in the LST and TST of the Es_{3U}. Three slope break belts limited the distribution of subaqueous gravity flow deposition in the eastern sag (Figure 10e,f). The western slope break belt with a high drop mainly controls the formation of the Es_{3L} of the Qinnan depression, which provides a favorable geological setting for the development of four slope fans filling up the low-lying lands in the eastern sag and of three near-shore subaqueous fans filling up the middle low-lying region (Figures 3 and 10). Additionally, there exists interconnected slope fans sourced from the Shijiutuo uplift (Figure 10f).

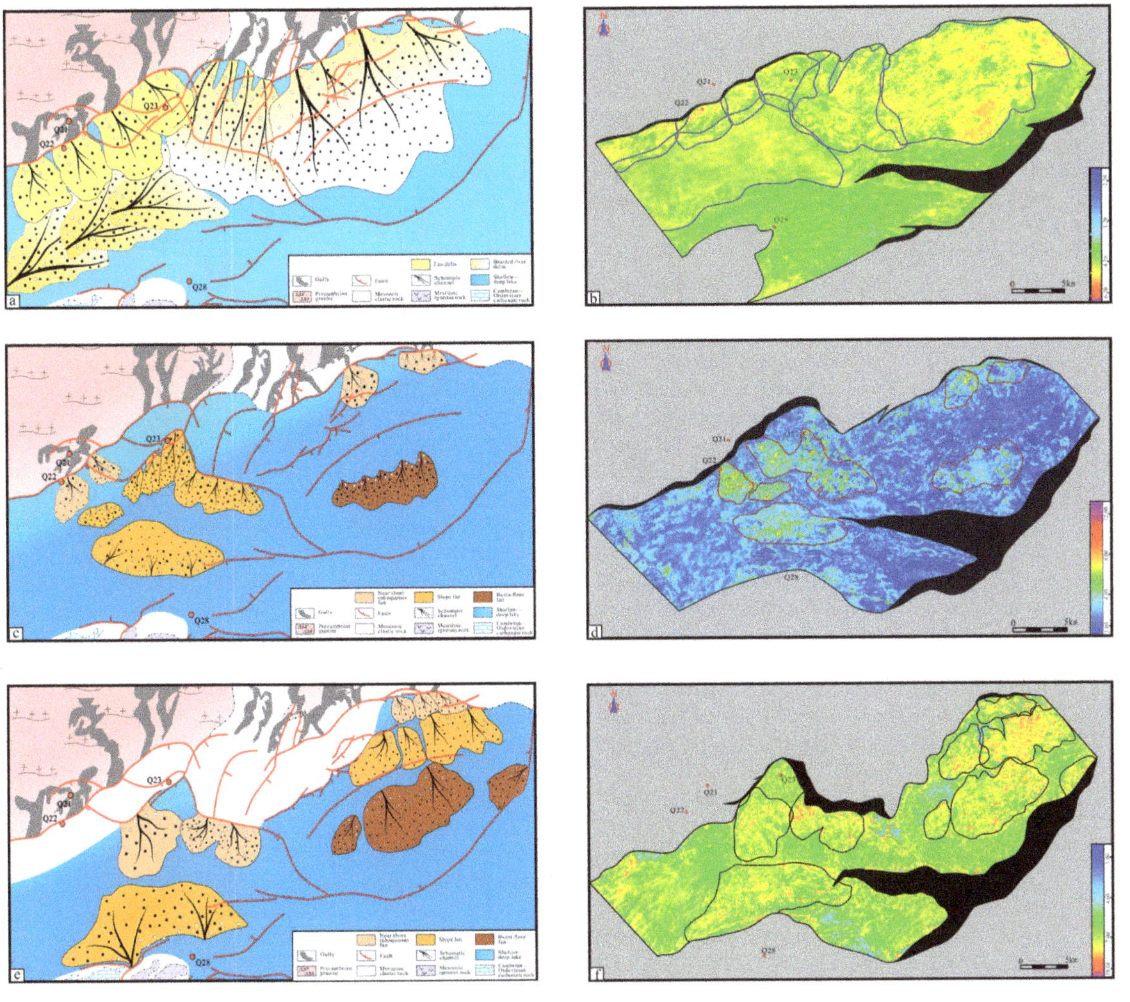

Figure 10. Sedimentary facies distribution and interpretative seismic amplitude extraction maps of the Es_3 in Qinnan depression. (**a**) The distribution of sedimentary facies in the HST of the Es_{3U} sequence. (**b**) Seismic maximum-peak amplitude extraction maps corresponding to the distribution of delta deposition in the HST of the Es_{3U} sequence. (**c**) The distribution of sedimentary facies in the LST and TST of the Es_{3U} sequence. (**d**) Seismic maximum-peak amplitude extraction maps corresponding to the distribution of subaqueous gravity flow deposition in the LST and TST of the Es_{3U} sequence. (**e**) The distribution of sedimentary facies in the Es_{3L} sequence. (**f**) Seismic maximum-peak amplitude extraction maps corresponding to the distribution of subaqueous gravity flow deposition in the Es_{3L} sequence.

5. Discussion

5.1. Advantages

The CAIM method can resolve the difficulty in accurately defining the sedimentary characteristics and distributions with inadequate data attained during reservoir exploration. Exploiting the compatible geological model established through the analysis of paleogeomorphology and the stratigraphic framework to conduct seismic interpretation, the CAIM method allows for the geological analysis and geophysical analysis to complement each other.

With geological principles added to the process of geophysical analysis, the error rate of the analysis results is reduced. The geological analysis centering on the interpretation of the seismic data can be viewed as non-digital seismic inversion, which provides a novel graphical analysis route for geophysical analysis. Meanwhile, the utilization of seismic forward modeling (geophysical analysis) as a supplement to the geological analysis can achieve the digitalization of geological analysis, by which the correctness and uniqueness of the geological analysis results are ensured.

In addition, the CAIM method exhibits a wide range of applicability. On the one hand, the four sections of the method, i.e., paleogeomorphology construction, stratigraphic pattern analysis, seismic data interpretation and seismic forward modeling, have all been widely applied to the exploration and development of oil and gas in basins, which proves that there is little restriction to the application of the CAIM method. On the other hand, this method can be utilized when there are only 2D seismic data obtained, which reduces the dependence on data acquisition. It is worth noting that the obtainment of rocky *P*-wave velocities in a basin lacking wells is regarded as a main difficulty for the utilization of the CAIM method. In basins short of wells, a sedimentary velocity model can be established based on the measured *P*-wave velocities of the outcrop [37].

It is also interesting that the CAIM method features strong inclusiveness. During the process of seismic interpretation, not only can seismic facies and amplitude mentioned in this paper be used, but multiple seismic parameters can also be exploited. In fact, all interpretation methods of seismic data can be included in this method to facilitate the recognition of sedimentary characteristics and distributions.

All in all, the CAIM method, with a wide range of applicability and strong inclusiveness, combines geophysical analysis with geological analysis, which retains the role of geology in the process of geophysical analysis and eliminates the uncertainty and subjective effects of the results obtained via geological analysis alone.

5.2. Differences from Other Methods

Compared with other methods, the CAIM method highlights a closed-loop workflow and innovatively combines geological analysis and geophysics analysis so that the accuracy of the recognition of sedimentary characteristics and distributions in a basin with few or no wells is increased, and the success rate of reservoir exploration is greatly improved.

Many methods have been used to define sedimentary characteristics and distributions for reservoir exploration, which can be roughly divided into five categories: (1) geological analysis based on sedimentology, sequence stratigraphy and seismic sedimentology, which includes core or outcrop observation, a comprehensive analysis of seismic facies and log facies and an analysis of the source-to-sink system centering on lithology and geomorphology [51–53]; (2) geophysical analysis, which contains seismic forward modeling, seismic inversion and the reprocessing of seismic data [11]; (3) geochemical analysis including elemental analysis and organic geochemical analysis [47,49]; (4) hybrid analysis, which assembles results derived from different methods to acquire a comprehensive result [8,11,16,53]; and (5) the artificial intelligent method, which is used to conduct a comprehensive data analysis based on a large database [11]. Although the methods above have each their own advantages, they rarely have a closed-loop research process to ensure the correctness and uniqueness of the analysis results. By contrast, the closed-loop workflow of the CAIM method allows for the results of geological analysis and geophysical analysis to verify each other and ensures the correctness of the results. Additionally, as opposed to other methods that require a lot of logging data or experimental data, the application of the CAIM method can tap into only 2D seismic sections and rocky Vp at different depths to complete the study. Therefore, the CAIM method can be applied to study areas with few wells for the exploration of potential reservoirs.

5.3. Significances

This study not only introduces the CAIM method by elaborating on its principle, workflow and technologies, but also applies the method to explore potential reservoirs in the Es_3 of the Qinnan depression. The sedimentary characteristics and distributions defined via the CAIM method imply the distribution of potential reservoirs and, in turn, promote the exploration of oil and gas reservoirs in the Es_3 of the Qinnan depression.

Since the development of mudstone caprocks is a necessary condition for the accumulation of oil and gas and the deposition in the Es_3 of the Qinnan depression features a high sandstone proportion (Figure 10), sandstones in the Es_{3L} and in the LST and TST of the Es_{3U} under the MFS are deemed as potential reservoirs. And based on the sedimentary distribution in transgressive sequences, the potential reservoirs are inferred to distribute along slope break belts. The average P-wave velocity of sand-rich sedimentary bodies is lower than that of muddy sediments at the pinch-out sites, which implies the development of lithologic pinch-out hydrocarbon reservoirs.

Overall, the sedimentary characteristics and distributions in the Es_3 of the Qinnan depression indicate that sand-rich near-shore fan deposition, slope fan deposition and basin-floor fan deposition are potential reservoirs, of which the distribution is controlled by local accommodation change induced by the development of slope break belts.

6. Conclusions

Exhibiting a wide range of applicability and strong inclusiveness, the CAIM method, with a closed-up workflow, consists of paleogeomorphology construction, stratigraphic pattern analysis, seismic data interpretation and seismic forwarding modeling. The application of the CAIM method is of great significance for the studies on basins with little drilling data attained in the early exploratory stage, and can ensure the validity of the study results.

The sedimentary characteristics and distributions in the Es_3 of the Qinnan depression are defined through the CAIM method. The study results show that the deposition with a high proportion of sandstone in the study area is composed of braided river deltas and fan deltas in the HST as well as subaqueous gravity flow deposition (near-shore subaqueous fans, slope fans and basin-floor fans) in the LST and TST sequences. The fan deltas and braided river deltas develop in the western sag with a steep slope, and develop in eastern sag with a gentle slope, respectively. The distribution of near-shore subaqueous fans, slope fans and basin-floor fans are, respectively, delimited by the first-order, second-order and third-order slope break belts. And the results suggest that sand-rich subaqueous gravity flow depositions are the potential reservoirs.

Author Contributions: Conceptualization, Z.Z. and H.W.; methodology, Z.Z. and H.W.; software, C.X., H.L., K.L., Y.S. and W.Z.; validation, Z.Z., C.X., C.W., H.L. and W.Z.; formal analysis, Z.Z., C.W. and C.X.; investigation, Z.Z.; resources, H.W., Z.Z., C.X., H.L. and C.W.; data curation, C.X., H.L., K.L. and C.W.; writing—original draft preparation, Z.Z.; writing—review and editing, Z.Z. and H.W.; visualization, Z.Z.; supervision, C.X. and H.W.; project administration, H.W.; funding acquisition, H.W. All authors have read and agreed to the published version of the manuscript.

Funding: This research was funded by the National Science and Technology Major Project, grant number 2016ZX05033.

Data Availability Statement: The data presented in this study are available upon request from the corresponding authors.

Acknowledgments: We acknowledge the financial support from the China National Offshore Oil Corporation. Thanks to the China National Offshore Oil Corporation Limited, Tianjin Branch, for providing the seismic and logging data.

Conflicts of Interest: The authors declare no conflict of interest.

References

1. Lu, Z.; He, Z.; Ma, S.; He, Y. Sedimentary characteristics and sand-body distributions in the Lower Permian He 8 Member, Ordos Basin, China. *Interpretation* **2022**, *10*, 223–236. [CrossRef]
2. Leila, M.; Yasser, A.; El Bastawesy, M.; El Mahmoudi, A. Seismic stratigraphy, sedimentary facies analysis and reservoir characteristics of the Middle Jurassic syn-rift sediments in Salam Oil Field, north Western Desert, Egypt. *Mar. Petrol. Geol.* **2022**, *136*, 105466. [CrossRef]
3. Yina, Z.; Wenjie, C.; Songling, Y.; Ke, Z.; Jingyang, C. Sedimentary characteristics of the Jurassic shelf-edge delta and oil and gas exploration in the Papuan Basin. *Earth Sci. Front.* **2021**, *28*, 167–176.
4. Cheng, Z.; Xu, H.; Wang, D.; Zhang, L.; Shi, Y.; Lu, Z.; Zhao, X. Research on sedimentary facies of Guantao Formation in Shanjiasi area of Dongying depression, China. *Energy Rep.* **2021**, *7*, 822–841. [CrossRef]
5. Xu, X.; Liu, L.; Li, X.; Yang, W.; Cao, Y.; Ma, H.; He, A.; Wang, R.; Leng, H.; Zhu, Y.; et al. Sequence stratigraphy, sedimentary characteristics of barrier coastal sedimentary system of the Benxi Formation (Gaoqiao area, Ordos basin) and favorable reservoir distribution. *Energy Rep.* **2021**, *7*, 5316–5329. [CrossRef]
6. Dalhoff, F.; Chalmers, J.A.; Gregersen, U.; Nohr-Hansen, H.; Rasmussen, J.A.; Sheldon, E. Mapping and facies analysis of Paleocene-Mid-Eocene seismic sequences, offshore southern West Greenland. *Mar. Petrol. Geol.* **2003**, *20*, 935. [CrossRef]
7. Sangree, J.B.; Widmier, J.M.; Payton, C.E. Seismic stratigraphy and global changes of sea level; Part 9, Seismic interpretation of clastic depositional facies. *AAPG Bull.* **1978**, *62*, 165–184.
8. Brown, L.F.; Fisher, W.L.; Payton, C.E. Seismic-stratigraphic interpretation of depositional systems; examples from Brazilian rift and pull-apart basins. In *Seismic Stratigraphy—Applications to Hydrocarbon Exploration;* Payton, C.E., Ed.; American Association of Petroleum Geologists: Tulsa, OK, USA, 1997; Volume 26, pp. 213–248.
9. Posamentier, H.W.; Kolla, V. Seismic geomorphology and stratigraphy of depositional elements in deep-water settings. *J. Sediment. Res.* **2003**, *73*, 367–388. [CrossRef]
10. Brown, A.R. *Interpretation of Three-Dimensional Seismic Data;* American Association of Petroleum Geologists: Tulsa, OK, USA, 2011; pp. 1–309.
11. Xu, G.; Haq, B.U. Seismic facies analysis: Past, present and future. *Earth-Sci. Rev.* **2022**, *224*, 103876. [CrossRef]
12. Posamentier, H. Seismic Geomorphology and Depositional Systems of Deep-Water Environments; Observations from Offshore Nigeria, Gulf of Mexico, and Indonesia. In Proceedings of the AAPG Annual Convention Program, Florence, Italy, 27 May 2002.
13. Yu, T.; Liu, H.; Liu, B.; Tang, S.; Tang, Y.; Yin, C. Restoration of karst paleogeomorphology and its significance in petroleum geology—Using the top of the Middle Triassic Leikoupo Formation in the northwestern Sichuan Basin as an example. *J. Petrol. Sci. Eng.* **2022**, *208*, 109638. [CrossRef]
14. Cheng, Y.F.; Dong, Y.L.D.; Zhu, X.M.; Yang, D.Q.; Wu, W.; Yang, K.; Su, B.; Zhao, R.X.; Qiao, C.K. Cretaceous paleogeomorphology restoration and its controlling mechanism on sand-bodies in Chunguang exploration area, Junggar Basin. *J. Palaeogeogr.* **2020**, *6*, 1127–1142.
15. Vail, P.R.; Audemard, F.; Bowman, S.A.; Eisner, P.N.; Perez-Cruz, G. *The Stratigraphic Signatures of Tectonics, Eustacy and Sedimentology: An Overview*; Einsele, G., Ricken, W., Seilacher, A., Eds.; Springer: Berlin/Heidelberg, Germany, 1991; pp. 617–659.
16. Leila, M.; Moscariello, A. Seismic stratigraphy and sedimentary facies analysis of the pre- and syn-Messinain salinity crisis sequences, onshore Nile Delta, Egypt: Implications for reservoir quality prediction. *Mar. Petrol. Geol.* **2019**, *101*, 303–321. [CrossRef]
17. Cai, Q.; Hu, M.; Ngia, N.R.; Hu, Z. Sequence stratigraphy, sedimentary systems and implications for hydrocarbon exploration in the northern Xujiaweizi Fault Depression, Songliao Basin, NE China. *J. Petrol. Sci. Eng.* **2017**, *152*, 471–494. [CrossRef]
18. Zimmer, E.H.; Howell, J.A.; Schofield, N.; Gawthorpe, R.L. Seismic geomorphology linked to sequence stratigraphy of an Eocene delta in the Outer Moray Firth, UKCS. *Mar. Petrol. Geol.* **2019**, *104*, 150–167. [CrossRef]
19. Batchelor, C.L.; Dowdeswell, J.A.; Pietras, J.T. Seismic stratigraphy, sedimentary architecture and palaeo-glaciology of the Mackenzie Trough: Evidence for two Quaternary ice advances and limited fan development on the western Canadian Beaufort Sea margin. *Quat. Sci. Rev.* **2013**, *65*, 73–87. [CrossRef]
20. Catuneanu, O.; Abreu, V.; Bhattacharya, J.P.; Blum, M.D.; Dalrymple, R.W.; Eriksson, P.G.; Fielding, C.R.; Fisher, W.L.; Galloway, W.E.; Gibling, M.R.; et al. Towards the standardization of sequence stratigraphy. *Earth-Sci. Rev.* **2009**, *92*, 102991. [CrossRef]
21. Matenco, L.C.; Haq, B.U. Multi-scale depositional successions in tectonic settings. *Earth-Sci. Rev.* **2020**, *200*, 102991. [CrossRef]
22. Catuneanu, O. Model-independent sequence stratigraphy. *Earth-Sci. Rev.* **2019**, *188*, 312–388. [CrossRef]
23. Xu, G.; Zhang, L.; Pang, X.; Chen, M.; Xu, S.; Liu, B.; Zuo, Y.; Luo, S.; Hu, L.; Chen, H.; et al. New method for the reconstruction of sedimentary systems including lithofacies, environments, and flow paths: A case study of the Xisha Trough Basin, South China Sea. *Mar. Petrol. Geol.* **2021**, *133*, 105268. [CrossRef]
24. Miall, A.D. *Stratigraphy: A Modern Synthesis*; Springer International Publishing: Cham, Switzerland, 2016; pp. 1–442.
25. Chopra, S.; Marfurt, K.J. Evolution of seismic interpretation during the last three decades. *Lead. Edge* **2012**, *31*, 654–676. [CrossRef]
26. Coléou, T.; Poupon, M.; Azbel, K. Unsupervised seismic facies classification: A review and comparison of techniques and implementation. *Lead. Edge* **2003**, *22*, 942–953. [CrossRef]
27. Song, C.; Liu, Z.; Wang, Y.; Li, X.; Hu, G. Multi-waveform classification for seismic facies analysis. *Comput. Geosci.* **2017**, *101*, 1–9. [CrossRef]

28. Ao, Y.; Li, H.; Zhu, L.; Ali, S.; Yang, Z. Identifying channel sand-body from multiple seismic attributes with an improved random forest algorithm. *J. Petrol. Sci. Eng.* **2019**, *173*, 781–792. [CrossRef]
29. Kumar, P.C.; Omosanya, K.O.; Sain, K. Sill Cube: An automated approach for the interpretation of magmatic sill complexes on seismic reflection data. *Mar. Petrol. Geol.* **2019**, *100*, 60–84. [CrossRef]
30. Zeng, H. Frequency-dependent seismic-stratigraphic and fades interpretation. *AAPG Bull.* **2013**, *97*, 201–221. [CrossRef]
31. Xie, T.; Zheng, X.; Zhang, Y. Seismic facies analysis based on speech recognition feature parameters. *Geophysics* **2017**, *82*, 23–35. [CrossRef]
32. Farzadi, P. Seismic facies analysis based on 3D multi-attribute volume classification Dariyan Formation, SE Persian Gulf. *J. Petrol. Geol.* **2006**, *29*, 159–173. [CrossRef]
33. Di, H.; Alfarraj, M.; AlRegib, G. Three-dimensional curvature analysis of seismic waveforms and its interpretational implications. *Geophys. Prospect.* **2019**, *67*, 265–281. [CrossRef]
34. Johansen, S.E.; Arntsen, B.; Raknes, E.B.; Omosanya, K.O.L.; Harishidayat, D.; Henningsen, T. Seismic forward modelling of the Kvalhovden outcrop, Spitsbergen, Norway. *Mar. Petrol. Geol.* **2023**, *147*, 106000. [CrossRef]
35. Shang, W.; Xu, S.; Li, X.; Liang, F.; Wu, C.; Wang, J.; Li, Z.; Sun, Y.; Li, Y.; Li, M.; et al. Utilizing 2D seismic forward modeling to constrain the seismic response and plane distribution of grain shoal reservoir in the northern slope of Central Sichuan Paleo-uplift, Sichuan Basin. *Mar. Petrol. Geol.* **2023**, *152*, 106228. [CrossRef]
36. Tomassi, A.; Trippetta, F.; de Franco, R.; Ruggieri, R. From petrophysical properties to forward-seismic modeling of facies heterogeneity in the carbonate realm (Majella Massif, central Italy). *J. Petrol. Sci. Eng.* **2022**, *211*, 110242. [CrossRef]
37. Xiong, R.; Zheng, J.; Wang, X.; Huang, L.; Guo, L. 3D Outcrop geologic modeling and seismic forward modeling of mound-beach complexes. *Arab. J. Geosci.* **2022**, *15*, 1289. [CrossRef]
38. Wan, L.; Hurter, S.; Bianchi, V.; Li, P.; Wang, J.; Salles, T. The roles and seismic expressions of turbidites and mass transport deposits using stratigraphic forward modeling and seismic forward modeling. *J. Asian Earth Sci.* **2022**, *232*, 105110. [CrossRef]
39. Grasseau, N.; Grélaud, C.; López-Blanco, M.; Razin, P. Forward seismic modeling as a guide improving detailed seismic interpretation of deltaic systems: Example of the Eocene Sobrarbe delta outcrop (South-Pyrenean foreland basin, Spain), as a reference to the analogous subsurface Albian-Cenomanian Torok-Nanushuk Delta of the Colville Basin (NPRA, USA). *Mar. Petrol. Geol.* **2019**, *100*, 225–245.
40. Shuster, M.W.; Aigner, T. Two-dimensional synthetic seismic and log cross sections from stratigraphic forward models. *AAPG Bull.* **1994**, *78*, 409–431.
41. Wang, D.; Yu, H.; Wang, J.; Li, R.; Li, L. Key exploration techniques for stratigraphic-lithologic reservoirs and their application in Qinnan sag, Bohai sea. *China Offshore Oil Gas* **2015**, *27*, 16–24. (In Chinese with English Abstract)
42. Lai, W.; Xu, C.; Wang, X.; Wang, C.; Liu, F. A study on Paleogene sequence stratigraphy and sedimentary systems and a discussion on hydrocarbon exploration directions in Qinnan depression. *China Offshore Oil Gas* **2007**, *5*, 300–305. (In Chinese with English Abstract)
43. Cai, S.; Zhou, D.; Wang, D.; Zhang, J.; Zhang, Z.; Li, Y. Tectonic development characteristics and favorable exploration direction of Qinnan sag in Bohai Bay Basin. *Acta Petrol. Sin.* **2019**, *40*, 532–541. (In Chinese with English Abstract)
44. Zhang, Z.; Xu, C.; Guo, R.; Li, L.; Yang, C. Cenozoic fault system and tectonic evolution of Qinan Sag in Bohai Sea. *Fault-Block Oil Gas Field* **2019**, *26*, 158–161+167. (In Chinese with English Abstract)
45. Cai, S.; Lü, D.; He, D.; Zhang, J.; Yu, Y. Characteristics of tectonic migration in Qinnan sag of the Bohai Bay Basin and its impact on hydrocarbon accumulation. *Acta Petrol. Sin.* **2019**, *40* (Suppl. 2), 67–78. (In Chinese with English Abstract)
46. Liu, H. Study on the Paleogene system Formation Sequence stratigraphy and depositional system in Qinnan depression of Bohai sea. Master's Thesis, Chengdu University of Technology, Sichuan, China, 2014. (In Chinese with English Abstract).
47. Wang, Q.; Hao, F.; Xu, C.; Zhu, Y.; Sun, Z.; Zou, H. The origin and charging directions of Neogene biodegraded oils: A geochemical study of large oil fields in the middle of the Shijiutuo Uplift, Bohai Sea, China. *Mar. Petrol. Geol.* **2017**, *88*, 200–213. [CrossRef]
48. Zhao, J.; Liu, C.; Huang, L.; Mountney, N.; Han, S.; Liu, P.; Hu, J. Original Sedimentary Pattern of an Inverted Basin: A Case Study from the Bozhong Depression, Offshore Bohai Bay Basin. *Acta Geol. Sin.* **2016**, *90*, 2163–2181. [CrossRef]
49. Yang, C.; Guo, R.; Wang, F.; Zhang, Z.; Pei, X. Quantitative prediction of TOC in source rocks of the Member 3 of Shahejie Formation in Qinnan Depression, Bohai sea, China. *J. Chengdu Univ. Technol. (Sci. Technol.)* **2019**, *46*, 549–557. (In Chinese with English Abstract)
50. Veeken, P.C.H.; Moerkerken, B.V. *Seismic Stratigraphy and Depositional Facies Models*; Academic Press: Amsterdam, The Netherlands, 2013; pp. 15–410.
51. Xu, C.G.; Du, X.F.; Zhu, H.T. *Sand Control Principle and Application of Source to Sink System in Continental Rift Basin*; China Science Publishing and Media Ltd.: Beijing, China, 2020; pp. 27–54.

52. Huang, C.; Wang, H.; Wu, Y.; Wang, J.; Chen, S.; Ren, P.; Liao, Y.; Zhao, S.; Xia, C. Genetic types and sequence stratigraphy models of Palaeogene slope break belts in Qikou Sag, Huanghua Depression, Bohai Bay Basin, Eastern China. *Sediment. Geol.* **2012**, *261–262*, 65–75. [CrossRef]
53. Wang, C.C.; Xu, G.Q.; Zhuo, Y.H. *Fundamental for Seismic-Geologic Interpretation (Chinese and English Edition)*; Petroleum Industry Press: Beijing, China, 2015; p. 126.

processes

Article

Factors That Control the Reservoir Quality of the Carboniferous–Permian Tight Sandstones in the Shilounan Block, Ordos Basin

Jing Wang [1,2], Fawang Ye [1,*], Chuan Zhang [1] and Zhaodong Xi [2]

1 National Key Lab of Remote Sensing Information and Imagery Analysis, Beijing Research Institute of Uranium Geology, Beijing 100029, China
2 School of Energy Resource, China University of Geosciences (Beijing), Beijing 100083, China
* Correspondence: yfwbeijing2008@sina.com

Abstract: The Carboniferous–Permian, coal-bearing, sedimentary succession on the eastern edge of the Ordos Basin in the Shilounan Block contains large accumulations of hydrocarbon resources. During the exploration of coalbed methane and tight sandstone gas in the study area, multiple drilling wells in the tight sandstone reservoirs have yielded favorable gas logging results. The Benxi, Taiyuan, Shanxi, Shihezi, and Shiqianfeng formations contain multiple sets of sandstone reservoirs, and the reservoir quality and the controlling factors of its tight sandstones were affected by sedimentation, diagenetic alteration, and pore structure. This study comprehensively examines the sedimentary environment, distribution of sand bodies, and physical characteristics of tight sandstone reservoirs through drilling, coring, logging, and experimental testing. The results indicate that the Carboniferous–Permian tight sandstones are mainly composed of lithic sandstone and lithic quartz sandstone. The reservoir quality is relatively poor, with an average permeability of 0.705 mD and porosity of 6.20%. The development of reservoirs in the study area is primarily influenced by diagenesis and sedimentation. Compaction and cementation, which are destructive diagenetic processes, significantly reduced the porosity of the sandstone reservoirs in the study area. Compaction primarily causes a reduction in porosity and accounts for over 70% of the overall decrease in porosity. Dissolution, as a constructive diagenetic process, has a limited effect on porosity and is the primary reason for the relatively tight nature of these reservoirs. The macroscopic and microscopic characteristics of tight sandstone reservoirs were used to establish the evaluation and classification criteria, after which the sandstone reservoirs in the study area were divided into three types. The poor quality type II and type III reservoirs are predominant, while high quality type I reservoirs are primarily limited to the Shihezi Formation.

Keywords: tight sandstone gas; reservoir; pore structure; diagenesis; porosity

Citation: Wang, J.; Ye, F.; Zhang, C.; Xi, Z. Factors That Control the Reservoir Quality of the Carboniferous–Permian Tight Sandstones in the Shilounan Block, Ordos Basin. *Processes* **2023**, *11*, 2279. https://doi.org/10.3390/pr11082279

Academic Editor: Carlos Sierra Fernández

Received: 25 June 2023
Revised: 25 July 2023
Accepted: 26 July 2023
Published: 28 July 2023

1. Introduction

Tight sandstone gas is an important unconventional natural gas. Tight gas reservoirs typically have an effective permeability of less than 0.1 mD (or an absolute permeability of less than 1 mD) and a porosity of less than 10% [1–3]. A diverse array of tight sandstone gas reservoirs with significant economic potential occurs across China. Progress in unconventional oil and gas exploration technologies has allowed for the identification of basins such as the Ordos, the Tarim, and the Songliao, where geological conditions are also conducive to the formation of tight sandstone gas reservoirs [4]. The annual production of tight sandstone gas in the Ordos Basin surpassed 3×10^{10} m^3 in 2010, with proven reserves exceeding 2×10^8 m^3. Tight sandstone gas accounts for over one third of China's natural gas energy structure, making China one of the leading global producers of tight sandstone gas after the United States and Canada [4,5].

Tight sandstone gas reservoirs are extensively distributed in the Upper Paleozoic strata of the Ordos Basin and occur primarily within the Carboniferous–Permian Taiyuan, Shanxi, and Shihezi formations, where these formations serve as the principal layers for tight sandstone gas production. Despite the widespread occurrence of sandstone across the Ordos Basin, its compositional and structural maturity is relatively low. Under surface conditions, a significant portion of the samples have porosity levels below 10%, and approximately 90% of the reservoirs have a matrix permeability below 0.1 mD under overburdened pressure conditions [6,7]. The development of tight sandstone reservoirs is primarily influenced by sedimentation, diagenesis, and tectonism. Sedimentation plays the largest role in the formation of low-permeability reservoirs and determines the types of diagenesis that occur. Diagenesis is the key driver in creating ultra-low porosity and permeability reservoirs, while tectonic processes transform tight sandstone reservoirs [8–11]. Advances in research methods and technologies have allowed for substantial progress in studying the characteristics and controlling factors in the development of tight sandstone reservoirs. Advanced technologies that have been used in reservoir characterization include high-precision and high-resolution microscopic observation, fluid inclusion testing, mineralogical composition identification, genetic analysis, and pore system characterization [12–14]. These technologies and research methods provide a reliable foundation for analyzing pore genesis, diagenesis types, and diagenetic evolution, and shift the study of tight sandstone reservoir characteristics and their controlling factors from a qualitative description to a quantitative study.

The tight sandstone formation associated with the Carboniferous–Permian coal measure in the eastern margin of the Ordos Basin exhibits a stable horizontal thickness distribution and vertical multi-layer superposition. This formation offers promising prospects for the exploration and development of tight sandstone gas. This study used experimental methods, such as thin section petrographic examinations, mercury intrusion, and porosity and permeability testing, to determine the distribution and reservoir characteristics of the sandstone formation of the Shilounan Block. The influence of sedimentation, diagenesis, and pore structure on the physical properties of tight sandstone reservoirs has also been investigated and provides valuable guidance for the exploration and development of tight sandstone gas in the Shilounan Block.

2. Geological Setting

2.1. Structural Characteristics and Stratigraphy

The Shilounan Block is situated in the southern part of the Jinxi Flexural Belt within the Ordos Basin. It shows an overall structural pattern known as "two uplifts and two depressions". The "two uplifts" are the Shilou Anticline and the Xixian Anticline, with the eastern wing of the Shilou Anticline being intersected by a nearly north–south westward dipping thrust fault. This fault causes a steeper eastern wing when compared to the western wing. The "two depressions" consist of the Caocun Fold Belt and the Xixiandong Syncline. The structural characteristics of the study area primarily involve north–south and northeast folding and flexing, accompanied by a significant presence of faults in the region (Figure 1).

The Upper Paleozoic strata in the Shilounan Block are well developed, with the lowermost part of the sedimentary succession consisting of Middle Ordovician limestones of the Majiagou Formation. The Middle Carboniferous Benxi Formations, which are characterized by a parallel unconformity, overlay the Ordovician limestones. The coal-bearing strata within this block include the Middle Carboniferous Benxi Formation, the Upper Carboniferous Taiyuan Formation, and the Lower Permian Shanxi Formation. Additionally, the uppermost part of the sedimentary succession is represented by the Shihezi Formation and the Shiqianfeng Formation (Figure 2).

2.2. Sedimentary Facies and Sand Body Distribution

The Shilounan Block is extensively developed in the Upper Paleozoic sandstones. It consists primarily of the tidal flat sandstones from the Taiyuan Formation and the channel sandstones from the Shanxi, Shihezi, and Shiqianfeng formations, as well as beach bar sandstones. A depositional hiatus occurred during the Middle Ordovician and the Early Carboniferous. During the sedimentation period of the Benxi Formation in the Middle Carboniferous, the Ordos Basin gradually underwent subsidence as seawater infiltrated from the east and southeast. The sediments from the Shilounan Block in the Ordos basin were deposited in an epicontinental sea. The sedimentary characteristics of the Taiyuan Formation are similar to the Benxi period, where the Shilounan Block received sediment from areas northeast and northwest of the Ordos basin. This caused the development of tidal flats, barrier islands, and lagoon deposits (Figure 3a). The sand bodies in this formation consist of gray fine-grained sandstones and light gray siltstones, and have a maximum sedimentary thickness of 14.8 m. The Shanxi Formation was predominantly deposited in the delta plain subfacies and is characterized by distributary channel sands that form the main sand body (Figure 3b). The sand body exhibits a northwest–southeast distribution trend and ranges in thickness from 5 to 30 m.

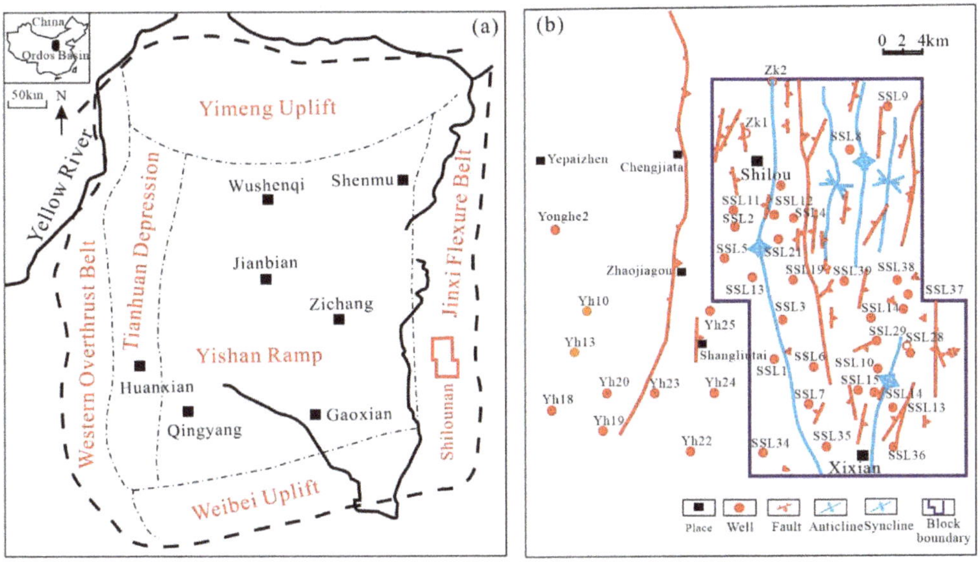

Figure 1. (**a**) The tectonic divisions of the Ordos Basin and the location of the Shilounan Block (irregular polygon colored in red); (**b**) structural map of the Shilounan Block.

The climate transitioned from warm and humid to arid and hot during the sedimentation of the Xiashihezi Formation, which significantly reduced the vegetation. A series of gray-white to yellow-green terrigenous debris, which developed the delta front subfacies, was deposited during this period. The main sedimentary microfacies comprised underwater distributary channels and underwater interdistributary bays, with the presence of underwater distributary channel sand bodies (Figure 3c). The sediments of the Shangshihezi Formation are characterized by purplish-red and yellowish-green hues. These colors indicate a lithofacies paleogeographic environment where delta and shallow lake conditions coexisted. Within the Shilounan Block, the delta front subfacies are characterized by the presence of underwater distributary channel sand bodies (Figure 3d). A series of purplish-red clastic rocks were deposited during the sedimentation of the Shiqianfeng Formation. This area mainly exhibits the development of delta front subfacies and shore-shallow lake facies, with the presence of underwater distributary channel sand bodies (Figure 3e).

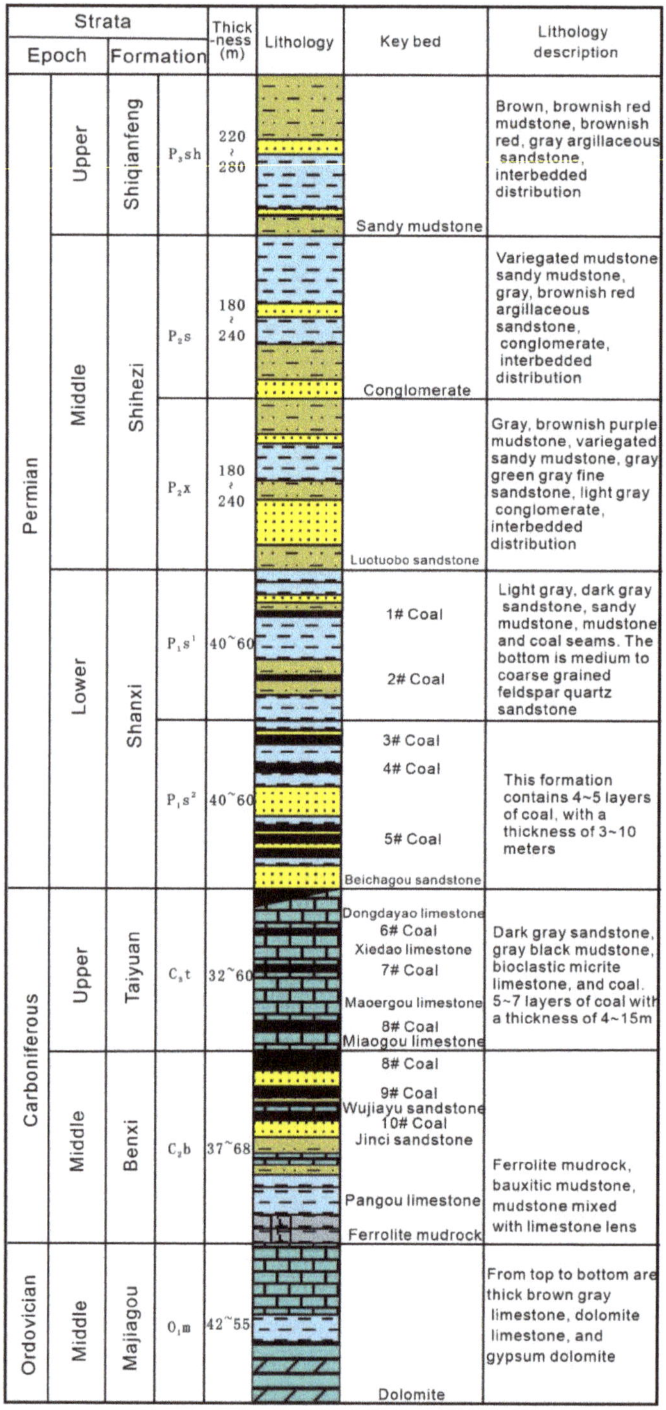

Figure 2. Comprehensive lithostratigraphic column of the Shilounan Block.

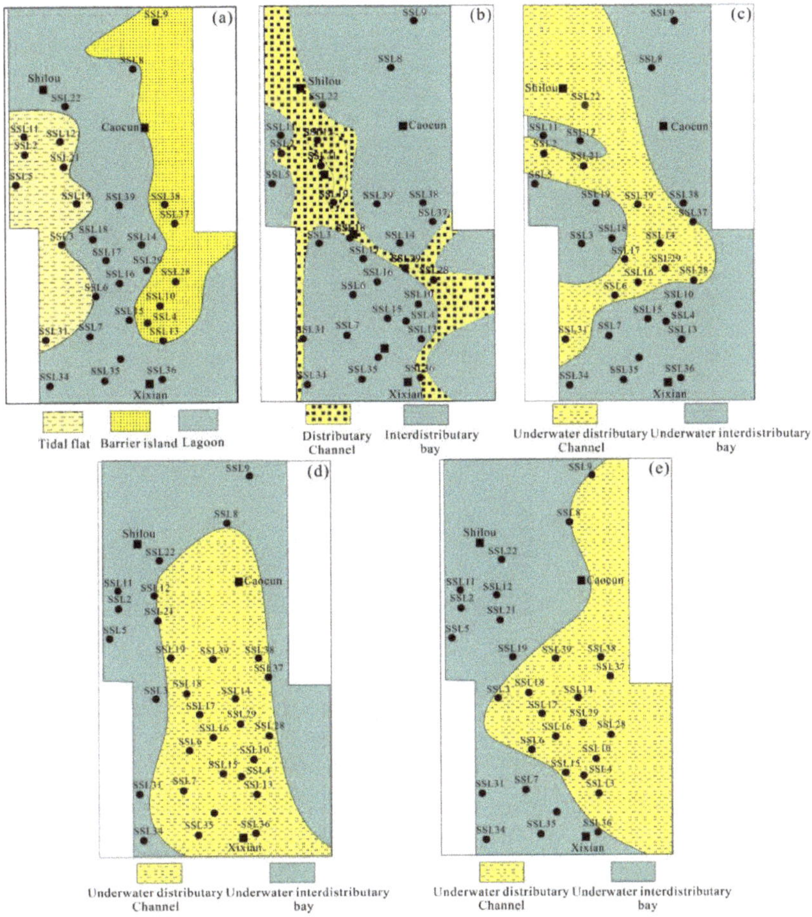

Figure 3. Distribution of the sedimentary facies and sand bodies within the study area (the Shilounan Block): (**a**) Taiyuan Formation, (**b**) Shanxi Formation, (**c**) Xiashihezi Formation, (**d**) Shangshihezi Formation, (**e**) Shiqianfeng Formation. Dots represent well locations and rectangles represent cities.

3. Sampling and Methodology

More than 280 samples were collected from 28 drilling wells in the Shilounan Block, with the main sampling layers including the Benxi, Taiyuan, Shanxi, Shihezi, and Shiqianfeng formations. Figure 1b shows the distribution of the well locations for sampling. The samples were analyzed via thin section petrographic examinations, SEM, porosity and permeability measurements, XRD analysis, and mercury intrusion tests. The specific details of the experimental methods and detailed procedures are based on previous papers published by our team [15,16].

4. Results

4.1. Petrological Characteristics

The sandstones from the Taiyuan and Shanxi formations range primarily from dark gray to gray. In contrast, the sandstones from the Shihezi Formation range from gray to gray-white to gray-green. The sandstones of the Shiqianfeng Formation, however, showcase a more diverse range of colors that range from mainly brown to grayish-green and purple-red. There is a notable color variation in the sandstones from the Benxi Formation and the

Shiqianfeng Formation, where the formations transition from gray and black to green and grayish-brown. These color changes indicate a shift from reducing to oxidizing conditions and a transition in climate from humid to arid in the sedimentary environment during the late Paleozoic (Figure 4).

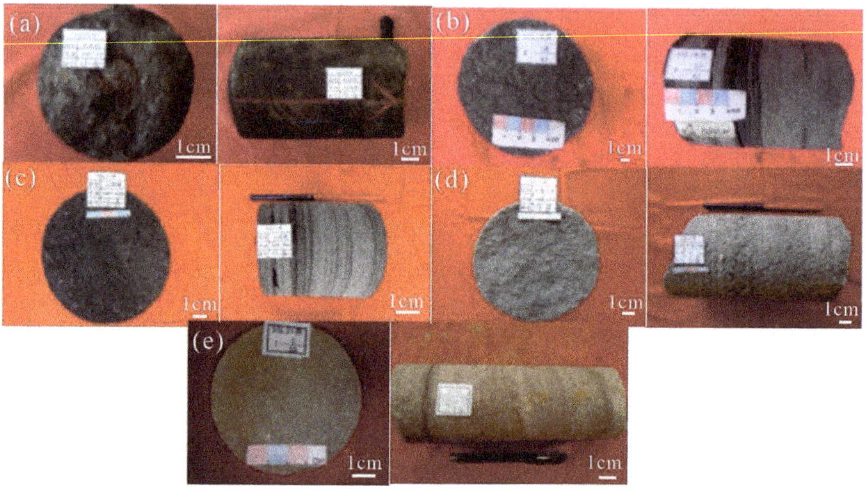

Figure 4. Core characteristics of the Carboniferous–Permian sandstones in the study area: (**a**) dark gray siltstone of the Benxi Formation; (**b**) gray medium-grained sandstone of the Taiyuan Formation; (**c**) gray fine-grained sandstone with carbonaceous and argillaceous material and muscovite fragments of the Shanxi Formation; (**d**) gray-green sandstone of the Shihezi Formation; (**e**) grayish brown fine-grained sandstone of the Shiqianfeng Formation.

The Shilounan Block contains several types of sandstones, such as lithic sandstone, lithic feldspathic sandstone, lithic quartz sandstone, quartz sandstone, and feldspathic lithic sandstone (Figure 5). Within the Taiyuan and Shanxi formations, three main types of sandstone are present—lithic sandstone, lithic quartz sandstone, and quartz sandstone. The quartz content varies from 10% to 96%, with an average quartz content of 53.2%. The rock debris content ranges from 1% to 70%, with an average rock debris content of 28.9%. The feldspar content in the Taiyuan and Shanxi formations is relatively low and is present as fine to medium feldspar grains.

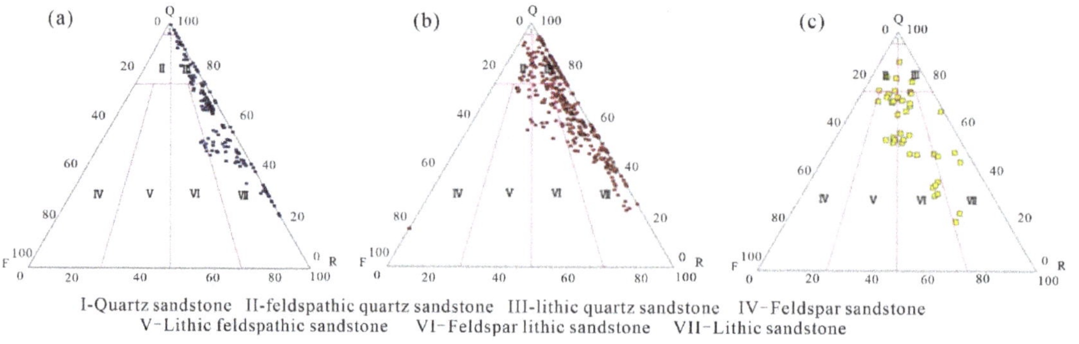

Figure 5. Sandstone types identified in the Carboniferous–Permian sediments from the Shilounan Block study area: (**a**) Taiyuan and Shanxi formations, (**b**) Shihezi Formation, (**c**) Shiqianfeng Formation. Q = quartz; F = feldspar; R = rock fragments.

The Shihezi Formation primarily comprises lithic sandstone, commonly known as lithic quartz sandstone, feldspathic lithic sandstone, and quartz sandstone. The quartz content within this formation ranges from 14% to 89%, while the rock debris content ranges from 2% to 63%, with an average content of 26.1%. In contrast, the sandstones of the Shiqianfeng Formation are predominantly composed of lithic feldspathic sandstone and feldspathic lithic sandstone, and are occasionally accompanied by lithic sandstone and feldspathic quartz sandstone. The feldspar content in the Shihezi Formation is relatively high. Quartz is the predominant mineral in the sandstone, with the highest content being observed in the Taiyuan Formation and the lowest content being observed in the Shiqianfeng Formation.

4.2. Porosity and Permeability

Porosity and permeability tests were conducted on 245 sandstone samples from the Shilounan Block. The samples have an average porosity of 6.2% and an average permeability of only 0.705 mD. The reservoir, therefore, has low porosity and ultra-low permeability (Table 1).

Table 1. Porosity and permeability of the sandstones from the study area.

Formation	Shangshihezi	Xiashihezi	Shanxi	Taiyuan	Benxi
Porosity/%	8.4 (101)	7.1 (94)	3.99 (11)	6.88 (2)	4.9 (4)
Permeability/mD	0.95 (97)	0.29 (93)	0.11 (11)	0.77 (2)	2.86 (4)

Note: The numbers in the parentheses indicate the number of samples.

The Shihezi Formation has the highest porosity, with an average of 7.7%. The Shangshihezi Formation has an average porosity of 8.4%, while the Xiashihezi Formation has an average porosity of 7.1%. The sandstones in the Xiashihezi Formation display a maximum porosity of 15.4% and a minimum porosity of 1.3%. Porosities that exceed 10% account for only 10% of the samples (Figure 6a). The sandstone in the Shangshihezi Formation ranges from a maximum porosity of 16.1% to a minimum porosity of 1.6%. Here, approximately 50% of the samples exhibit porosity exceeding 10%, indicating favorable reservoir characteristics (Table 1 and Figure 6a). Even though the Taiyuan Formation has a porosity close to 7%, the sample size is relatively small and shows higher variability. The Shanxi and Benxi formations have average porosities of 4.0% and 4.9%, respectively.

The limited number of samples from the Benxi and Taiyuan formations exhibit relatively high permeabilities. The average permeability is 0.11 mD in the Shanxi Formation, 0.29 mD in the Xiashihezi Formation, and 0.95 mD in the Shangshihezi Formation. The permeability range of the Shihezi Formation is quite extensive, with the Xiashihezi Formation displaying a maximum permeability of 6.14 mD and a minimum of only 0.0013 mD (Figure 6b). Similarly, the Shangshihezi Formation has a maximum permeability of 14.18 mD and a minimum of 0.0033 mD. The peak permeability of the Xiashihezi Formation ranges from 0.1 to 0.5 mD, while the peak permeability of the Shangshihezi Formation ranges from 0.1 to 0.5 mD and exceeds 2 mD (Figure 6b).

Even though the data points on the porosity and permeability of the sandstone reservoirs in the study area are relatively scattered, they exhibit a certain degree of correlation (Figure 7). This correlation suggests that the physical properties are influenced by multiple factors rather than a single geological factor. The variations in the sedimentary background, the burial depth, the lithology of the roof and floor, and the heterogeneity of diagenesis, all contribute to the weak correlation between the porosity and the permeability [17,18]. Cai (2015) also observed a similar weak positive correlation between the porosity and the permeability when investigating the Chang 8 reservoir in the Jiyuan area of the Ordos Basin [19].

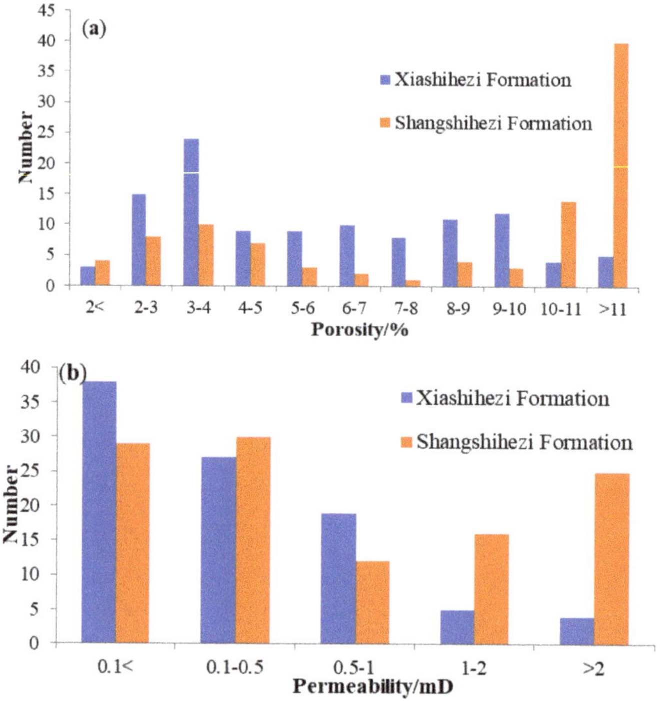

Figure 6. Distribution of porosity and permeability in the sandstone reservoirs from the Shihezi Formation in the study area: (**a**) Distribution of porosity, and (**b**) Distribution of permeability.

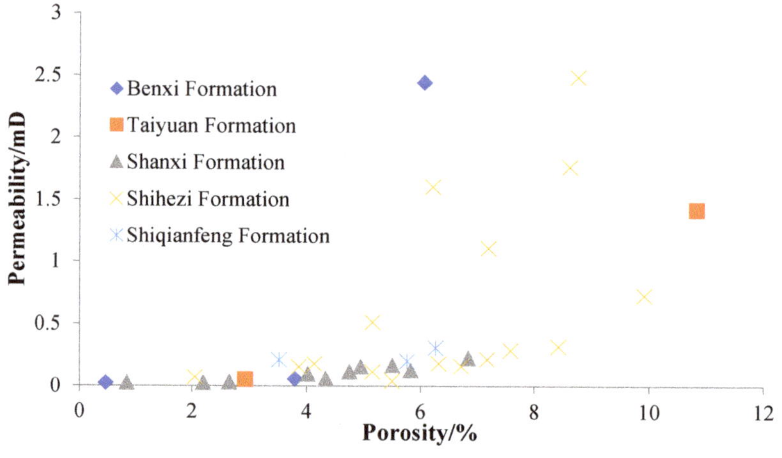

Figure 7. Relationship between porosity and permeability in Carboniferous–Permian sandstone samples.

4.3. Pore System

4.3.1. Pore Type

Petrographical examinations and statistical analysis of 117 sandstone cast thin sections, combined with scanning electron microscopy (SEM) analysis, were used to classify the pore types into residual intergranular pores, dissolution pores, and intercrystalline pores.

Residual intergranular pores: the primary intergranular pores are almost imperceptible, while the residual intergranular pores are predominantly observed. The residual intergranular pores are mainly formed due to compaction and cementation, and exhibit triangular or polygonal shapes with fairly straight pore boundaries. There is limited connectivity and significant heterogeneity among these isolated pores because the distribution of the residual intergranular pores is primarily isolated (Figure 8).

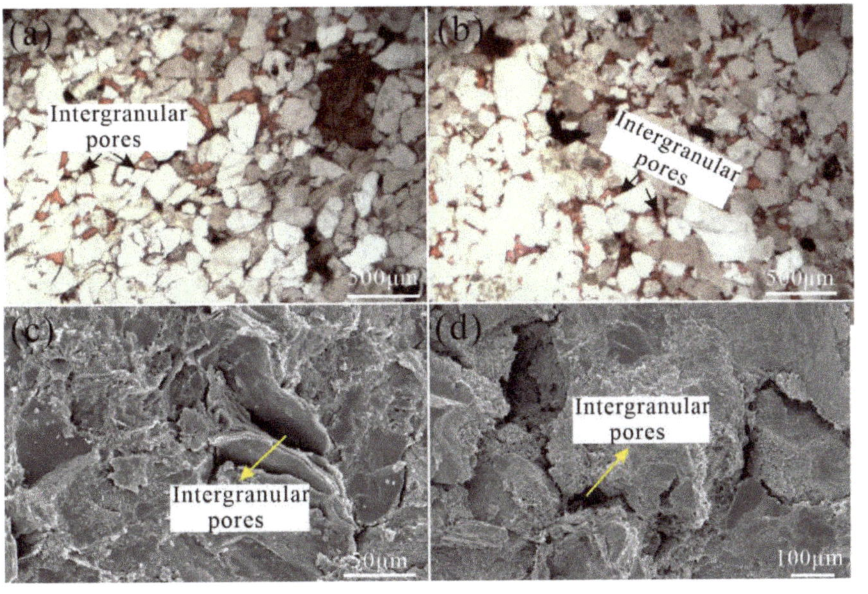

Figure 8. Microscopic images of polished thin sections and high-resolution SEM images of residual intergranular pores: (**a**) medium-grained feldspathic sandstone, with small residual intergranular pores; (**b**) medium-grained feldspathic sandstone, with small residual intergranular pores; (**c**) particles that are bent due to compaction, causing the irregular morphology of the intergranular pores; (**d**) kaolinite aggregates and illite crystals between particles, with development of residual intergranular pores within these aggregates.

Dissolved pores: the dissolved pores consist of intergranular pores and intragranular pores (Figure 9a–d). The feldspar, rock debris, and clay minerals are susceptible to dissolution. The intergranular pores are large due to the partial or complete erosion of larger minerals such as feldspars. The intergranular pores tend to have good connectivity and an irregular morphology, and feature multilateral and harbor-shaped structures. The intragranular pores, in contrast, are predominantly circular and elliptical, with relatively limited connectivity. These pores are mainly present within the rock debris particles.

Intercrystalline pores: the clay minerals kaolinite and illite are commonly found in this rock. Kaolinite typically forms through feldspar alteration and often exhibits a book-like or plate-like structure. Illite is predominantly fibrous and filamentous, with numerous pores developed within its aggregates (Figure 9e,f).

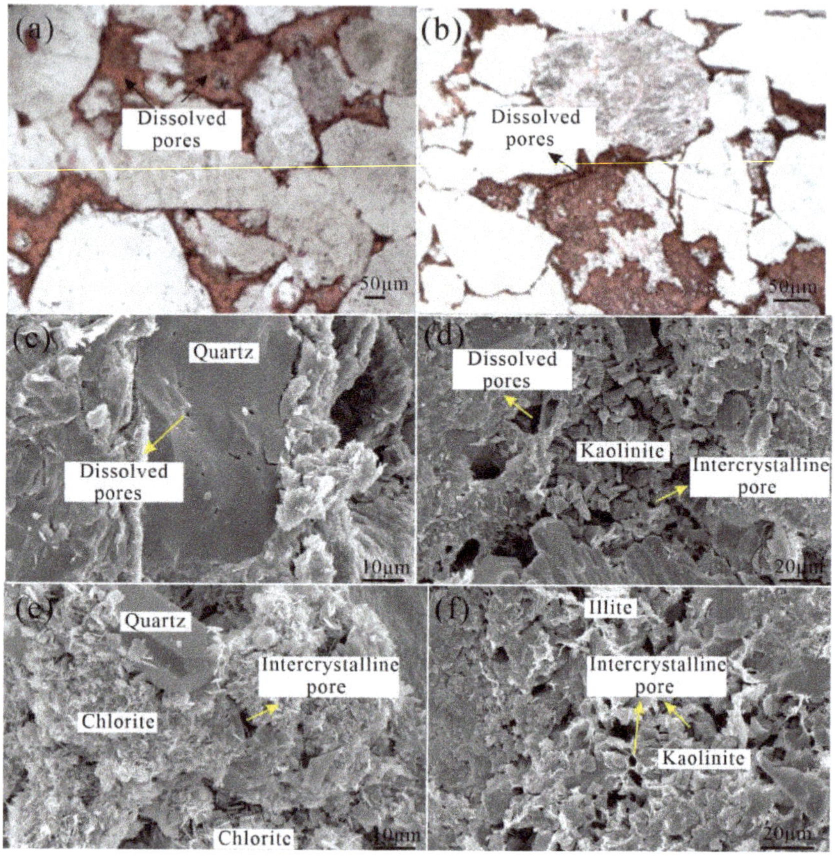

Figure 9. Microscopic images of polished thin sections and high-resolution SEM images of the dissolved pores and intercrystalline pores: (**a**) fine- to medium-grained feldspathic sandstone, with well-developed pores that formed due to feldspar dissolution; (**b**) coarse-grained lithic quartz sandstone exhibiting significant feldspar that was corroded to form well-developed dissolved pores; (**c**) illite-wrapped quartz particle surfaces with limited dissolved pores; (**d**) intergranular pores filled with kaolinite aggregates, along with developed intercrystalline micropores and dissolved pores; (**e**) chlorite aggregate covering particle surfaces, intermixed with authigenic quartz crystals, and displaying developed intergranular micropores; (**f**) intergranular pores filled with kaolinite and illite aggregates, accompanied by developed intercrystalline micropores.

4.3.2. Pore Structure

The pore structure of the sandstone in the Shilounan Block is categorized into three types based on the pore structure parameters and the capillary pressure curve derived from the mercury intrusion measurements of the 29 sandstone samples, the observations of the cast thin sections, and the SEM (Table 2 and Figure 10).

Table 2. Pore structure parameters and classification of pore structure types based on mercury intrusion testing.

Pore Type	Sample ID	Formation	Sorting Coefficient	Structural Coefficient	Coefficient of Mean Value	Average Pore Throat Diameter/μm	Displacement Pressure/MPa	Maximum Mercury Saturation/%	Efficiency of Mercury Ejection/%	Porosity /%
I	S6-5	Xiashi hezi	2.89	3.25	0.16	0.97	0.23	69.03	37.59	3.1
	S6-18	Xiashi hezi	0.87	2.3	0.46	0.37	0.99	81.08	43.38	8.2
	S6-19	Xiashi hezi	1.12	3.42	0.22	0.44	0.79	80.09	38.76	7.2
	S7-6	Xiashi hezi	1.52	1.59	0.3	0.59	0.61	86.85	51.68	12.4
	S7-7	Xiashi hezi	1.62	1.67	0.32	0.62	0.59	84.59	42.85	10.8
	S1-2	Xiashi hezi	0.23	0.08	/	2.48	0.7	84.6	39.8	7.7
	S1-3	Xiashi hezi	0.2	0.17	/	0.31	0.7	91.54	35.92	7
	S1-4	Xiashi hezi	0.21	0.15	/	0.31	0.5	88	42.02	8.6
	S2-22	Shang shihezi	0.21	0.13	/	0.33	0.3	86.89	46.63	16.8
	S2-38	Shang shihezi	0.28	0.19	/	0.41	0.3	79.63	45.26	10.6
	S2-47	Shang shihezi	0.26	0.03	/	0.48	0.3	84.33	35.91	13.8
	Average		0.86	1.18	0.29	0.66	0.55	83.33	41.8	9.65
II	S12-44	Tai yuan	11.2	9.58	0.35	3.66	0.12	85.4	20.08	7.9
	S12-45	Tai yuan	13.2	14.5	0.32	4.26	0.11	75.89	16.19	8.4
	S2-17	Shang shihezi	0.12	0.01	/	0.02	7	90.65	22.28	4
	S2-30	Shang shihezi	0.2	0.2	/	0.38	0.5	86.69	30.17	12.2
	S6-2	Shanxi	0.63	1.06	0.33	0.28	1.65	81.67	26.28	3.6

Table 2. *Cont.*

Pore Type	Sample ID	Formation	Sorting Coefficient	Structural Coefficient	Coefficient of Mean Value	Average Pore Throat Diameter/μm	Displacement Pressure/MPa	Maximum Mercury Saturation/%	Efficiency of Mercury Ejection/%	Porosity /%
II	S12-13	Shang shihezi	2.2	5.09	0.24	0.77	0.38	76.89	33.41	9.9
	S12-14	Shang shihezi	5.2	40.8	0.23	1.64	0.19	61.57	21.18	12.6
	S1-1	Xiashi hezi	0.48	0.05	/	0.11	1	63.57	33.38	2.5
	S2-9	Shanxi	0.52	0.1	/	0.17	0.8	62.45	31.29	4
	S2-14	Shanxi	0.69	0.01	/	0.02	7	51.08	31.73	3.2
	Average		3.43	7.14	0.29	1.13	1.87	73.58	26.6	6.83
III	S6-3	Tai yuan	0.52	1.04	0.39	0.27	1.12	37.42	32.92	3.2
	S6-4	Tai yuan	1.92	0.87	0.24	0.72	0.45	38.98	36.1	5.6
	S6-9	Tai yuan	0.04	0.08	0.6	6.88	6.88	30.79	35.6	2.7
	S2-2	Tai yuan	1.05	0.13	/	0.14	1.5	37.4	44.52	5.2
	S2-52	Xiashi hezi	0.55	0.05	/	0.09	2	59.42	38.08	4
	S2-62	Xiashi hezi	1.66	0.03	/	0.06	3	22.3	33.61	5.7
	S2-67	Xiashi hezi	1.06	0.04	/	0.05	3	36.75	36.2	3.7
	S2-75	Xiashi hezi	1.05	0.13	/	0.14	1.5	37.4	44.52	5.2
	Average		0.98	0.3	0.41	1.04	2.43	37.56	37.69	4.4

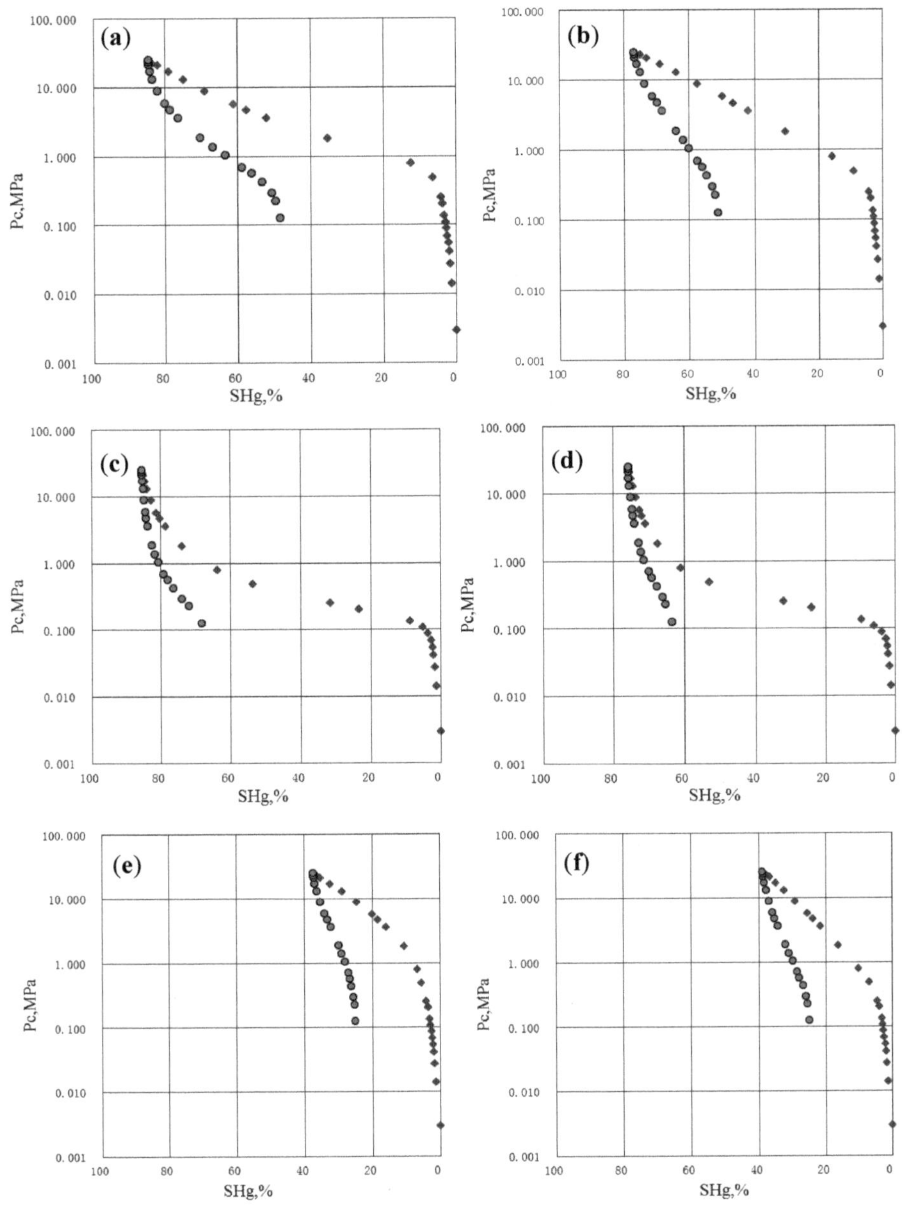

Figure 10. Mercury intrusion curves of three types of sandstone samples in the study area. Porosity type I: (**a**,**b**); type II: (**c**,**d**); type III: (**e**,**f**). The blue circles indicate inject mercury quantity curve, and the red circles indicate withdrawal mercury quantity curve.

Type I: the displacement pressure of a reservoir with the type 1 pore structure ranges from 0.3 to 0.99 MPa, with an average of 0.55 MPa. The average pore throat size ranges from 0.31 to 2.48 μm, indicating relatively small pore throats, while the sorting coefficient is relatively low. The average maximum mercury saturation reached 85%, with over 40% of the mercury being efficiently ejected (with an average mercury injection porosity of 9.65%). The capillary pressure curve has a gentle slope, deviating towards the bottom left of the graph.

It was initially only possible to inject a small amount of mercury, suggesting the absence of well-developed fractures and large pores. However, at 0.1 MPa pressure, a significant amount of mercury entered the pores, indicating good pore connectivity. The pore morphology is predominantly cylindrical and parallel plate-shaped (Figure 10a,b). Sandstone reservoirs with such pore structures generally exhibit favorable physical properties.

Type II: the displacement pressure of a reservoir with the type II pore structure is relatively high, ranging from 0.11 to 7 MPa, with an average of 1.87 MPa. The average pore throat size ranges from 0.02 to 4.26 μm. The sorting coefficient of the type II pore structure is relatively high, and the average maximum mercury saturation reaches 73%. However, the efficiency of mercury ejection is low at only 26.6%, yielding an average mercury injection porosity of 6.83%. The volume of mercury increases slowly when the pressure reaches 0.1 MPa, but rapidly increases when the pressure reaches 1 MPa (Figure 10c,d). These results indicate the presence of relatively well-developed small-sized pores within which a significant amount of mercury gets trapped. Sandstone reservoirs with the type II pore structure have a relatively high porosity but poor connectivity.

Type III: the mercury intrusion curve of a reservoir with the type III pore structure exhibits a steep slope and upward convexity, leaning towards the upper right part of the figure (Figure 10e,f). Both the mercury saturation and efficiency of mercury ejection are below 40%, indicating limited available pore space for mercury intrusion. The average porosity of this type of reservoir is 4.4%, suggesting poor suitability for gas storage and gas seepage.

The Shihezi Formation mainly has sandstone reservoirs with type I and type II pore structures. These reservoirs exhibit fairly well-developed dissolution pores and residual intergranular pores, resulting in higher porosity and better connectivity. These pore structures are primarily formed in a delta front environment. Sandstone reservoirs with a type III pore structure, however, have low porosity and limited connectivity. The Taiyuan Formation consists of many of these reservoirs.

5. Discussion

5.1. Sedimentation Controls on Reservoir Quality

Sedimentary facies play a crucial role in controlling the development and distribution characteristics of sandstone reservoirs. The Taiyuan Formation is primarily composed of tidal flat and lagoon deposits, while the Shanxi Formation consists predominantly of delta plain deposits. The Shihezi Formation is characterized by delta front deposits, while the Shiqianfeng Formation is dominated by shore shallow lake and delta facies. The physical parameters of the sandstone reservoirs vary across the different sedimentary facies zones (Table 3). The sandstone reservoirs in the Shihezi Formation exhibit favorable reservoir properties, such as high porosity and good connectivity. The barrier sandstone reservoirs in the Taiyuan Formation and the distributary channel sandstone reservoirs in the Shanxi Formation, however, exhibit low porosity and permeability. The low porosity and permeability are due to the frequent marine invasion and regression events during their sedimentation period that have led to the interbedding of sand and mudstone [20]. The pore structures have become complex, while the pore tortuosity increased. It is important to note that even sandstone reservoirs that formed in the same sedimentary facies can exhibit significant differences in their properties [21,22]. Sandstone reservoirs that developed in the central areas of river channels, such as those in the Shihezi Formation, generally have better porosity and permeability compared to those developed in the inter-channel or two-wing regions of the river channel.

Table 3. Porosity and permeability of sandstone reservoirs in different sedimentary facies in the study area.

Formation	Sedimentary Facies	Porosity (%)		Permeability (mD)	
		Range	Average	Range	Average
Shangshihezi	Delta Front	0.51–15.34	8.4	0.06–2.31	0.95
Xiashihezi	Delta Front	0.23–14.34	7.1	0.04–0.79	0.29
Shanxi	Distributary Channel	1.43–5.89	3.99	0.007–0.45	0.11
Taiyuan	Tidal Flat Lagoon	4.56–9.2	6.88	0.002–3.64	0.77

Sandstones that formed under different sedimentary conditions can exhibit significant variations in their mineral composition, thickness, distribution, particle size, sorting, and rounding. These further increase the substantial differences in the reservoir properties [23,24]. A positive correlation is observed between quartz content and porosity, indicating that porosity tends to increase with higher quartz content (Figure 11a). The sediments in the study area have experienced deep burial (>1500 m) and significant compaction and pressure dissolution. Since quartz is a rigid mineral, it enhances the rock's ability to withstand compression and preserves some primary intergranular pores [25,26]. Quartz also serves as the material basis for the formation of secondary dissolution pores, and there is a positive correlation between the porosity and the quartz content of the sediment. There is, however, a negative correlation between rock fragments and porosity since higher rock fragment content decreases the porosity (Figure 11c). Unlike quartz, rock fragments have smaller particle sizes which can easily block pores. Sandstone-type reservoirs with higher quartz content (for example quartz sandstone) usually have higher porosities compared to samples with a higher rock fragment content, such as lithic sandstone. No significant correlation was observed between the content of quartz and rock fragments and the permeability (Figure 11b,d). This means that quartz and rock fragments are not the primary controlling factors that influence permeability.

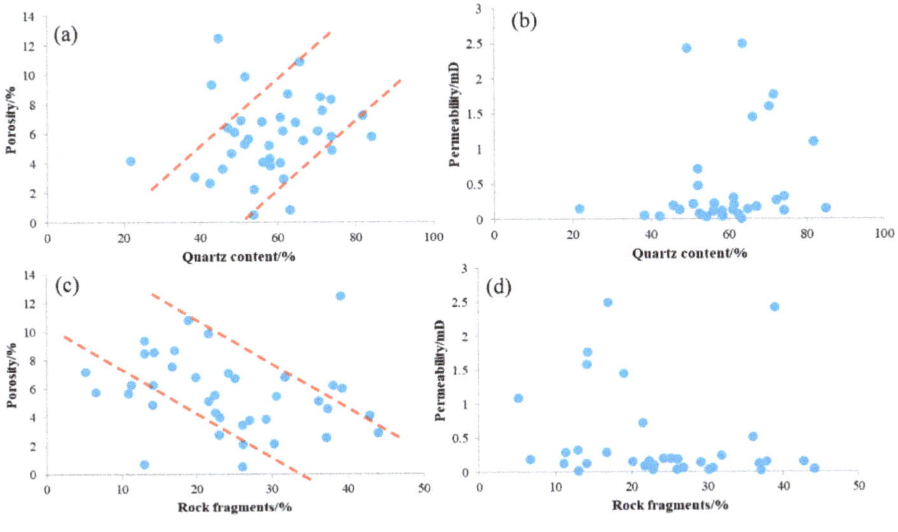

Figure 11. The relationship between the quartz content, the rock fragment content, the porosity, and the permeability of the sandstone samples in the study area: (**a**) The relationship between the quartz content and porosity, and the red dotted lines indicate a positive trend. (**b**) The relationship between the quartz content and permeability. (**c**) The relationship between the rock fragment content and porosity, and the red dotted lines indicate a negative trend. (**d**) The relationship between the rock fragment content and permeability.

5.2. Diagenetic Controls on Reservoir Quality

5.2.1. Diagenesis Type

Compaction: the pressure that is exerted by the overlying strata on the target layer increases as the burial depth increases. This causes intensified compaction [27,28]. The particles primarily experience linear and concave–convex contact during compaction, causing a significant reduction in the pore content and a deterioration in the reservoir quality. Compaction can decrease the porosity of sandstone reservoirs by over 50% and shale gas reservoirs by approximately 70% [29]. In the study area, sandstone reservoirs with favorable porosity and permeability are predominantly found in the relatively shallow-buried Shiqianfeng Formation and Shihezi Formation. The porosity of the sandstone in the study area generally decreases with increasing burial depth, while there is no significant correlation between the permeability and burial depth (Figure 12), indicating that compaction affects porosity. Even though there is a correlation between burial depth and porosity, the data points also exhibit variability, suggesting that the intensity of compaction is influenced by factors such as sandstone composition and other factors such as cementation. A sandstone that is predominantly composed of rigid particles like quartz may exhibit less compaction even at greater burial depths, indicating the effect of composition on compaction and porosity [26].

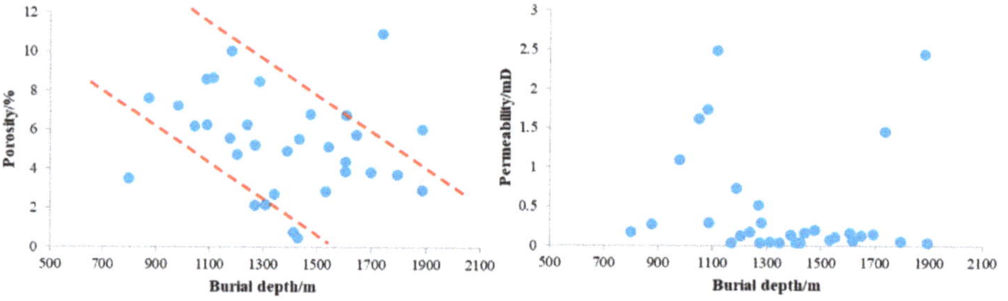

Figure 12. The relationship between burial depth and the porosity and permeability of sandstone reservoirs. The red dotted lines indicate a negative trend.

Cementation: the main types of cementation in the study area include clay mineral cementation, siliceous cementation, and calcite cementation (Figure 13). Illite, which forms during the middle-late diagenetic stage, is a common clay mineral that occurs in the rocks from the study area. It primarily occurs in hair-like, needle-like, and fibrous forms, where it coats other particles. Kaolinite occurs predominantly as plate-like, book-like, or vermicular structures. Plate-like kaolinite forms due to feldspar dissolution in acidic water bodies during the early diagenetic stage. Kaolinite, which exhibits well-developed crystal forms, primarily fills the dissolution pores or intergranular pores and is mainly formed during the late diagenetic stage. Similar to illite, chlorite partially fills the pores. Siliceous cementation in sandstone reservoirs commonly appears in two forms: quartz overgrowth and microcrystalline quartz aggregates (Figure 13e,f). Calcite cementation is also prevalent in the Shihezi Formation. During the early diagenetic stage, microcrystalline calcite primarily grows in a ctenoid form around clastic particles. During the late diagenetic stage, it is susceptible to replacement by carbonate cement (Figure 13d). Late-stage calcite cement typically forms during intense compaction in the late diagenetic stage.

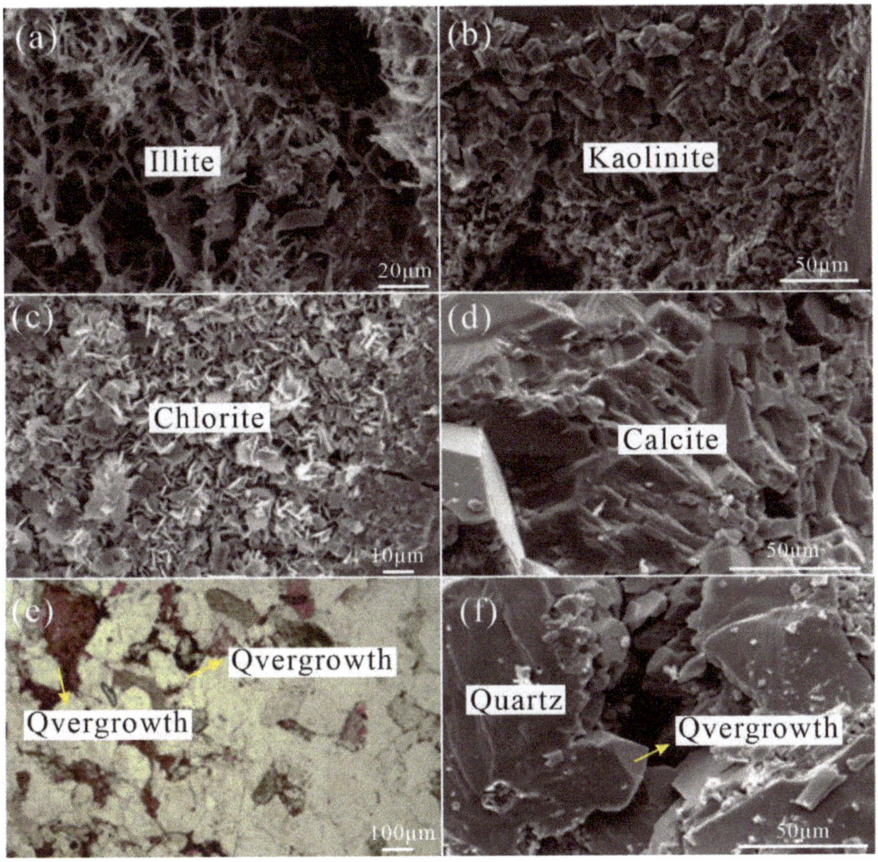

Figure 13. The occurrence and characteristics of cement observed under optical microscopy and SEM: (**a**) lamellar and hair-like authigenic illite; (**b**) book-like kaolinite aggregate; (**c**) needle-like chlorite aggregate; (**d**) authigenic calcite; (**e**,**f**) quartz overgrowth.

Cementation can occur during all the stages of diagenesis. Since the cement mainly fills the pores, a higher cement content usually leads to poorer reservoir properties [30,31]. The cements identified in the study area rocks mainly include authigenic clay minerals such as kaolinite, carbonate minerals such as calcite and dolomite, and siliceous minerals such as quartz overgrowth. The carbonate mineral content shows a negative correlation with porosity; as the carbonate mineral content increases, porosity decreases (Figure 14a). Even though carbonate cement can be partially dissolved to form dissolution pores in an acidic environment, the overall correlation between the carbonate mineral content and the reservoir's physical parameters is negative, indicating relatively underdeveloped dissolution pores. There is no significant correlation between the clay mineral content and the porosity and permeability (Figure 14c,d). Kaolinite, illite, and chlorite have high plasticity and are susceptible to compaction, causing a reduction in the primary pores. The intercrystalline pores of these clays are, however, well developed, and the micropores within the clay mineral aggregates are easily preserved [32]. The influence of clay minerals on reservoir properties is relatively complex and can have both positive and negative effects.

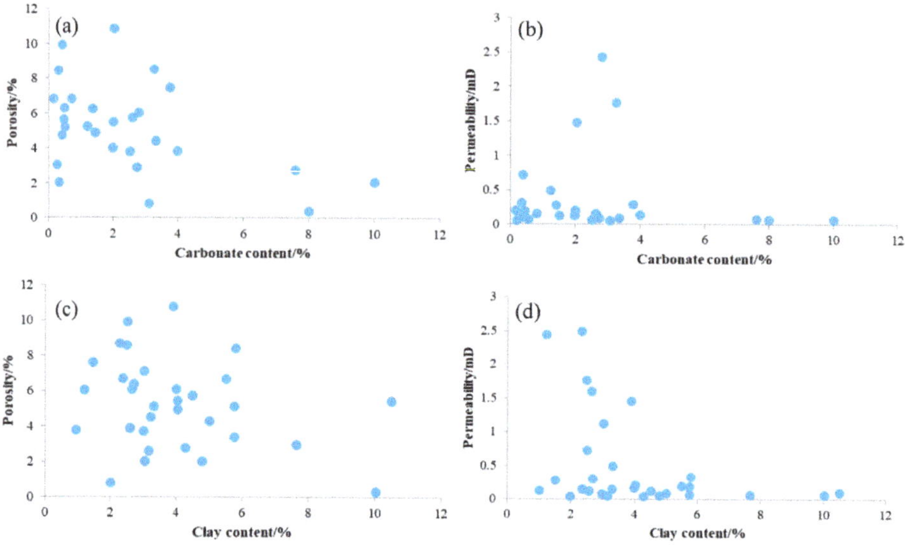

Figure 14. The relationship between the cement content and the porosity and permeability of the sandstone hydrocarbon reservoirs: (**a**) The relationship between the carbonate content and the porosity, (**b**) the relationship between the carbonate content and the permeability, (**c**) the relationship between the clay content and the porosity, and (**d**) the relationship between the clay content and the permeability.

Dissolution: dissolution is a constructive diagenetic process that can enhance the physical properties of hydrocarbon reservoirs. Chemically unstable debris particles, cement, and matrices can undergo dissolution under specific conditions, causing the development of secondary pores [33]. The sandstone reservoirs in the study area experienced varying degrees of dissolution (Figure 9a–d). The thermal maturity of source rocks increases as the burial depth increases, causing the production of organic acids during hydrocarbon generation. These acidic fluids can infiltrate the sandstone reservoir, causing the dissolution of feldspar and carbonate minerals. The resulting dissolution of pores can significantly enhance the pore connectivity of the reservoir [34,35]. When the carbonate mineral content is less than 2% (Figure 14a,b), certain sandstone samples exhibit high porosity and permeability, which could be attributed to the dominant role that dissolution and limited cementation play in improving the reservoir quality.

5.2.2. Influence of the Diagenesis Type on the Porosity

The physical properties of sandstone reservoirs in the study area are influenced by both destructive diagenesis, such as compaction and cementation, and constructive diagenesis, namely dissolution. To quantify these processes, the pore evolution was calculated using a quantitative method [36,37]. The porosity loss resulting from compaction and cementation during the burial of sandstone reservoirs in the Taiyuan, Shanxi, Shihezi, and Shiqianfeng formations in the Shilounan area was calculated, along with the porosity increase caused by dissolution.

The initial porosity can be determined using the following equations, which are based on the relationship between the initial porosity and sorting coefficient of sandstone reservoirs as proposed by Beard and Weyl (1973) [36]:

$$\varphi_1 = 20.91 + 22.90/S_0 \quad (1)$$

$$S_0 = (d_{75}/d_{25})^{1/2}, \tag{2}$$

where φ_1 represents the initial porosity in %; S_0 denotes the sorting coefficient; d_{75} is the particle diameter in the cumulative curve corresponding to 75% cumulative content; and d_{25} represents the particle diameter in the cumulative curve corresponding to 25% cumulative content. The value of d_{75} and d_{25} were obtained through particle size analysis experiments.

The formulae for calculating the porosity loss caused by compaction and cementation, as well as the porosity increase caused by dissolution, are as follows [38]:

$$\varphi_2 = C + (P_1 \times P_0/P_t) \tag{3}$$

$$\varphi_3 = P_1 \times P_0/P_t \tag{4}$$

$$\varphi_4 = P_2 \times P_0/P_t, \tag{5}$$

where φ_2 represents the porosity reduced by compaction in %; φ_3 denotes the porosity reduced by cementation in %; φ_4 is the porosity increased by dissolution in %; C represents the cement content in %; P_0 denotes the measured porosity in %; P_1 is the intergranular porosity in %; P_2 represents the dissolution porosity in %; and P_t denotes the total porosity in %. The values of C, P_1, P_t, and P_2 were obtained through observation and statistical analysis of cast thin sections.

The primary porosity in the study area exhibits a relatively narrow range, varying from 31.3% to 38.9% (Figure 15), which is comparable to the predominant tight sandstone gas formations in the Ordos Basin. Compaction and cementation have notably reduced the porosity, with compaction accounting for a porosity decrease ranging from 22% to 28.9%, while cementation contributed to a porosity decrease ranging from 5.7% to 7.2%. The Xiashihezi Formation shows the smallest porosity decrease due to compaction, while the Shanxi Formation exhibits the largest porosity decrease due to compaction, and the Benxi Formation shows the largest porosity decrease due to cementation. Compaction primarily causes a reduction in porosity, accounting for over 70% of the porosity decrease. Dissolution has also positively affected the physical properties of the reservoir, causing a porosity increase ranging between 2.9% and 4.8%. The Shangshihezi Formation exhibits the most developed dissolution process and the highest porosity.

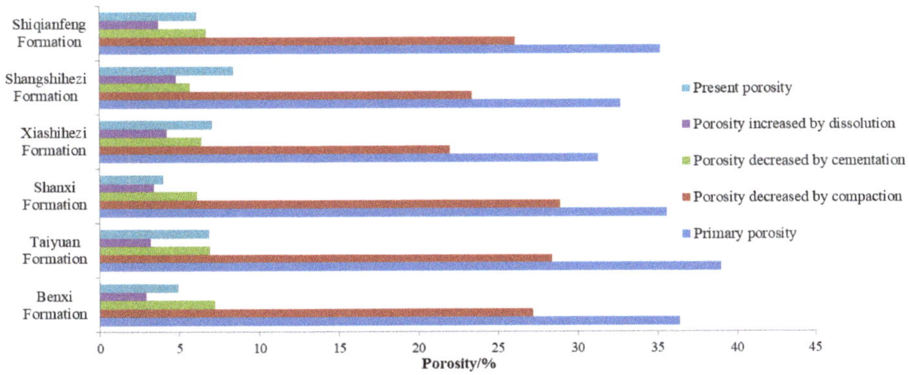

Figure 15. Statistical analysis of the quantitative parameters for the porosity evolution in the Carboniferous–Permian sedimentary succession from the study area.

As the sediments experience an increasing burial depth during the early diagenetic stage, compaction becomes the dominant diagenetic process. This caused the closure of numerous primary intergranular pores, causing a rapid decrease in porosity. During the early diagenetic stage, porosity reduction due to compaction can be as high as 20%. As the burial depth continues to increase during the later stages of diagenesis, porosity reduction due to compaction is around 10%. The gradual rise in the formation temperature also causes the generation of authigenic minerals like quartz, kaolinite, and illite. These minerals further fill the pore spaces and consequently decrease porosity. The Carboniferous–Permian coal measures are characterized by the widespread development of source rocks, including coal and black shale. The abundant organic matter in these rocks produced organic acid as it underwent hydrocarbon generation [32]. These acids dissolve feldspar, rock fragments, and other minerals, thereby improving the physical properties of the hydrocarbon reservoir and increasing the porosity by approximately 5%. The Carboniferous–Permian strata experienced gradual uplift in the study area during the late diagenetic stage, and the occurrence of micro-cracks due to tectonic processes was relatively limited. The sandstone reservoirs in the study area, therefore, exhibit tight characteristics (Figure 16).

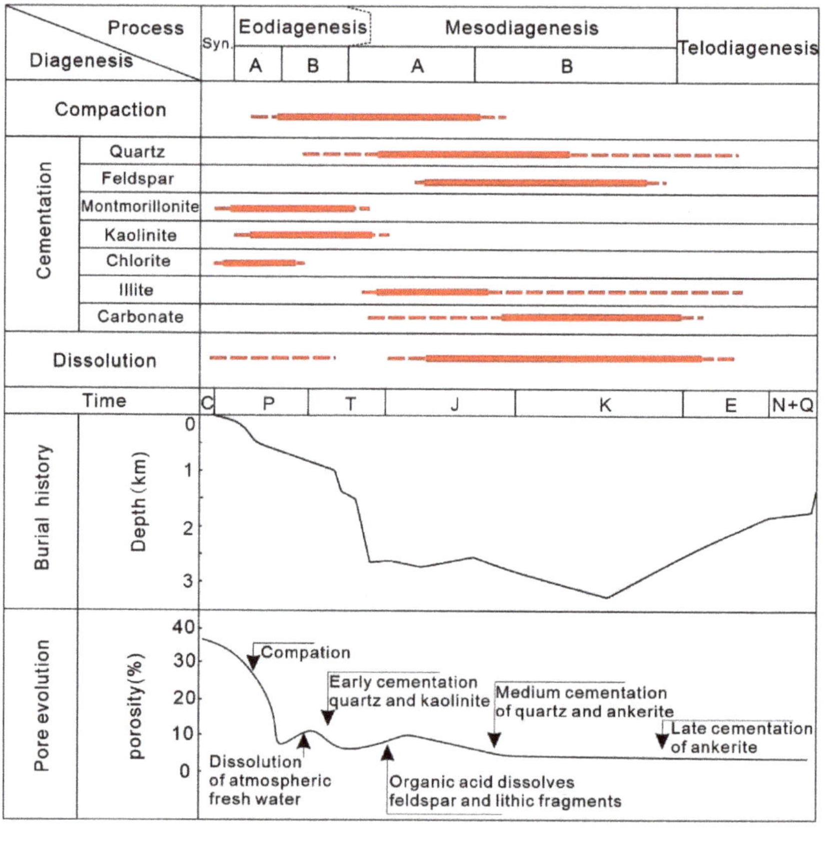

Figure 16. Diagenetic evolution of sandstone reservoirs in the study area and its effect on porosity.

5.3. Pore Structure Controls on the Reservoir Quality

Using mercury intrusion experiments, various parameters reflecting pore structure characteristics were obtained (Table 2). There is a weak positive correlation between the porosity and the sorting coefficient, indicating that porosity tends to increase with an increase in the sorting coefficient (Figure 17a). SEM and cast thin section analyses revealed that the Shihezi Formation has relatively well-developed pores, with intergranular pores and dissolution pores being the main type. These dissolution pores exhibit diverse morphologies, and the pore throat sizes have an uneven distribution. Samples with higher sorting coefficients tend to have larger pore sizes and smaller throats. Sandstone reservoirs with larger pores have higher permeability, while micro throats facilitate gas occurrence and storage [39,40].

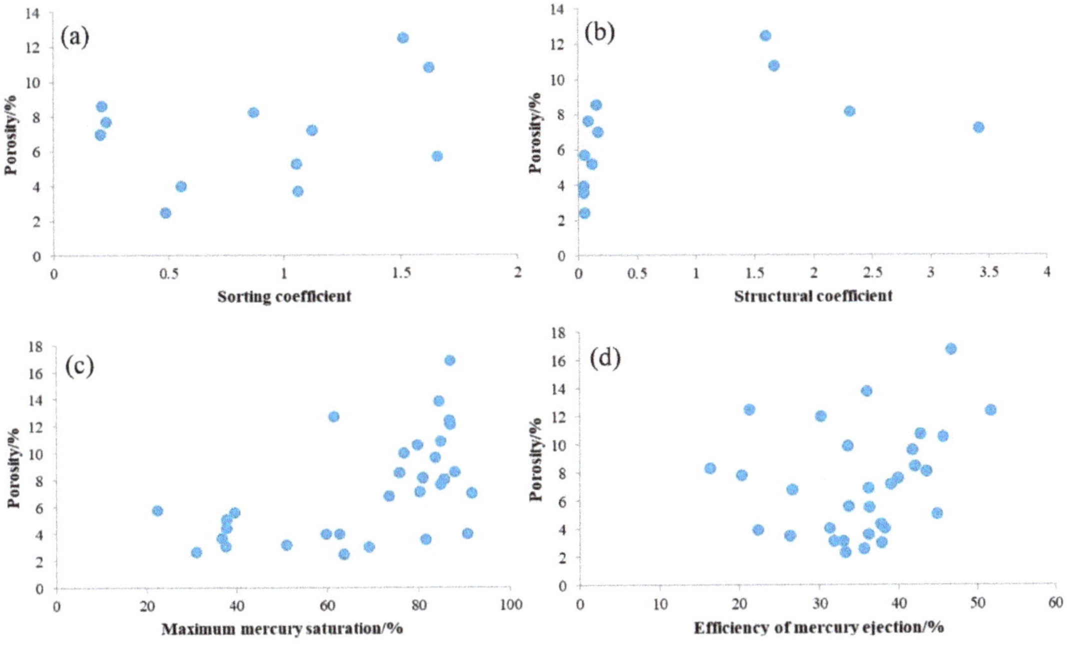

Figure 17. Relationship between the pore structure parameters and the porosity of the sandstone reservoirs in the study area: (**a**) The relationship between the sorting coefficient and the porosity, (**b**) the relationship between the structural coefficient and the porosity, (**c**) the relationship between the maximum mercury saturation and the porosity, and (**d**) the relationship between the efficiency of mercury ejection and the porosity.

Unlike the sorting coefficient, there is no significant correlation between the structural coefficient and the porosity (Figure 17b), suggesting that the tortuosity of the pore throat has a minimal impact on porosity. There may, however, be a distinct relationship between the structural coefficient and permeability. Higher tortuosities in the pore throat generally make seepage more challenging. The parameters related to mercury saturation include maximum mercury saturation and efficiency of mercury ejection. There is a positive correlation between the maximum mercury saturation and the porosity, while the efficiency of mercury ejection shows no significant correlation with porosity (Figure 17c,d). This means that the pore structure is favorable for the accumulation of sandstone gas but not for gas seepage.

5.4. Evaluation of Sandstone Reservoirs

Within the Carboniferous–Permian sedimentary succession, the characteristics of sandstone reservoirs in the study area vary significantly. The sandstone reservoirs in the Shilounan Block were classified into three types based on factors such as sedimentary facies, sandstone distribution, physical properties, and the influence of sedimentation, diagenesis, and pore structure on reservoir quality [41–43] (Table 4):

Type I: the type I reservoir is predominantly found in the primary areas of the superimposed multiphase distributary channels within the study area. The type I sandstone reservoir exhibits relatively high porosity and permeability, with quartz sandstone being the primary lithology [44]. The grain size of the type I reservoir ranges from medium to coarse and displays good sorting. The main types of cement are siliceous minerals and kaolinite. The pore structure of the type I reservoirs consists of a combination of dissolution pores and residual intergranular pores. The pore structure characteristics are relatively straightforward, with a displacement pressure that is typically below 1 MPa and an average throat radius of 0.4 μm. Type I reservoirs, which are mainly developed in the Shihezi Formation, should be considered as a primary focus for future hydrocarbon exploration and development efforts.

Type II: the type II reservoir occurs in the principal sand bodies of the subaqueous distributary channels. The rock particle size ranges from medium to fine, with an increased presence of interstitial materials. Compaction and cementation are more pronounced in the type II reservoirs compared to the type I reservoirs, causing a significant reduction in the primary intergranular pores. The porosity generally ranges between 5% and 10%, with a moderate displacement pressure. The pore development of the type II reservoirs is noticeably lower compared to the type I reservoirs. The predominant pore types are intergranular pores and intercrystalline pores, with an average throat size of 0.2 to 0.4 μm. This type of reservoir can serve as an alternative target for exploration and development within the Shilounan Block [45,46]. The type II reservoir is primarily developed in the Shiqianfeng Formation.

Type III: the type III reservoir is mainly developed in the underwater distributary channels and consists predominantly of fine-grained lithic sandstones, siltstones, and argillaceous siltstones. This reservoir exhibits relatively compact characteristics. It has a high presence of interstitial material, and compaction and cementation significantly affect the reservoir properties [47,48]. The particles often exhibit concave–convex and mosaic-like contacts, which causes a substantial reduction in the primary pores and limited development of the secondary dissolution pores. The porosity typically falls below 5%. The predominant pore type is intercrystalline pores, with an average throat size of less than 0.2 μm. The displacement pressure of type III reservoirs is relatively high, and larger-sized pores and fractures are generally scarce. The type III reservoirs are primarily developed in the Shanxi Formation and the Taiyuan Formation.

Table 4. Comprehensive evaluation parameters for sandstone reservoirs in the study area.

	Parameters	Type I	Type II	Type III
Depositional Feature	Single-layer sandstone thickness/m	>8	3–8	3<
	Sandstone type	Coarse- to medium-grained quartz sandstone, lithic quartz sandstone, with low interstitial content	Medium- to fine-grained quartz lithic sandstone, lithic sandstone, medium to low interstitial content, kaolinite is the predominant clay mineral	Fine-grained lithic sandstone, with high content of fillings, mainly carbonates and clay minerals
	Sand body type	Channel sand body	Channel sand body, underwater distributary channel sands	Underwater distributary channel sands
Physical Property	Porosity/%	>10	5–10	5<
	Permeability/mD	>0.5	0.1–0.6	0.1<
Pore Structure	Pore type	Intergranular and dissolution pores	Intergranular and dissolution pores	Intercrystalline pores
	Pore-Throat	Mesopore-fine throat, moderately sorted	Fine-micro throat, poorly sorted	Micro-throat, poorly sorted
	Displacement pressure	Low	Medium	High
	Average throat size/μm	>0.4	0.2–0.4	0.2<
Reservoir Properties		High quality	Good	Average and lower

6. Conclusions

(1) The predominant sandstone type in the Shilounan Block is lithic sandstone, which has a low porosity and ultra-low permeability reservoir. The primary types of pores present are intergranular pores, intercrystalline pores, and dissolution pores. The Shihezi Formation exhibits relatively good porosity and pore connectivity, characterized by a larger average throat size. In contrast, the Taiyuan and Shanxi formations have poor pore connectivity and have smaller throats.

(2) The primary controlling factors influencing the sandstone reservoirs in the study area are diagenesis and sedimentation. Sedimentation determines the distribution characteristics of the sand bodies and their petrological features, including the composition and grain size. Compaction and cementation caused the loss of a significant number of primary pores, leading to unfavorable reservoir properties. Compaction has notably reduced the porosity, accounting for over 70% of the overall porosity decrease. Dissolution processes have been relatively limited in the Shilounan Block, contributing to the relatively tight nature of the reservoirs in this region.

(3) A comprehensive evaluation system has been established to assess tight sandstone reservoirs in the Shilounan Block. This system is based on sedimentary characteristics such as sandstone thickness, sandstone type, and sand body type, while the primary discriminant criteria are porosity, permeability, and pore structure parameters. Using this evaluation system, it is determined that the high-quality reservoirs in the study area are primarily developed in the Shihezi Formation.

Author Contributions: Methodology, J.W.; writing—original draft preparation, J.W., F.Y. and Z.X.; writing—review and editing, J.W., F.Y. and C.Z. All authors have read and agreed to the published version of the manuscript.

Funding: This research received no external funding.

Data Availability Statement: Data is unavailable due to privacy or ethical restrictions.

Acknowledgments: The authors would like to thank Fundamental Research Funds for the Central Universities (grant no. 2652019106). We also thank the editors and reviewers very much for their critical comments and valuable suggestions, which were very helpful in improving the manuscript.

Conflicts of Interest: The authors declare no conflict of interest.

References

1. Nelson, P.H. Pore-throat sizes in sandstones, tight sandstones, and shales. *AAPG Bull.* **2009**, *93*, 329–340. [CrossRef]
2. Zou, C.N.; Zhu, R.K.; Wu, S.T.; Yang, Z.; Tao, S.Z.; Yuan, X.J.; Hou, L.H.; Yang, H.; Xu, C.C.; Li, D.H.; et al. Types, characteristics, genesis and prospects of conventional and unconventional hydrocarbon accumulations: Taking tight oil and tight gas in China as an instance. *Acta Pet. Sin.* **2012**, *33*, 173–187.
3. Nabawy, B.S.; Abd El Aziz, E.A.; Ramadan, M.; Shehata, A.A. Implication of the micro- and lithofacies types on the quality of a gas-bearing deltaic reservoir in the Nile Delta, Egypt. *Sci. Rep.* **2023**, *13*, 8873. [CrossRef] [PubMed]
4. Zou, C.N.; Yang, Z.; Tao, S.Z.; Yuan, X.J.; Zhu, R.K.; Hou, L.H.; Pang, Z.L. Continuous hydrocarbon accumulation over a large area as a distinguishing characteristic of unconventional petroleum: The Ordos Basin, North-Central China. *Earth Sci. Rev.* **2013**, *126*, 358–369. [CrossRef]
5. Qu, X.; Sun, W.; Lei, Q.; Huang, H.; Huo, L. Study on saturation of movable fluid in the low-permeability sandstone reservoirs of Huaqing Oilfeld and its influencing factors. *J. Xi'an Shiyou Univ. (Nat. Sci. Ed.)* **2016**, *31*, 93–98.
6. Liu, M.J.; Xiong, C. Diagenesis and reservoir quality of deep-lacustrine sandy-debris-flow tight sandstones in Upper Triassic Yanchang Formation, Ordos Basin, China: Implications for reservoir heterogeneity and hydrocarbon accumulation. *J. Pet. Sci. Eng.* **2021**, *202*, 108548. [CrossRef]
7. Yang, H.; Fu, J.H.; Liu, X.S.; Meng, P.L. Accumulation conditions and exploration and development of tight gas in the Upper Paleozoic of the Ordos Basin. *Pet. Explor. Dev.* **2012**, *39*, 295–303. [CrossRef]
8. Schmid, S.; Worden, R.H.; Fisher, Q.J. Diagenesis and reservoir quality of the sherwood sandstone (triassic), corrib feld, slyne basin, west of Ireland. *Mar. Pet. Geol.* **2004**, *21*, 299–315. [CrossRef]
9. Lai, J.; Wang, G.; Ran, Y.; Zhou, Z.; Cui, Y. Impact of diagenesis on the reservoir quality of tight oil sandstones: The case of Upper Triassic Yanchang Formation Chang 7 oil layers in Ordos Basin, China. *J. Pet. Sci. Eng.* **2016**, *145*, 54–65. [CrossRef]

10. Abdel-Fattah, M.I.; Sen, S.; Abuzied, S.M.; Abioui, M.; Radwan, A.E.; Benssaou, M. Facies analysis and petrophysical investigation of the late miocene abu madi sandstones gas reservoirs from offshore baltim east feld (nile delta, Egypt). *Mar. Pet. Geol.* **2022**, *137*, 105501. [CrossRef]
11. Radwan, A.E.; Husinec, A.; Benjumea, B.; Kassem, A.A.; Abd El Aal, A.K.; Hakimi, M.H.; Thanh, H.V.; Abdel-Fattah, M.I.; Shehata, A.A. Diagenetic overprint on porosity and permeability of a combined conventional-unconventional reservoir: Insights from the Eocene pelagic limestones, Gulf of Suez, Egypt. *Mar. Pet. Geol.* **2022**, *146*, 105967.
12. Dong, S.; Zeng, L.; Lyu, W.; Xia, D.; Liu, G.; Wu, Y.; Du, X. Fracture identification and evaluation using conventional logs in tight sandstones: A case study in the Ordos Basin, China. *Energy Geosci.* **2020**, *1*, 115–123. [CrossRef]
13. Wei, J.X.; Chen, S.; Gu, Y.; Zhang, H.; Yin, S.; Yuan, H. Diagenesis and reservoir classifcation criteria for Jurassic continental sandstone oil reservoir in the western Ordos Basin, China. *Geol. J.* **2021**, *56*, 3868–3882. [CrossRef]
14. Zhao, Z.; Liu, Z.; He, F.Q.; Zhang, W.; Li, M.; Hou, Y.J.; Fu, S.; Zhu, M.L. An improved time–depth dual porosity evolution model and a new parameter for tight sandstone reservoir quality evaluation. *J. Asian Earth Sci.* **2023**, *252*, 105684. [CrossRef]
15. Tang, S.L.; Tang, D.Z.; Liu, S.M.; Li, S.; Tang, J.C.; Wang, M.F.; Zhang, A.B.; Pu, Y.F. Multiscale pore characterization of coal measure reservoirs and gas storage and transport behavior in Yanchuannan gas field of China. *AAPG Bull.* **2022**, *106*, 2387–2415. [CrossRef]
16. Zhang, A.B.; Chen, S.D.; Tang, D.Z.; Tang, S.L.; Zhang, T.Y.; Pu, Y.F.; Sun, B. The Study on Diagenetic Characteristics of Coal Measures Sandstone Reservoir in Xishanyao Formation, Southern Margin of the Junggar Basin. *Energies* **2022**, *15*, 5499. [CrossRef]
17. Worden, R.H.; Mayall, M.; Evans, I.J. The effect of ductile-lithic sand grains and quartz cement on porosity and permeability in Oligocene and lower Miocene clastics, South China Sea: Prediction of reservoir quality. *Am. Assoc. Pet. Geol. Bull.* **2000**, *84*, 345–359.
18. Liu, K.; Wang, R.; Shi, W.Z.; Travéb, A.; Martín-Martín, J.D.; Baqués, V.; Qi, R.; Lin, J.W.; Ye, H. Diagenetic controls on reservoir quality and heterogeneity of the Triassic Chang 8 tight sandstones in the Binchang area (Ordos Basin, China). *Mar. Pet. Geol.* **2022**, *146*, 105974. [CrossRef]
19. Cai, Y. Study on the Relationship between Reservoir Evolution and Oil Accumulation Process of Chang 8 Tight Sandtone Oil Reservoir in Jiyuan Area, Ordos Basin. Ph.D. Thesis, Chang'an University, Xi'an, China, 2015.
20. Liang, J.T.; Huang, W.H.; Wang, H.L.; Blum, M.J.; Chen, J.; Wei, X.L.; Yang, G.Q. Organic geochemical and petrophysical characteristics of transitional coalmeasure shale gas reservoirs and their relationships with sedimentary environments: A case study from the Carboniferous-Permian Qinshui Basin, China. *J. Pet. Sci. Eng.* **2020**, *184*, 106510. [CrossRef]
21. Mansurbeg, H.; De Ros, L.F.; Morad, S.; Ketzer, J.M.; El-Ghali, M.A.K.; Caja, M.A.; Othman, R. Meteoric-water diagenesis in late Cretaceous canyou-fll turbidite reservoirs from the Espirito Santo Basin, eastern Brazil. *Mar. Pet. Geol.* **2012**, *37*, 7–26. [CrossRef]
22. Yang, T.; Cao, Y.; Friis, H.; Liu, K.; Wang, Y.; Zhou, L.; Yuan, G.; Xi, K.; Zhang, S. Diagenesis and reservoir quality of lacustrine deep-water gravity-flow sandstones in the Eocene Shahejie Formation in the Dongying sag, Jiyang depression, eastern China. *AAPG Bull.* **2020**, *104*, 1045–1073. [CrossRef]
23. Zhou, X.; He, Y.; Wang, J.; Li, S.; Ling, A. Characteristics of sandy debris flow reservoir from Chang 6 Formation in Ordos Basin. *Sci. Technol. Eng.* **2014**, *14*, 216–220, (In Chinese with English Abstract).
24. Liao, J.; Xi, A.; Li, Z.; Liu, H.; Li, X.; Wanyan, R. Microscopic characterization and formation mechanisms of deepwater sandy-debris-flow and turbidity-current sandstones in a lacustrine basin: A case study in the Yanchang Formation of the Ordos Basin, China. *Pet. Sci.* **2018**, *15*, 28–40. [CrossRef]
25. Xi, Z.D.; Tang, S.H.; Li, J.; Zhang, Z.Y.; Xiao, H.Q. Pore characterization and the controls of organic matter and quartz on pore structure: Case study of the Niutitang Formation of northern Guizhou Province, South China. *J. Nat. Gas Sci. Eng.* **2019**, *61*, 18–31. [CrossRef]
26. Xi, Z.D.; Tang, S.H.; Lash, G.G.; Ye, Y.P.; Lin, D.L.; Zhang, B. Depositional controlling factors on pore distribution and structure in the lower Silurian Longmaxi shales: Insight from geochemistry and petrology. *Mar. Pet. Geol.* **2021**, *130*, 105114. [CrossRef]
27. Lundegard, P.D. Sandstone porosity loss–a "big picture" view of the importance of compaction. *J. Sediment. Res.* **1992**, *62*, 250–260. [CrossRef]
28. Paxton, S.T.; Szabo, J.O.; Ajdukiewicz, J.M.; Klimentidis, R.E. Construction of an intergranular volume compaction curve for evaluating and predicting compaction and porosity loss in rigid-grain sandstone reservoirs. *Am. Assoc. Pet. Geol.* **2002**, *86*, 2047–2067.
29. Ehrenberg, S.N. Assessing the relative importance of compaction processes and cementation to reduction of porosity in sandstones: Discussion; compaction and porosity evolution of Pliocene sandstones, Ventura basin, California: Discussion. *AAPG Bull.* **1989**, *73*, 1274–1276.
30. Taylor, K.G.; Gawthorpe, R.L.; Curtis, C.D.; Marshall, J.D.; Awwiller, D.N. Carbonate cementation in a sequence-stratigraphic framework: Upper Cretaceous sandstones, Book Cliffs, Uttah-Colorado. *J. Sediment. Res.* **2000**, *70*, 360–372. [CrossRef]
31. Dutton, S.P. Calcite cement in Permian deep-water sandstones, Delaware Basin, west Texas: Origin, distribution, and effect on reservoir properties. *AAPG Bull.* **2008**, *92*, 765–787. [CrossRef]
32. Xi, Z.D.; Tang, S.H.; Wang, J.; Yang, G.Q.; Li, L. Formation and development of pore structure in marine-continental transitional shale from northern China across a maturation gradient: Insights from gas adsorption and mercury intrusion. *Int. J. Coal Geol.* **2018**, *200*, 87–102. [CrossRef]

33. Ma, B.; Cao, Y.; Jia, Y. Feldspar dissolution with implications for reservoir quality in tight gas sandstones: Evidence from the Eocene Es4 interval, Dongying Depression, Bohai Bay Basin, China. *J. Pet. Sci. Eng.* **2017**, *150*, 74–84. [CrossRef]
34. Mahmi, O.; Dypvik, H.; Hammer, E. Diagenetic influence on reservoir quality evolution, examples from Triassic conglomerates/arenites in the Edvard Grieg feld, Norwegian North Sea. *Mar. Pet. Geol.* **2018**, *93*, 247–271. [CrossRef]
35. Zhang, Y.; Jiang, S.; He, Z.; Wang, Y.; Chen, G. Characteristics of heterogeneous diagenesis and modifcation to physical properties of Upper Paleozoic tight gas reservoir in eastern Ordos Basin. *J. Pet. Sci. Eng.* **2022**, *208*, 109243. [CrossRef]
36. Beard, D.C.; Weyl, P.K. Influence of texture on porosity and permeability of unconsolidated sand. *AAPG Bull.* **1973**, *57*, 349–369.
37. Yang, J.Q.; Ji, Y.L.; Wu, H.; Meng, L.J. Diagenesis and Porosity Evolution of Deep Reservoirs in the Nanpu Sag: A case study of Sha 1 Member of the Paleogene in No. 3 structural belt. *Acta Sedimentol. Sin.* **2022**, *40*, 203–216.
38. Wilson, J.C.; McBride, E.F. Compaction and porosity evolution of Pliocene sandstones, Ventura Basin, California. *AAPG Bull.* **1989**, *73*, 664–681.
39. Wang, R.; Shi, W.; Xie, X.; Zhang, W.; Qin, S.; Liu, K.; Busbey Arthur, B. Clay mineral content, type, and their effects on pore throat structure and reservoir properties: Insight from the Permian tight sandstones in the Hangjinqi area, north Ordos Basin, China. *Mar. Pet. Geol.* **2020**, *115*, 104281. [CrossRef]
40. Law, B.E. Basin-centered gas systems. *AAPG Bull.* **2002**, *86*, 1891–1919.
41. Bell, D.; Kane, I.A.; Pontén, A.S.M.; Flint, S.S.; Hodgson, D.M.; Barrett, B.J. Spatial variability in depositional reservoir quality of deep-water channel-fll and lobe deposits. *Mar. Pet. Geol.* **2018**, *98*, 97–115. [CrossRef]
42. Fic, J.; Pedersen, P.K. Reservoir characterization of a "tight" oil reservoir, the middle jurassic upper shaunavon member in the whitemud and eastbrook pools, SW saskatchewan. *Mar. Pet. Geol.* **2013**, *44*, 41–59. [CrossRef]
43. Liu, D.; Sun, W.; Ren, D.; Li, C. Quartz cement origins and impact on storage performance in Permian Upper Shihezi Formation tight sandstone reservoirs in the northern Ordos Basin, China. *J. Pet. Sci. Eng.* **2019**, *178*, 485–496. [CrossRef]
44. Ghanizadeh, G.; Clarkson, C.R.; Aquino, S.; Ardakani, O.H.; Sane, H. Petrophysical and geomechanical characteristics of Canadian tight oil and liquid-rich gas reservoirs: I. Pore network and permeability characterization. *Fuel* **2015**, *153*, 664–681. [CrossRef]
45. Li, Z.; Wu, S.; Xia, D.; He, S.; Zhang, X. An investigation into pore structure and petrophysical property in tight sandstones: A case of the Yanchang Formation in the southern Ordos Basin, China. *Mar. Pet. Geol.* **2018**, *97*, 390–406. [CrossRef]
46. Sakhaee-Pour, A.; Bryant, S.L. Effect of pore structure on the producibility of tightgas sandstones. *AAPG Bull.* **2014**, *98*, 663–694. [CrossRef]
47. Christopher, B.; Kuiwu, L.; Oswald, G. Diagenesis and reservoir properties of the permian ecca group sandstones and mudrocks in the eastern cape province, South Africa. *Minerals* **2017**, *7*, 88.
48. Enayati-Bidgoli, A.; Saemi, E. Effects of late diagenesis on primary reservoir quality of a quartz arenite unit: A case study from the lower Cretaceous successions of SW Iran. *Pet. Sci.* **2019**, *16*, 267–284. [CrossRef]

processes MDPI

Article

Matrix Compressibility and Its Controlling Factors of the Marine Shale Gas Reservoir: A Case Study of the Ning228 Well in the Southwest Sichuan Basin, China

Jiaming Chen [1,2], Yongkai Qiu [1,2,*], Yujing Qian [1,2] and Xianglong Fang [1,2]

1 School of Energy Resources, China University of Geosciences, Beijing 100083, China; 2006220001@email.cugb.edu.cn (J.C.); 2106200031@email.cugb.edu.cn (Y.Q.); xianglong.fang@email.cugb.edu.cn (X.F.)
2 Beijing Key Lab of Geological Evaluation and Development of Unconventional Natural Gas, Beijing 100083, China
* Correspondence: qyk@cugb.edu.cn

Abstract: Exploring the compressibility of the deeply buried marine shale matrix and its controlling factors can help achieve efficient petroleum production. Taking ten sets of deeply buried marine shale core samples from Ning228 wells in the Yanjin area as an example, the matrix compressibility and pore characteristics of deeply buried marine shale reservoirs were investigated by applying mercury intrusion porosimetry (MIP) and nitrogen adsorption/desorption isotherms at a low temperature of 77 K. Mathematical models (based on MIP and nitrogen adsorption/desorption isotherms) were established to analyze the effects of TOC, mineral components, and pore structure on matrix compressibility. The relationship between the compressibility coefficient and the brittleness index was also established. The results show that the compressibility of the shale matrix is significant when the mercury injection pressure ranges from 8.66 to 37 MPa. For deeply buried marine shale, the matrix compressibility is in the range of 0.23×10^{-4}–22.03×10^{-4} MPa^{-1}. The influence of TOC and minerals on matrix compressibility is mainly reflected in the control effect of pore structure. High TOC content decreases the overall shale elastic modulus, and high clay mineral content enhances shale stress sensitivity, resulting in a significant matrix compressibility effect. For the effect of pore structure on compressibility, the pore content in shale has a positive effect on matrix compressibility. In addition, the pore-specific surface area is critical to the effective variation of shale matrix compressibility, indicating that the complexity of the shale pore structure is a key factor affecting matrix compressibility.

Keywords: marine shale; matrix compressibility; mercury intrusion porosimetry; low-temperature nitrogen adsorption; brittleness index

Citation: Chen, J.; Qiu, Y.; Qian, Y.; Fang, X. Matrix Compressibility and Its Controlling Factors of the Marine Shale Gas Reservoir: A Case Study of the Ning228 Well in the Southwest Sichuan Basin, China. *Processes* **2023**, *11*, 2136. https://doi.org/10.3390/pr11072136

Academic Editors: Qingbang Meng and Albert Ratner

Received: 20 April 2023
Revised: 15 June 2023
Accepted: 13 July 2023
Published: 17 July 2023

1. Introduction

Shale is a kind of heterogeneous and organic rock composed of matrix and pore fractures [1,2]. Its matrix has a solid particle structure with many micro-pores, and there is a pore fracture system within or between the matrices. The large number of pore spaces serves as the main bearer of shale gas fugacity, which is closely related to shale gas generation and storage capacity. Since shale gas reservoirs are mostly dense porous media with low porosity and ultra-low permeability, conventional natural gas extraction methods cannot achieve commercial exploitation of shale gas. Therefore, hydraulic fracturing technology is often used to artificially fracture shale reservoirs to form more fracture networks in the reservoirs [3,4], thus increasing the transport channels for shale gas seepage. The pore structure and seepage capacity of shale are both highly stress sensitive. As part of the shale gas exploitation process, changes in stress will lead to changes in shale reservoir and matrix volume, which in turn will cause changes in matrix permeability, influencing the effect of hydraulic fracturing and shale gas production [4–6]. Due to the complexity of shale reservoirs, such changes in volume data are usually not directly available but can be estimated by matrix compressibility.

Two effects can be attributed to the change in volume of the shale reservoir matrix due to stress: the pressure induced effect characterized by matrix compressibility and the desorption effect characterized by shrinkage of the matrix [5,7]. There have been many studies on the effects of matrix shrinkage. However, relatively few studies have been conducted with pressure-induced matrix compressibility in shale reservoirs. Mercury intrusion porosimetry (MIP) is commonly applied to verify pore structure characteristics of shale reservoir, including pore distribution and pore volume [8–10]. In the process of the MIP experiment, the higher mercury injection pressure will cause the destruction of the primary pores of the shale and the compression of the matrix, resulting in a larger amount of incoming mercury than the actual number of pores in the shale, causing a large error [11,12]. Many scholars tend to use the matrix compressibility to calibrate the MIP data [12–14]. Within a certain pressure range, the matrix deformation caused by compression belongs to elastic deformation [10,15], and the deformation caused by compression is linearly correlated with the pressure increment in MIP experiments. Based on this principle, scholars have deduced the calculation formula of the matrix compressibility coefficient for the quantitative study of reservoir matrix compressibility [11,14,16]. Many remarkable and effective results have been achieved in the research process. Previous studies have shown that when the injection pressure of mercury is higher than 10 MPa [14,16,17], the reservoir matrix can be compressed, which also indicates that the data measured by MIP in the high pressure section contain information about the compressibility of rock samples. Li et al. proposed and verified the feasibility of using fractal dimensions to distinguish the matrix compressibility from the pore filling effect in the high-pressure stage of MIP [11]. Guo et al. showed that the effect of different pressure levels on the matrix compressibility is quite small and rather negligible [13]. Han et al. indicated that the matrix compressibility and pore filling effect existed simultaneously at pressures of 10–206 MPa, with the matrix compressibility dominating [10]. Cai et al. determined that the compressibility of the coal matrix varies as the rank of coal changes by applying MIP with the N_2 adsorption isotherm at 77 K [14]. Shale reservoir matrix compression involves lithology, microscopic pore-fracture structure, and fluid characteristics. Previous studies in this area have tended to focus on coal of different ranks and different regions. The quantitative characterization and control factors of the matrix compressibility of marine shale have not been as well covered, and the current status of research in this area is still relatively weak. In particular, deeply buried marine shale has become an important target for shale gas exploration and development in China [18]. Therefore, based on the experience and results of previous studies on the compressibility of the coal matrix, we apply the kerogen vitrinite reflectance test, the total organic carbon (TOC) test, X-ray diffraction whole-rock quantitative analysis, mercury intrusion porosimetry (MIP), and low-temperature liquid nitrogen adsorption experiments to achieve quantitative characterization and qualitative analysis of the matrix compressibility of deeply buried marine shale and its influencing factors by using marine shale core samples from the Ning228 well in the southwestern Sichuan basin, China.

This paper proposes a theoretical method for assessing the matrix compressibility of shale reservoirs and establishes the interrelationship between organic matter, mineral fraction, pore structure, and shale matrix compressibility, with the goal of providing some theoretical basis for efficient shale gas production. Distinguishing features of this work include the following: (1) the evaluation of matrix compressibility of deep marine shale; (2) the microscopic pore structure assessment of shale using MIP combined with the N_2 adsorption isotherm at 77 K; (3) factors that influence the compressibility of the deep marine shale matrix.

2. Samples and Methodology

2.1. Regional Overview and Sample Analysis

Yanjin is a key area for the exploration and exploitation of shale gas in Yunnan Province. It lies on the southwestern edge of the Sichuan Basin, in a restricted–semi-confined paleogeographic pattern formed by the combination of the Chuanzhong Uplift, the Qianzhong Uplift, and the Kangdian Archicontinent (Figure 1a). Under this pattern, transgression and regression events have occurred successively; then, the black shale of the Lower Silurian Longmaxi Formation, which has a broad distribution, and is thick and rich in organic matter, has been studied [19,20]. The Longmaxi Formation is distributed in the area north of Zhenxiong and Yiliang counties, and its stratigraphy has been subjected to different degrees of uplift and denudation due to late tectonic movements. In general, the Longmaxi Formation in the Daguan–Yanjin–Suijiang region has a relatively small denudation area, more outcrop distribution, and less and sparse distribution of faults and folds; it is an essential area for research and development of shale gas reservoirs [20–22] (Figure 1b). The Yanjin area focuses on the development of two series of marine shale from the Lower Paleozoic, the Lower Cambrian Qiongzhusi Formation, and the Upper Ordovician Wufeng Formation–Lower Silurian Longmaxi Formation. The main formation discussed in this paper is the Longmaxi Formation of Lower Silurian in the Yanjin Ning228 well.

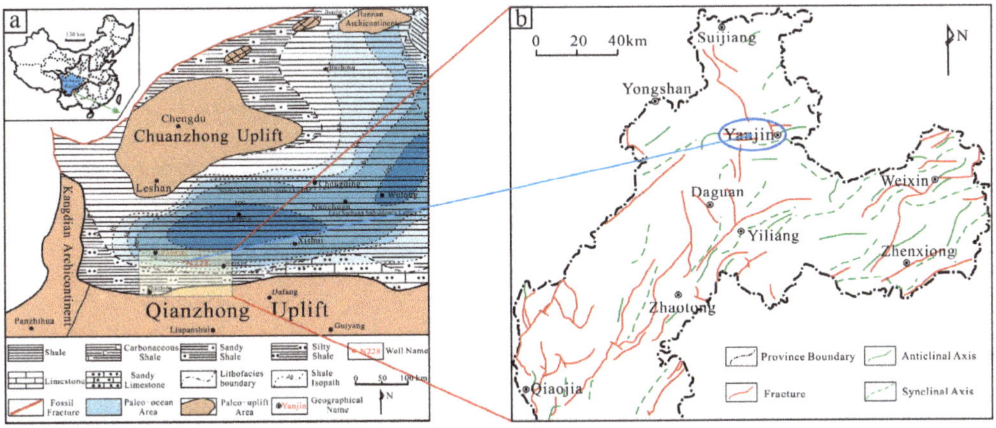

Figure 1. Palaeogeography map [19] (reproduced with permission from author, Journal of Palaeogeography, 2021) (**a**) of the Late Ordovician–Early Silurian in southwestern Sichuan Basin and the structure distribution map [22] (reproduced with permission from author, Natural Gas Geoscience, 2015) (**b**) of the Yanjin block.

The samples were collected from the Ning228 well in the Yanjin block, and ten sets of shale core samples were selected from the Longmaxi Formation in the depth range of 3365 m to 3465 m. According to the SYT 5124-2012 standard procedure [23], the GB/T 19145-2022 standard procedure [24], and the SY/T 5163-2010 standard procedure [25], the basic experimental tests, including the kerogen vitrinite reflectance ($R_{o,m}$) test, the total organic carbon (TOC) test, and X-ray diffraction (XRD) whole-rock quantitative analysis were conducted, respectively. The $R_{o,m}$ test was conducted by using the vitrinite reflectance tester model QDI302 produced by Craic. The kerogen was first prepared from the original shale core; then, the vitrinite reflectance analysis was conducted with the measurement point of asphalt. The TOC test is based on the principle of the combustion method using C-744 carbon and sulfur analyzers. Before starting, the shale sample was crushed to a particle size of less than 0.2 mm, and after removing the inorganic carbon from the sample with dilute hydrochloric acid, the sample was burned and oxidized at high temperature to convert the organic carbon into CO_2. Then, the CO_2 volume was detected with a thermal conductivity detector, and the TOC value of shale samples can be converted. The X-ray diffraction whole-rock quantitative analysis was performed with a Rigaku Ultima IV X-ray diffractometer with an operating voltage of 20 kV–40 KV and a scanning current of 10 mA–40 mA, which were mainly used to determine the mineral species based on the crystal structure of the scanned samples and to determine the level of phase content based on the strength of the diffraction peaks. Here, the basic information of the samples was obtained, as shown in Table 1.

Table 1. Sample information and basic parameters of the selected Longmaxi shale.

Sample No.	Depth (m)	$R_{o,m}$ (%)	TOC (%)	Mineral Content (%)					
				Quartz	Feldspar	Calcite	Dolomite	Pyrite	Clay Minerals
N228-1	3365.10	4.45	1.456	27.10	9.00	20.00	3.90	0.60	39.40
N228-13	3376.81	4.10	1.46	25.00	6.30	28.00	3.50	0.90	36.30
N228-24	3387.99	4.18	2.055	28.30	4.30	3.60	0.00	1.00	62.70
N228-35	3398.71	4.40	1.729	30.10	6.90	18.10	4.20	0.80	39.90
N228-46	3409.93	4.20	1.973	35.60	5.40	8.90	0.00	1.70	48.40
N228-57	3420.96	4.38	1.531	36.20	5.80	6.10	0.00	0.60	51.50
N228-67	3431.09	4.18	1.783	35.60	10.90	10.90	0.00	0.80	41.70
N228-78	3442.25	4.60	1.642	42.40	11.40	10.30	1.60	0.80	33.30
N228-88	3452.79	4.32	4.248	34.00	4.40	9.00	0.00	1.40	51.10
N228-100	3465.08	4.60	4.233	63.30	3.90	11.40	2.40	1.40	17.60

2.2. *Mercury Intrusion Porosimetry and Nitrogen Adsorption/Desorption*

Mercury intrusion porosimetry (MIP) is a conventional pore characterization technique that is widely used to determine the size distribution of pores in porous materials. It is suitable for conventional and unconventional reservoirs, which can obtain pore data such as pore volume, pore size distribution, and pore throat connectivity of samples [10,14,26]. The MIP analysis was conducted using an American Autopore 9500 Instrument (Micrometrics, US) in accordance with the standard procedure of GB/T 29171-2012 [27] at Sichuan Koyuan Engineering and Technology Testing Center. The experimental pressure can reach up to 228 MPa, and the measurable pore size range is 5 nm to 1000 μm. At each pressure point, when the capillary pressure equilibrium is reached in the rock samples, the injected pressure and the inlet mercury data are recorded at the same time; then, the sample pressure–mercury curves can be plotted [28]. The MIP curves of ten groups of shale samples were obtained, as shown in Figure 2. Based on the Washburn equation [29], it is possible to calculate the pore size of shale samples by combining mercury injection parameters. Here, we take the contact angle between the sample surface and the mercury vapor to be 140 °C, and the surface tension of mercury is fixed at 0.48 N/m in the experiment. In addition, the low-temperature N_2 adsorption/desorption experiment is a more common technique for testing pore structure both domestically and internationally. Different from the mercury pressure experiment, the former tests a broad range of pore sizes, which is suitable for reflecting the distribution of medium and large pores, while the low-temperature N_2 adsorption/desorption experiment can measure pores with a smaller pore size range and is suitable for characterizing the structural parameters of micropores. The low-temperature (77 K) N_2 adsorption/desorption experiment was performed using the Micromeritics ASPA 2460 system at Sichuan Keyuan Engineering Technology Testing Center, following the standard GB/T 19587-2017 procedure [30]. The range of pore sizes tested is 1.7 to 300 nm. Its basic principle is that under the condition of constant temperature, the adsorption capacity V is a function of nitrogen relative pressure (equilibrium pressure P/saturation vapor pressure P_0) when nitrogen reaches adsorption equilibrium on the shale surface. Therefore, the amount of nitrogen adsorption can be determined by the magnitude of the relative pressure, so the pore structure parameters such as pore volume and pore specific surface area of all samples are known. The N_2 adsorption/desorption curves of all samples at 77 K were acquired as shown in Figure 3. Compared with conventional reservoirs, the pore size of the shale reservoir is extremely small, with more complex pore size distribution and pore genesis. Its pore structure also has strong non-homogeneous characteristics, so there is still no unified standard terminology system for the classification of shale gas micro-reservoir space. In this paper, the shale reservoir space is divided into micropores (<2 nm), mesopores (2–50 nm), and macropores (>50 nm) by IUPAC (1982) [31].

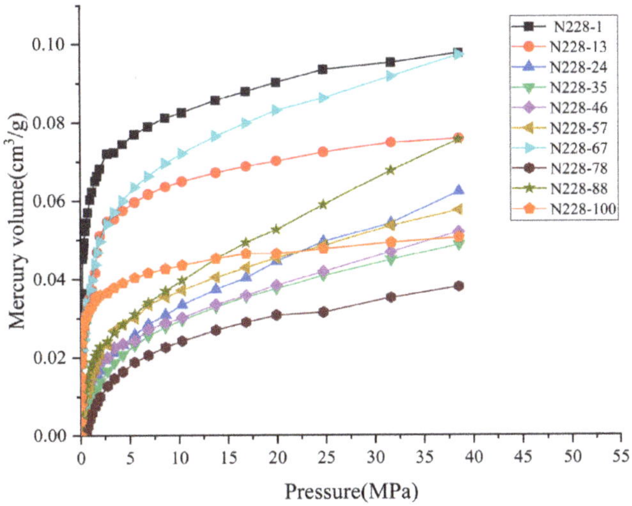

Figure 2. MIP curves of different buried depths.

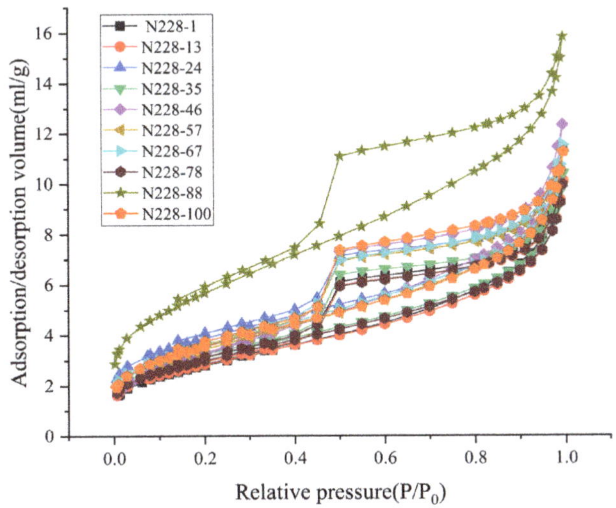

Figure 3. N_2 adsorption/desorption at low temperature (77 K) of different buried depths.

2.3. *Shale Matrix Compressibility*

For rocks, mercury is a fluid that is non-wetting phase. It is only when the pressure applied to mercury is greater than or equal to the capillary pressure of the pore throat that mercury is able to overcome the capillary resistance to penetrate the pore [28]. When applying the Washburn equation [29] to obtain the pore radius, it is generally assumed that the pore shape of the sample is cylindrical. The calculation is shown in Equation (1):

$$P_c = -\frac{2\sigma\cos\theta}{r_c} \quad (1)$$

where σ represents the surface tension of mercury (N/m); θ is the wetting contact angle from the sample to mercury (°); P_c is the inlet pressure of mercury (MPa); and r_c represents the radius of the capillary (μm) at the respective pressure. Here, we set σ to 0.48 N/m and θ to 140°. The equation can be simplified as

$$P_c = \frac{0.735}{r_c} \quad (2)$$

Based on Equation (2), the corresponding pore radius can be obtained from the mercury injection pressure. Then, the pore volume of the corresponding pore diameter can be analyzed according to the amount of mercury injection.

If mercury compressibility is neglected, the shale matrix compressibility can be defined as [10,11]

$$C_m = \frac{dV_m}{V_m dP} \quad (3)$$

where dV_m/dP is the volume change of the shale matrix as a function of pressure, V_m is the shale matrix volume, and P is the fluid pressure.

The shale matrix volume V_m can be defined as

$$V_m = \frac{1}{\rho} - V_T \quad (4)$$

where ρ is the density of shale samples (g/cm^3), which represents the reciprocal of the total volume of the shale sample; and V_T (cm^3/g) is the BJH pore volume [32], which represents the total pore volume of shale samples and can be calculated from low-temperature nitrogen adsorption experimental data. The BJH theory is based on the theory of multilayer adsorption by BET and the phenomenon of gas-phase capillary coalescence in porous materials [32]. In order to obtain the BET pore area, BJH pore volumes, and pore size distributions of the shale samples, nitrogen adsorption and desorption isotherms were measured over the relative pressure range of 0.01–0.99 MPa.

In MIP, with increasing mercury intrusion, significant compression can be detected in shale. In the case of compressible porous solids, a relation like Equation (5) exists:

$$\Delta V_{mercury} = \Delta V_{pore} + \Delta V_{compaction} \quad (5)$$

where $\Delta V_{mercury}$ is the mercury injection volume, ΔV_{pore} is the pore volume, and $\Delta V_{compaction}$ is the matrix compression volume. This formula indicates that the observed mercury volume increment comes from pore filling ΔV_{pore} and solid compression $\Delta V_{compaction}$, respectively.

Drawing on the experience and achievements of previous studies on the compressibility characteristics of the coal matrix [11,14,16], the pressure is set in the range of 8.66 MPa to 37 MPa, corresponding to the interstitial interval of 170 nm to 40 nm (Equation (2)). It can be found that when the experimental pressure is 8.66–37 MPa, the mercury injection volume and pressure show a better linear relationship that is approximately linear by observing the curve of cumulative mercury volume at different pressures (Figure 2). The shale pore fracture space is filled with incompressible mercury. Prior studies have demonstrated that for shale samples with a pore size of diameters < 200 nm, neither sample size nor particle size has a significant effect on the evaluation of pore structure. Therefore, we can assume that the ratio of mercury injection volume to pressure is constant N in this pressure section, while the volume change of pore-filled mercury at this stage can be approximated by the pore volume with the pore diameter ranging from 40 to 170 nm. Therefore, the change in mercury injection volume observed at this stage is equal to the sum of the compressed volume of the shale matrix and the pore-filled volume with a pore diameter of 40–170 nm. Thus, we can approximate Equation (6):

$$\frac{\Delta V_{\text{compaction}}}{\Delta P} = N - \frac{\sum_{40\text{nm}}^{170\text{nm}} \Delta V_{\text{pore}}}{\Delta P} \tag{6}$$

When the pores of the sample remain constant during the compression process, $\Delta V_{\text{mercury}}/\Delta P$ or $\Delta V_{\text{compaction}}/\Delta P$ is valid. Assuming that $\Delta V_{\text{mercury}}/\Delta P$ or $\Delta V_{\text{compaction}}/\Delta P$ is independent of pressure, we can replace $\mathrm{d}V_{\text{m}}/\mathrm{dP}$ with $\Delta V_{\text{compaction}}/\Delta P$ and combine Equations (3) and (6). The equation for the shale matrix compression coefficient can be obtained as

$$C_{\text{m}} = \frac{1}{V_{\text{m}}}\left(N - \frac{\sum_{40\text{nm}}^{170\text{nm}} \Delta V_{\text{pore}}}{\Delta P}\right) \tag{7}$$

2.4. *Brittleness Index Calculation and Lithofacies Classification*

2.4.1. Shale Brittleness Index

The brittleness characteristics of the shale reservoirs have significant implications for shale gas development. Typically, the brittleness index is used to quantitatively calculate and describe the brittleness characteristics of shale. There are numerous methods for calculating the brittleness index, while mineral components are often used to evaluate rock brittleness in China [33–35]. Not only do the type and content of brittle minerals such as quartz, feldspar, and dolomite affect the brittleness of the shale, but they are also related to the ease of fracturing at the final extraction step [36,37]. The better the brittleness of the shale reservoir, the easier a fracture network can be formed during fracturing and the higher the shale gas productivity. On the contrary, the worse the brittleness is, the more obvious the plasticity of the shale will be, which will lead to poor fracturing effects. The evaluation of brittleness using rock mineral fraction characteristics is a method proposed and developed by the Weatherford Company [38], which mainly determines the sample mineral fraction using the whole-rock XRD experiment. Taking quartz as the main brittle mineral, the rock brittleness index can be calculated using XRD data. The calculation formula is shown in Equation (8). This practical method is relatively narrow, and it was later promoted and extended by scholars using the ratio of quartz, feldspar, and brittle carbonate minerals to the total amount of minerals as the brittleness index. Its calculation is shown in Equation (9). This type of method is more comprehensive in characterizing the brittleness characteristics in shale with more complex mineral composition. Therefore, Equation (9) is used to calculate the brittleness index of shale samples in this study.

$$BRIT = \frac{V_{\text{quartz}}}{V_{\text{quartz}} + V_{\text{calcite}} + V_{\text{clay minerals}}} \tag{8}$$

where *BRIT* is the brittleness index (%) and *V* is the mineral content (%).

$$BRIT = \frac{V_{\text{quartz}} + V_{\text{feldspar}} + V_{\text{carbonate minerals}}}{V_{\text{total minerals}}} \tag{9}$$

where V_{feldspar} is the sum of potassium feldspar and plagioclase mineral content (%) and $V_{\text{carbonatite minerals}}$ is the sum of calcite and dolomite mineral content (%).

2.4.2. Shale Lithofacies Classification

Shale lithofacies are important for studying shale reservoirs. Scholars often analyze the depositional environment and depositional processes of sedimentary rocks by lithofacies classification. This paper mainly uses lithofacies classification to characterize lithological characteristics of shale. There is still no unified standard for shale lithofacies classification. This work mainly draws on the three-terminal element method of shale formation lithofacies classification proposed by Wang et al. to petrographically delineate the shale samples of the Longmaxi Formation in the Ning228 well [39]. This method mainly uses data from the whole-rock XRD experiment to establish a triangular plate composed of quartz + feldspar, carbonate (e.g., calcite + dolomite), and clay. Then, based on sedimentary petrological classification criteria, the shale lithofacies are divided into six petrographic zones according to the principle of equal probability of the same pattern, e.g., siliceous shale, clay shale, calcareous shale, clay siliceous mixed shale, clay calcareous mixed shale, and calcareous siliceous mixed shale. The latter three are all mixed shale facies. The specific criteria for the division are shown in Table 2.

Table 2. Classification scheme of lithofacies types of marine shale.

Lithofacies Types		**Percentage of Shale Mineral Components (%)**		
		Quartz + Feldspar	**Carbonate**	**Clay**
	Siliceous shale	50~75	<30	10~50
	Clay shale	25~50	<30	50~75
	Calcareous shale	<30	50~75	25~50
Mixed shale facies	Clay siliceous mixed shale	30~50	<33	30~50
	Clay calcareous mixed shale	<33	30~50	30~50
	Calcareous siliceous mixed shale	30~50	30~50	<33

3. Results

3.1. Shale Characteristics

The ten sets of shale core samples from the Ning228 well in the Yanjin area are all marine over-mature hydrocarbon source rocks, and their maximum vitrinite reflectance is greater than 4%. Previous studies have shown that the maturity of shale in the Lower Paleozoic in southern China is generally high, with $R_{o,m}$ varying from 2% to 5%. Even though their thermal evolution is high, they still maintain a certain degree of porosity [40]. Samples with high maturity can still be used for matrix compressibility studies. The organic matter abundance of all samples is large, with a TOC content between 1.456% and 4.248%, in a mean of 2.211%. With the increase of burial depth, it shows that the typical characteristics in the lower section are high, and in the upper section, they are low, which is mainly related to the sedimentary background of the Longmaxi Formation in the Yanjin area [20,41,42]. The mineral fraction varies widely, with the highest content of clay minerals ranging from 17.6% to 62.7% with a mean of 42.19%; the quartz content is the second highest, ranging from 25% to 63.3%, with an average of 35.76%; calcite and feldspar (potassium feldspar and plagioclase) are less abundant, ranging from 3.6% to 28% and 3.9% to 11.4%, respectively, accounting for 12.63% and 6.63% of the total proportion. In addition, there is some dolomite and pyrite.

Both quartz and pyrite show a good dependence on TOC from the vertical upward. Figure 4 shows that the TOC content increases significantly with the increase of quartz and pyrite content. Meanwhile, the quartz and pyrite contents show an overall increasing trend with increasing burial depth. Closer to the bottom of the Longmaxi Formation, the quartz and pyrite contents increase steeply with larger TOC values to reach the peak. Previous studies have shown that both silica-rich and pyrite-rich mineralogical characteristics indicate a reductive deep-water depositional environment favorable for organic matter enrichment [43], suggesting that a set of high-quality silica-rich hydrocarbon source rocks were deposited in the lower part of the Longmaxi Formation in the Ning228 well.

3.2. Pore Structure Characteristics from Mercury Intrusion Porosimetry

The experimental analysis shows that mesopores are the main component of the pores in the Longmaxi Formation shale of the Ning228 well (Figure 5). The volume fraction of mesopores ranges from 29.64% to 73.75%, with a mean of 55.33%. The average pore volume fraction of macropores is 44.67%. The size of shale pores tends to decrease with increasing burial depth, with the number of mesopores gradually increasing and the number of macropores gradually decreasing. This is mainly related to the special 'three high' formation environment of deep reservoirs, namely, high ground stress, high formation temperature, and high reservoir pressure [44]. This special geological environment has a great influence on the pore-fracture structure of shale. The results of mercury injection pore structure analysis of all shale samples at different burial depths are shown in Table 3. The porosity of the shale samples at different burial depths is found to be highly variable, between 1.04% and 7.01%. The maximum mercury saturation and mercury withdrawal efficiency also vary widely, ranging from 7.63% to 31.58% and 25.64% to 77.65%, respectively. This is obviously related to the development of macropores and microfractures in shale. Previous research has revealed that the mercury withdrawal efficiency can effectively reflect the connectivity of pores and fractures, thus affecting the permeability of the shale reservoir. In general, the higher the mercury withdrawal efficiency, the more uniform the pore structure and the better the connectivity, which is conducive to the diffusion and infiltration of the shale gas.

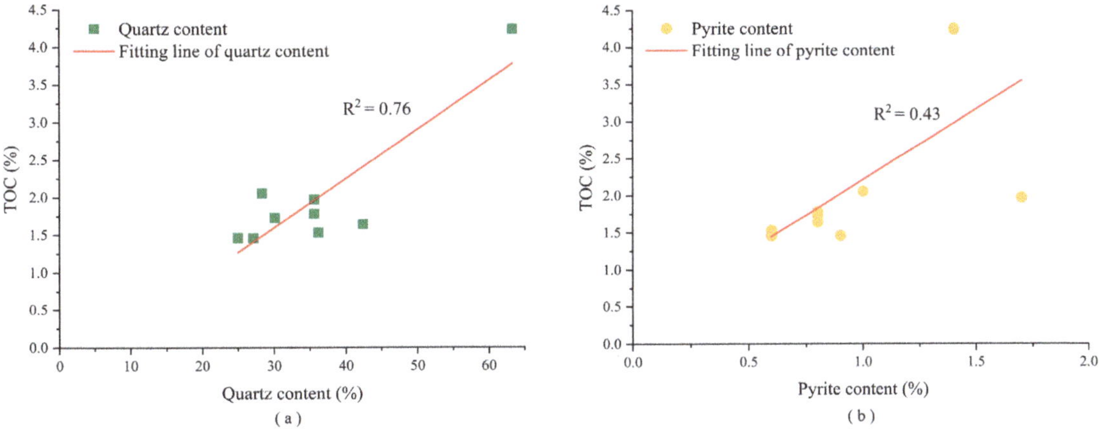

Figure 4. Variation of TOC with quartz and pyrite content. (**a**) Significant positive linear correlation between TOC and quartz content. (**b**) Positive linear correlation between TOC and pyrite content.

Table 3. Pore structure characteristics of shale with different buried depths by MIP.

Sample No.	Porosity (%)	Maximum Mercury Saturation (%)	Mercury Withdrawal Efficiency (%)	Total Pore Volume ($\times 10^{-3}$ cm^3/g)	Percentage Content of Pores in Different Pore Sizes (%)			Mercury Injection Curve Type
					<2 nm	2–50 nm	>50 nm	
N228-1	2.01	19.39	37.52	7.74	0.00	37.72	62.28	I
N228-13	1.04	26.00	25.64	3.95	0.00	29.64	70.36	I
N228-24	2.00	21.85	72.82	7.66	0.00	70.44	29.56	II
N228-35	2.64	10.56	64.86	10.25	0.00	62.22	37.78	III
N228-46	1.81	20.53	71.67	6.98	0.00	70.76	29.24	II
N228-57	1.86	18.28	67.56	7.13	0.00	62.84	37.16	II
N228-67	1.36	31.58	51.54	5.20	0.00	47.62	52.38	II
N228-78	1.17	18.87	57.57	4.49	0.00	62.91	37.09	II
N228-88	7.01	7.63	77.65	29.68	0.00	73.75	26.25	III
N228-100	1.22	15.28	31.62	4.83	0.00	35.40	64.60	I

In addition, the mercury intrusion curves for all samples are highly variable (Figure 6). Previous studies have shown that the different mercury intrusion curve patterns reflect the variability of pore connectivity and pore size distribution in the shale reservoir. The mercury intrusion curves for all shale samples presented in this paper are further subdivided into three typical types based on the shape of the mercury injection curve.

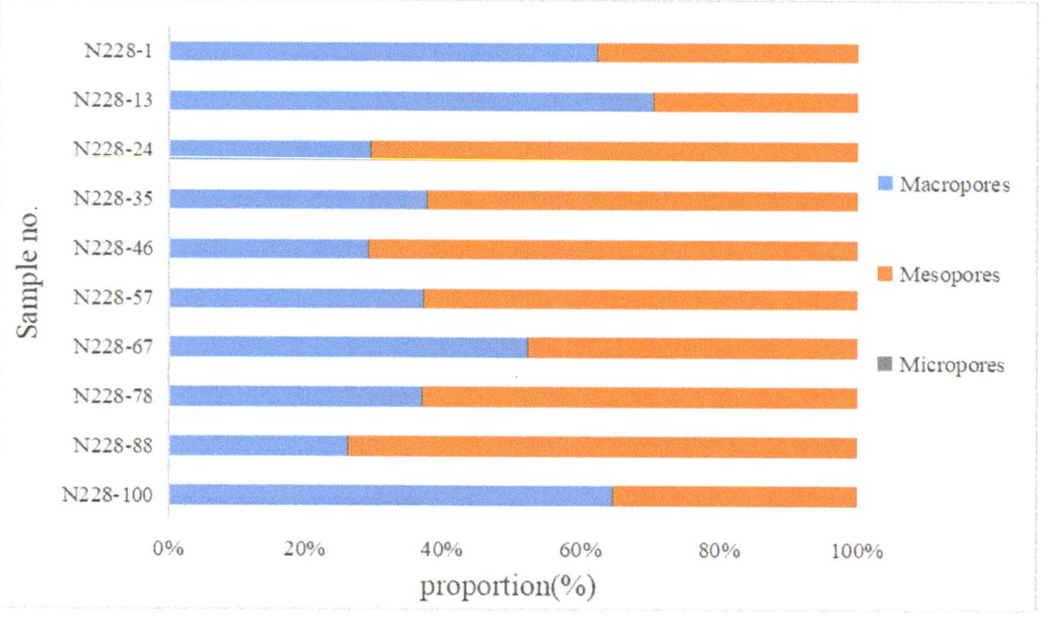

Figure 5. Pore volume distribution of selected shale samples at different sizes from MIP data.

There are three samples of the type (I) curve, accounting for 30% of the total number of samples. The mercury injection curve is a three-stage distribution of "front steep–middle slow–back steep". When the pressure is less than 1 MPa, there is a steeper oblique section; when the pressure is between 1 and 2 MPa, it is a slower oblique section; when the pressure is greater than 2 MPa, there is a steeper oblique segment. Its maximum mercury saturation is between 15.28% and 26.00%, and the mercury withdrawal efficiency is generally about 31.58%. In addition, when the pressure is less than 0.05 MPa, the curve is approximately a straight line, indicating that mercury is almost not introduced at this stage. The pore structure of the shale reservoir represented by the type (I) curve is extremely unevenly developed, and the connectivity between pores is common.

There are five samples of the type (II) curve, accounting for 50% of the overall number of samples. The mercury injection curve is a two-segment type of "steep at the front and slow at the back". This means that when the pressure is below 30 MPa, there is a steeper obliquity segment, and when the pressure is greater than 30 MPa, it shows a gentle slash segment. The maximum mercury saturation is 18.28–31.58%, and the mercury withdrawal efficiency is generally about 64.23%. When the pressure is less than 1 MPa, the curve is approximately linear, indicating almost no mercury is introduced at this stage. The pore structure of the shale reservoir represented by the type (II) curve is better developed, with better connectivity. Its micropore content is significantly higher compared with the previous type of mercury intrusion curve.

There are only two samples of the type (III) curve, accounting for 20% of the overall number of samples. The overall mercury injection curve is a nearly smooth curve, indicating that the mercury injection is slow and stable. Its maximum mercury saturation is 7.63%–10.56%, in general, and the mercury withdrawal efficiency is about 71.26%. The pore development of shale samples with such curves is uneven, and their micropores are more developed with good pore connectivity.

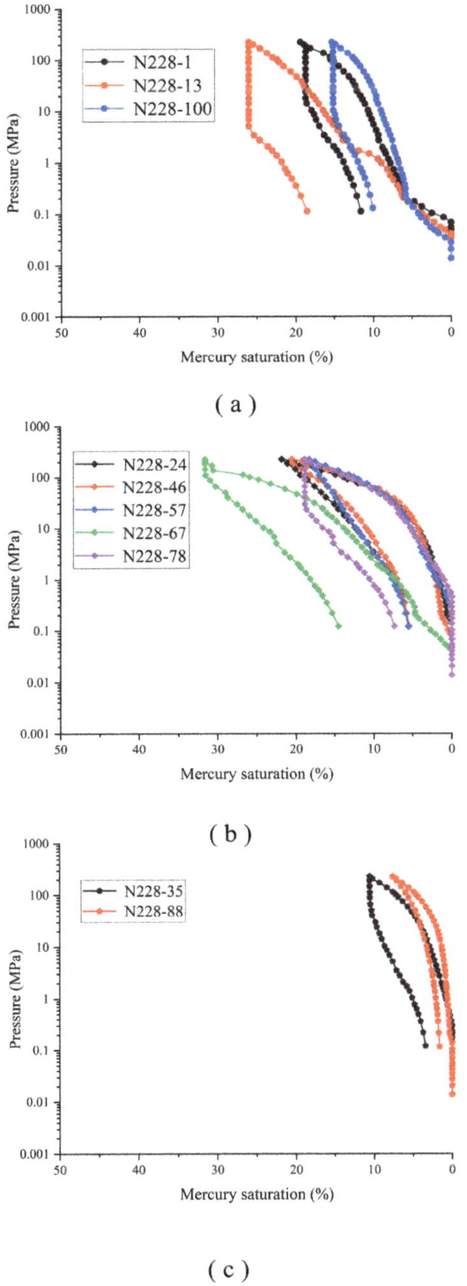

Figure 6. Three different types of mercury intrusion curves of selected shale samples. (**a**) The type (I) curve is a three-stage distribution of "front steep–middle slow–back steep", accounting for 30% of the total number of samples. (**b**) The type (II) curve is a two-segment type of "steep at the front and slow at the back", accounting for 50% of the overall number of samples. (**c**) The type (III) curve is a nearly smooth curve, accounting for 20% of the overall number of samples.

3.3. Pore Structure Characteristics from N_2 Adsorption/Desorption at 77 K

Figure 3 shows that the selected shale samples in this experiment all form hysteresis loops. There is no overlap between the adsorption and desorption curves of the samples in the $P/P_0 > 0.5$ relative pressure part. The formation of hysteresis loops denotes that the pores in the Longmaxi Formation shale reservoir within the study area are mostly open pores, which are mainly cylindrical pores, conical pores, and parallel plate-like pores with both ends open [45–47]. The adsorption curves have the characteristics of both the type H3 loop and the type H4 loop recommended by IUPAC, which show that the adsorption curves are steep near the saturation vapor pressure and the desorption curves are steep at medium pressure. The rising rate of the adsorption curves is closely related to the openness of the open pores, and the steeper adsorption curve indicates a greater degree of pore opening. None of the samples reaches saturation adsorption at relative pressures close to 1, indicating that capillary condensation occurs [48]. At a relative pressure of 0.5, there is a clear inflection point on the adsorption curve, indicating the presence of a certain amount of fine-necked vial pores in the sample.

The results from the low-temperature N_2 adsorption experiment show that the pore-specific surface area of the selected samples ranges from 10.0153 to 20.5537 m^2/g, with an average specific surface area of 12.4607 m^2/g. With the increase of buried depth, there is a tendency for the specific surface area of the shale samples to increase and then decrease. Shale samples vary in pore volume from 0.0163 to 0.0263 cm^3/g, with a large variation in pore volume among different samples. Figure 7 shows that the specific surface areas of mesopores and micropores account for 81.99% and 17.23% of the total specific surface area, respectively. As shown in Figure 8, the pore volumes of mesopores and macropores account for 79.28% and14.54% of the total pore volume, respectively. Mesopores provide the majority of the shale gas storage space, which means that mesopores are the main component of the pores in the Longmaxi Formation shale of the Ning228 well. However, the average pore volume of mesopores measured by MIP is 55.33%, which is 23.95% different from the measured value of the low-temperature N_2 adsorption experiment. This is a big difference. The main reason is that the matrix pores of shale reservoirs are easily damaged under high pressure due to the matrix compressibility, resulting in lower accuracy of pore characterization below 50 nm in mercury intrusion porosimetry [49]. The low-temperature nitrogen adsorption has no destructive effect on pores, and it has an excellent characterization effect on nano-scale pores. This shows that the MIP and low-temperature N_2 adsorption have their own advantages and complementarity in shale pore structure characterization.

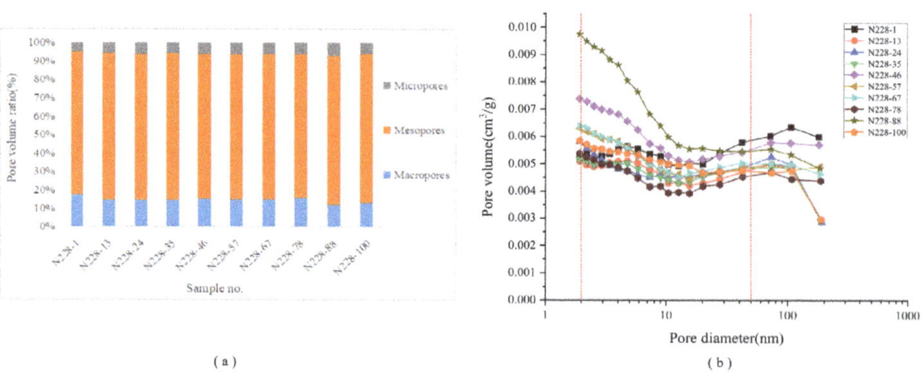

Figure 7. Specific surface area distribution of selected shale samples with different buried depths. (**a**) Histogram of pore volume distribution of shale samples with different pore types. (**b**) Pore volume distribution of shale samples with different pore size ranges.

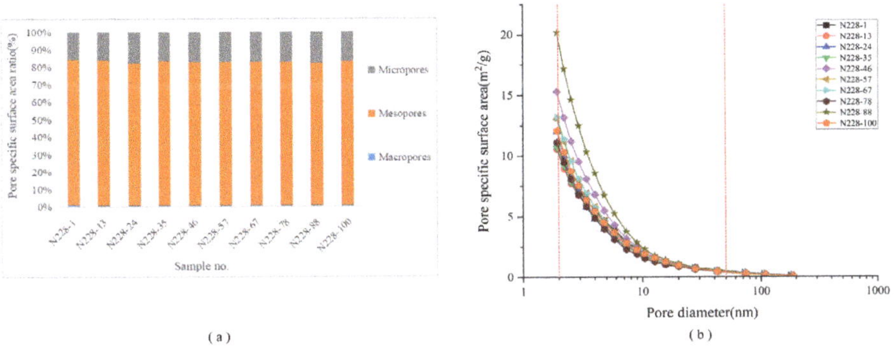

Figure 8. Pore volume distribution of selected shale samples with different buried depths. (a) Histogram of pore specific surface area distribution of shale samples with different pore types. (b) Distribution of pore specific surface area of shale samples with different pore size ranges.

3.4. Calculated Shale Matrix Compressibility

Consistent with the MIP and low-temperature N_2 adsorption/desorption data, the matrix compression coefficients of the ten shale samples with pore sizes of approximately 40 to 170 nm were calculated, as shown in Table 4. In the pressure interval of 8.66–37 MPa, the mercury injection volume of all samples shows an obvious linear relationship with the pressure. The values of the constant N can thus be obtained by fitting the slopes of the linear relation as presented in Table 4. Then, based on Equation (7), it is possible to acquire the shale matrix compressibility for all samples. The calculated shale matrix compressibility is within the range of 0.23×10^{-4} to 22.03×10^{-4} MPa^{-1}, which varies greatly with the buried depth.

Table 4. Calculation parameters of the shale matrix compressibility at different depths of Longmaxi Formation in the Ning228 well.

Sample No.	$R_{o,m}$ (%)	V_m (cm^3/g)	N ($\times 10^{-4}$)	PV ($\times 10^{-3}$, cm^3/g)	C_m ($\times 10^{-4}$, MPa^{-1})
N228-1	4.45	0.37	6.00	14.49	2.42
N228-13	4.10	0.36	4.00	11.10	0.23
N228-24	4.18	0.37	10.00	11.13	16.60
N228-35	4.40	0.37	7.00	10.92	8.48
N228-46	4.20	0.37	8.00	13.20	9.14
N228-57	4.38	0.36	7.00	11.27	8.28
N228-67	4.18	0.36	9.00	11.47	13.62
N228-78	4.60	0.37	5.00	10.46	3.57
N228-88	4.32	0.39	13.00	12.27	22.03
N228-100	4.60	0.38	5.00	10.82	3.14

Note: V_m—shale matrix volume; C_m—shale matrix compressibility; PV—pore volume.

The calculation results of the shale matrix compressibility indicate that in the pressure band from 8.66 to 37 MPa, the shale matrix has already started to produce a shrinkage phenomenon under the influence of pressure changes. Thus, matrix shrinkage is also bound to occur at pressures greater than 37 MPa. Therefore, it can be concluded that when the mercury injection pressure reaches 8.66 MPa or more, the shale produces a matrix shrinkage phenomenon in the mercury intrusion porosimetry. It makes the mercury intrusion volume greater than the original pore volume, causing the pore volume measured by mercury injection data to be larger than the actual original pore volume. Therefore, in order to guarantee the correctness of the experimental conclusions obtained later when considering pore structure as an influencing factor for correlation analysis, full-aperture pore data are needed in order to study the effect of pore structure on shale matrix compressibility. Based on previous research results and the quantitative analysis of the pore structure of shale samples above, we use low-temperature N_2 adsorption experimental data to quantitatively characterize micropores and mesopores. Quantitative characterization of macropores is performed using MIP data. On this basis, the influence of the shale pore structure on matrix compressibility is investigated below.

3.5. Shale Brittleness and Lithofacies Characteristics

Based on the results of XRD test analysis of all samples, the mineral brittleness index of the shale in the Ning228 well in the Yanjin area ranges from 36.2% to 81%, with an average of 56.78% by combining with the Equation (9). The empirical brittleness index in North America is 40%, which indicates that the brittleness of Longmaxi Formation shale in the Ning228 well in the Yanjin area is better than that in North America. This implies that the Longmaxi Formation shale in the Yanjin area can form complex network fractures rapidly during fracturing operations, which is conducive to shale oil and gas development.

According to the three-terminal element method of shale formation lithofacies classification, ten groups of shale samples can be divided into clay siliceous mixed shale, clay calcareous mixed shale, clay shale, and siliceous shale as shown in Figure 9. Among them, clay siliceous mixed shale is more distributed, which is more common in the study area. It is mainly distributed in the middle and upper parts of the selected Longmaxi Formation section, with an average TOC content of 1.4%. The mineral composition of this shale lithofacies is dominated by clay minerals, ranging from 36.3% to 48.4%, followed by siliceous minerals (including quartz and feldspar), ranging from 33.3% to 46.5%, and carbonate minerals (including calcite and dolomite) being the lowest. The clay calcareous mixed shale is mainly distributed in the upper part of the selected Longmaxi Formation section, with less distribution in the region. The clay shale is mainly distributed in the middle and lower parts of the selected Longmaxi Formation section, with a high clay mineral content, averaging 55.43%. The siliceous shale is mainly distributed in the lower part of the selected Longmaxi Formation section, with a high siliceous mineral content and abundant organic matter. Based on the above characteristics of shale samples, it can be found that there is a positive correlation between siliceous content and TOC content, indicating that the main siliceous body is biogenic and the early depositional environment of the Longmaxi Formation shale in the Ning228 well may be a deep-water shelf at a high sea level.

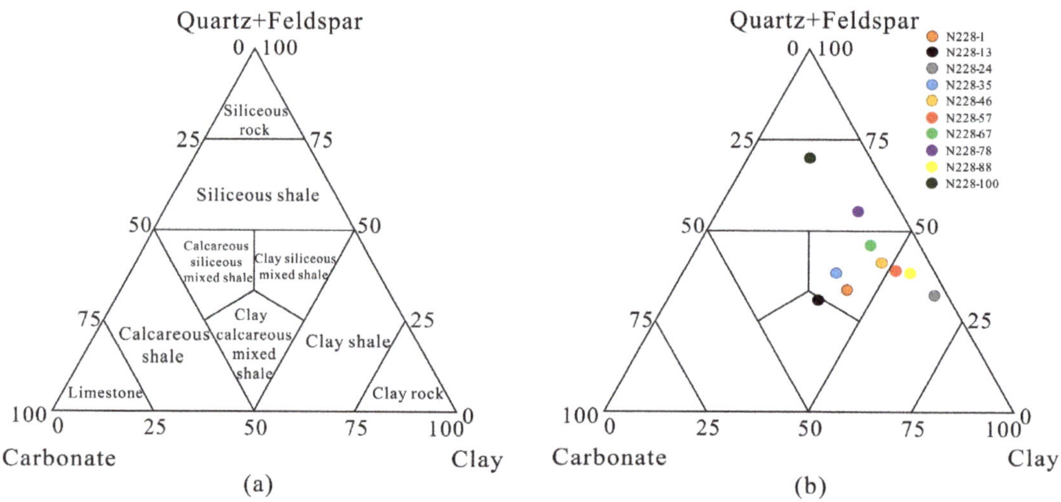

Figure 9. Lithofacies classification of Longmaxi Formation shale in the Ning228 well. (**a**) Three-terminal element method of shale formation lithofacies classification. (**b**) Lithofacies classification results for ten sets of shale samples.

4. Discussion

4.1. Effect of TOC on Shale Matrix Compressibility

Discarding the two excessive TOC anomalies, Figure 10 shows that the overall shale matrix compressibility is positively correlated with the increase of TOC content. We conclude that the effect of TOC in shale on the shale matrix compressibility can be regarded as a pore material filling problem. The high TOC content in shale enhances the softening effect of pore filling material on the shale skeleton, decreasing the overall shale modulus of elasticity and making the shale matrix susceptible to deformation due to pressure changes. Thus, it results in an increase in the shale matrix compressibility. Meanwhile, the high TOC content also indicates that the shale is rich in organic matter. Previous studies have shown that organic-rich shale often develops a large number of organic matter nanopores [50,51]. On the one hand, the rich organic matter provides a lot of space for the development of organic matter pores. On the other hand, the higher the organic matter content in the shale, the more organic acids and hydrocarbon gases will be generated during the maturation process, thus providing a large number of organic matter pores for the shale, which will lead to an increase in porosity and specific surface area. The rich pore space greatly enhances the compression effect of the shale matrix. Therefore, the effect of TOC on matrix compressibility mainly lies in the control of shale pore structure by organic matter.

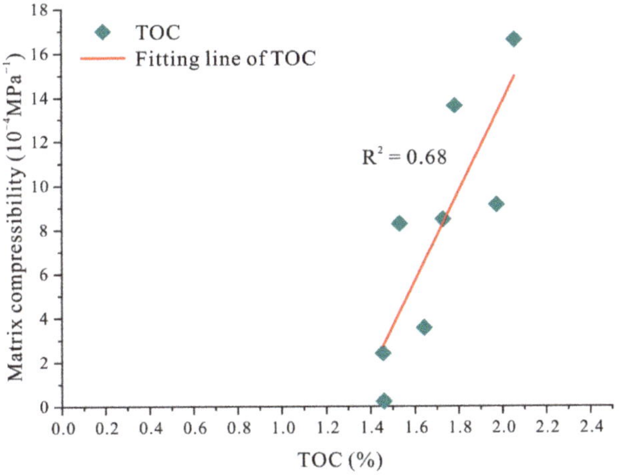

Figure 10. Relation between TOC content in shale and shale matrix compressibility.

4.2. Effect of Minerals on Shale Matrix Compressibility

As shown in Figure 11, the compressibility of the matrix is positively correlated with clay mineral content and negatively correlated with brittle mineral content and carbonate mineral content. These results agree with previous research that the higher the clay content, the stronger the corresponding shale stress sensitivity, causing the greater shale matrix compressibility. In addition, the relationship between shale lithofacies and matrix compressibility shows that clay shale and clay siliceous mixed shale with higher clay content tend to have higher matrix compressibility, while siliceous shale samples tend to correspond to lower matrix compressibility. It further validates the positive correlation between clay minerals and shale matrix compressibility. Meanwhile, previous studies have shown that clay minerals can be combined with organic matter to form organic clay complexes, in which a large number of organic matter pores are developed [50,52]. In the process of late diagenetic transformation, organic matter and clay mineral transformation promote each other, forming more intergranular pores of clay minerals. This will lead to the increase of porosity and specific surface area in shale and also affect its matrix compression characteristics. The role of minerals on the matrix compressibility of shale is also certainly determined by various factors, such as whether the minerals fill the pore space, the state of mineral fugacity within the pore space, etc., which all influence the control effect of minerals on matrix compressibility.

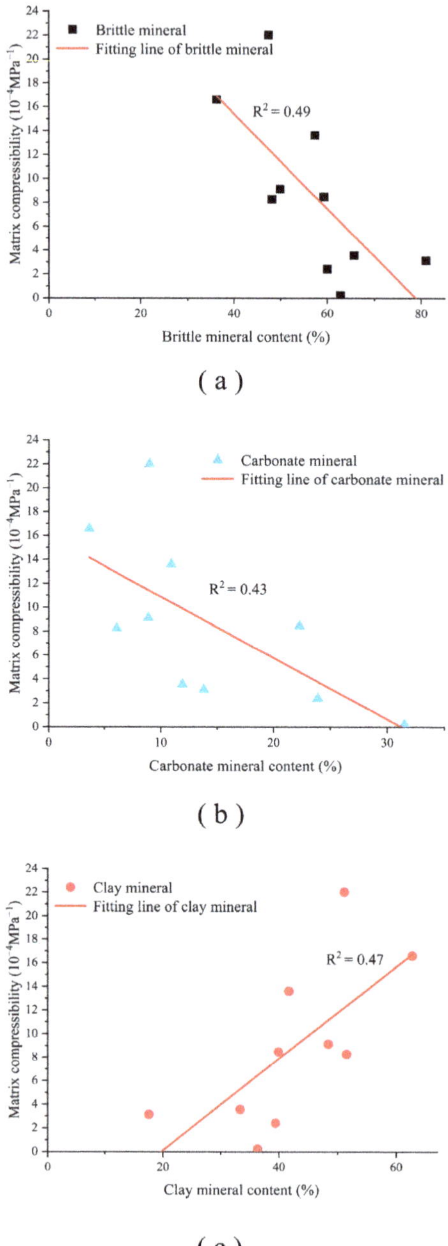

Figure 11. Relations between three types of mineral composition and shale matrix compressibility. **(a)** Brittle mineral and matrix compressibility. **(b)** Carbonate mineral and matrix compressibility. **(c)** Clay mineral and matrix compressibility.

4.3. Effect of Pore Structure on Shale Matrix Compressibility

The effect of pore structure on shale matrix compressibility is mainly reflected by porosity, specific surface area, total pore volume, and pore size in this work. Figure 12a shows that there is a positive correlation between shale matrix compressibility and porosity. The higher porosity of shale indicates that the matrix has more pore space. On the one hand, when the pressure increases, these pore spaces are more susceptible to compression. On the other hand, these empty pore spaces may be filled with organic matter or clay minerals, thus reducing the overall elastic modulus of shale and resulting in enhanced stress sensitivity of shale, which in turn leads to enhanced shale matrix compressibility. Figure 12b shows that the shale matrix compressibility overall has a more obvious positive correlation with specific surface area. The larger specific surface area indicates that the shale is more heterogeneous and the pore structure is more complex. This may imply that the shale with more complex pore structure has stronger matrix compressibility. The positive effect of specific surface area on shale matrix compressibility is stronger than porosity. Figure 12c shows a weak positive correlation between shale matrix compressibility and total pore volume. This means that specific surface area and porosity are the critical parameters that affect the compressibility of the deep marine shale matrix. Figure 12d indicates the effect of different pore sizes on matrix compressibility. The shale matrix compressibility has a significant positive correlation with micropore content, a positive correlation with mesopores, and no obvious relationship with macropore content. It can be synthesized to show that the pore content in shale has a positive effect on the matrix compressibility, especially the effect of micropore content. The larger micropore content in shale corresponds to larger porosity and specific surface area, indicating that the shale matrix is more easily compressed.

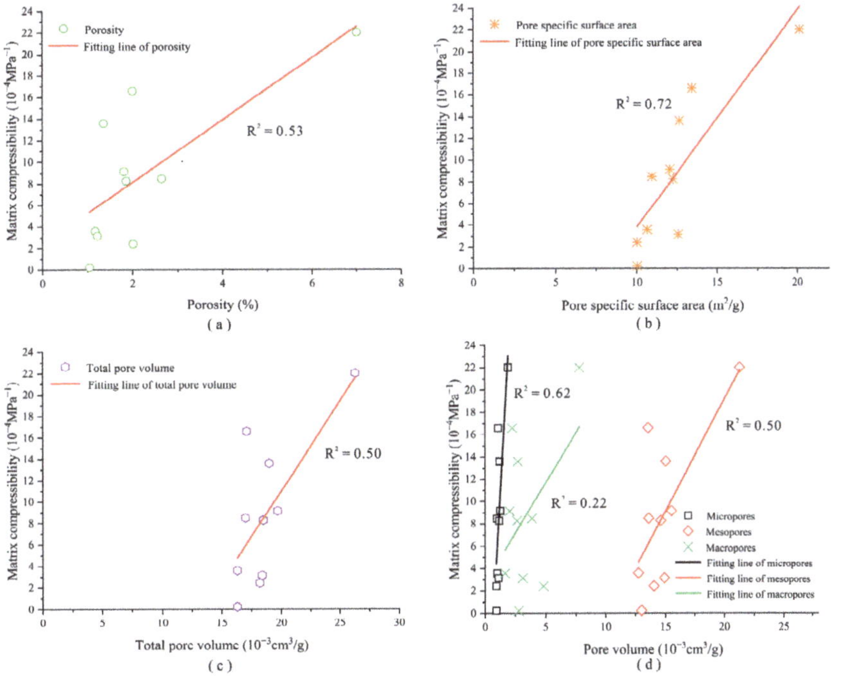

Figure 12. Relations between pore structure and shale matrix compressibility. (**a**) Porosity and matrix compressibility. (**b**) Total pore volume and matrix compressibility. (**c**) Specific surface area and matrix compressibility. (**d**) Pore size and matrix compressibility.

4.4. Relationship between Brittleness Index and Matrix Compressibility

Figure 13 shows that shale matrix compressibility is somewhat negatively correlated with the brittleness index. This indicates that with greater rock brittleness, the shale matrix is more resistant to volume compression caused by pressure change. Matrix compressibility is defined as the amount of change in matrix volume per unit mass sample at unit incremental pressure, which is an elastic deformation. As for oil and gas reservoirs, rock brittleness is generally defined as the ease of transient changes before rock fracture [53,54]. Macroscopically, the larger the brittleness index is, the easier it is for the rock to form fractures, making the microscopic rock matrix less susceptible to its internal compressibility. This view is consistent with our understanding of rock brittleness and matrix compressibility.

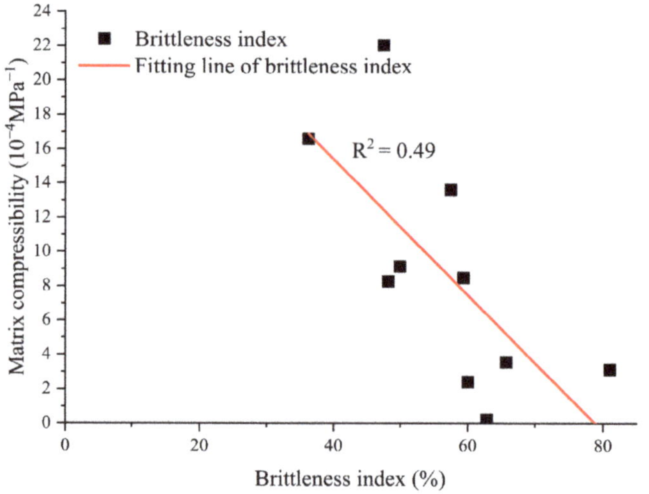

Figure 13. Relationship between brittleness index and shale matrix compressibility.

5. Conclusions

When applying the MIP combined with the low-temperature N_2 adsorption/desorption experiment to evaluate the pore structure of shale, it is necessary to consider the matrix volume compression changes caused by mercury intrusion into pore fissures. Factors that affect the shale matrix compressibility were also discussed herein. We determined that the matrix compressibility of shale from Ning228 in the Yanjin area varies in the range of 0.23×10^{-4} to 22.03×10^{-4} MPa^{-1} as the buried depth changes. The following conclusions were drawn.

(1) The shale of the Ning228 well in the Yanjin area consists of all marine, over-mature, hydrocarbon source rocks, but it still maintains a certain degree of porosity. Its organic matter content is typically high in the lower section and low in the upper section, with increasing burial depth, which is mainly related to the depositional background of the Longmaxi Formation in the Yanjin area. Mesopores are the main components of shale pores in the Longmaxi Formation of the Ning228 well in the Yanjin area, which is the main occurrence site for shale gas.

(2) The average mineral brittleness index of Ning228 well shale in the Yanjin area is 56.78%, which implies that the Longmaxi Formation shale in the study area can form complex mesh fractures rapidly during fracturing operation. The shale here can be divided into clay siliceous mixed shale, clay calcareous mixed shale, clay shale, and siliceous shale. Among these, clay siliceous mixed shale is more distributed in the study area. The main siliceous body of the Longmaxi Formation shale in the Ning228 well is biogenic, indicating that its early depositional environment may be a deep-water shelf with high sea level.

(3) The shale matrix compressibility is positively correlated with the increase of TOC. The high TOC content in shale can enhance the softening effect of pore filling material on the shale skeleton, decreasing the overall shale modulus of elasticity and making the shale matrix susceptible to deformation due to pressure changes. The matrix compressibility is positively correlated with clay mineral content and negatively correlated with brittle mineral content and carbonate mineral content. The higher the clay mineral content, the stronger the corresponding shale stress sensitivity, causing greater shale matrix compressibility. The influence of TOC and minerals on the compressibility of the shale matrix is mainly reflected in the control effect on the pore structure.

(4) The pore structures of shale—including porosity, specific surface area, total pore volume, and pore size—are all positively correlated with matrix compressibility. The pore content in shale has a positive effect on the matrix compressibility, and the pore specific surface area is critical for effective changes in shale matrix compressibility, indicating that the complexity of shale pore structure is the key factor affecting matrix compressibility.

Author Contributions: Conceptualization, J.C. and Y.Q. (Yongkai Qiu); methodology, J.C.; software, Y.Q. (Yujing Qian); validation, Y.Q. (Yujing Qian) and X.F.; formal analysis, J.C.; investigation, J.C.; resources, Y.Q. (Yongkai Qiu); data curation, Y.Q. (Yujing Qian) and X.F.; writing—original draft preparation, J.C.; writing—review and editing, J.C. and Y.Q. (Yongkai Qiu); visualization, X.F.; supervision, Y.Q. (Yongkai Qiu); project administration, Y.Q. (Yongkai Qiu); funding acquisition, Y.Q. (Yongkai Qiu) All authors have read and agreed to the published version of the manuscript.

Funding: This research was funded by the National Natural Science Foundation of China (Grant Nos. 42102227 and 42130806).

Data Availability Statement: Restrictions apply to the availability of these data. Data are available from the authors with the permission.

Acknowledgments: The authors all greatly appreciate Cai Yidong for his great help in the conception and writing of this paper. This work was supported by the National Natural Science Foundation of China (Grant Nos. 42102227 and 42130806).

Conflicts of Interest: The authors declare no conflict of interest.

References

1. Zhang, J.J.; Zeng, Q.Q.; Yin, X.Y.; Li, H.B.; Jiang, R.; Huang, J.Q.; Zhang, L.Q.; Zhang, G.Z.; Gu, Y.P. Multiple-porosity variable critical porosity model and pore structure characterization. *Chin. J. Geophys.* **2021**, *64*, 724–734.
2. Sun, L.D.; Liu, H.; He, W.Y.; Li, G.X.; Zhang, S.C.; Zhu, R.K.; Jin, X.; Meng, S.W.; Jiang, H. An analysis of major scientific problems and research paths of Gulong shale oil in Daqing Oilfield, NE China. *Pet. Explor. Dev.* **2021**, *48*, 527–540. [CrossRef]
3. Tang, J.P.; Qi, T.; Dai, S.H.; Pan, Y.S.; Lu, J.W. Experimental study on crack propagation of coal shale under stress transformation of cyclic water injection based on acoustic emission energy. *J. Exp. Mech.* **2020**, *35*, 639–649.
4. Zhuang, Z.; Liu, Z.L.; Wang, Y.L. Fundamental Theory and Key Mechanical Problems of Shale Oil Gas Effective Extraction. *Chin. Q. Mech.* **2015**, *36*, 11–25.
5. Cao, C.; Li, T.T.; Zhang, L.; Gao, C.; Wang, H. Shale gas dual porosity-dual permeability model with matrix shrinking. *Nat. Gas Geosci.* **2015**, *26*, 2381–2387. [CrossRef]
6. Liu, D.M.; Yao, Y.B.; Wang, H. Structural compartmentalization and its relationships with gas accumulation and gas production in the Zhengzhuang Field, southern Qinshui Basin. *Int. J. Coal Geol.* **2022**, *259*, 104055. [CrossRef]
7. Shi, J.T.; Zhang, L.; Li, Y.S.; Yu, W.; He, X.N.; Liu, N.; Li, X.F.; Wang, T. Diffusion and Flow Mechanisms of Shale Gas through Matrix Pores and Gas Production Forecasting. In Proceedings of the Forecasting, SPE Canadian Unconventional Resources Conference, Calgary, AL, Canada, 5–7 November 2013.
8. Zhang, N.; Wang, S.D.; Xun, X.J.; Wang, H.Y.; Sun, X.M.; He, M.C. Pore Structure and Fractal Characteristics of Coal-Measure Sedimentary Rocks Using Nuclear Magnetic Resonance (NMR) and Mercury Intrusion Porosimetry (MIP). *Energies* **2023**, *16*, 3812. [CrossRef]
9. Wang, X.M.; Jiang, Z.; Jiang, S.; Chang, J.Q.; Zhu, L.; Li, X.H.; Li, J.T. Full-Scale Pore Structure and Fractal Dimension of the Longmaxi Shale from the Southern Sichuan Basin: Investigations Using FE-SEM, Gas Adsorption and Mercury Intrusion Porosimetry. *Minerals* **2019**, *9*, 543. [CrossRef]
10. Han, B.B.; Qin, Y.; Zhang, Z.; Wang, G.; Yu, P. Study on coal compressibility and correction of compression amount based on compressibility of mercury injection test. *Coal Sci. Technol.* **2015**, *43*, 68–72.

11. Li, Y.H.; Lu, G.Q.; Rudolph, V. Compressibility and Fractal Dimension of Fine Coal Particles in Relation to Pore Structure Characterisation Using Mercury Porosimetry. *Part. Part. Syst. Charact.* **1999**, *16*, 25–31. [CrossRef]
12. Peng, S.; Zhang, T.W.; Loucks, R.G.; Shultz, J. Application of mercury injection capillary pressure to mudrocks: Conformance and compression corrections. *Mar. Pet. Geol.* **2017**, *88*, 30–40. [CrossRef]
13. Guo, X.Q.; Liu, D.M.; Yao, Y.B.; Cai, Y.D.; Li, J.Q. Influence of Pressure on Application of Mercury Injection Capillary Pressure for Determining Coal Compressibility. *Appl. Mech. Mater.* **2013**, *295*, 2726–2731. [CrossRef]
14. Cai, Y.D.; Li, Q.; Liu, D.M.; Zhou, Y.F.; Lv, D.W. Insights into matrix compressibility of coals by mercury intrusion porosimetry and N_2 adsorption. *Int. J. Coal Geol.* **2018**, *200*, 199–212. [CrossRef]
15. Liu, S.; Harpalani, S. Compressibility of sorptive porous media: Part 1. Background and theory. *AAPG Bull.* **2014**, *98*, 1761–1772. [CrossRef]
16. Guo, X.Q.; Yao, Y.B.; Liu, D.M. Characteristics of Coal Matrix Compressibility: An Investigation by Mercury Intrusion Porosimetry. *Energy Fuel* **2014**, *28*, 3673–3678. [CrossRef]
17. Toda, Y.; Toyoda, S. Application of mercury porosimetry to coal. *Fuel* **1972**, *51*, 199–201. [CrossRef]
18. Wang, Y.; Liu, L.; Zheng, S.; Luo, Z.; Sheng, Y.; Wang, X. Full-scale pore structure and its controlling factors of the Wufeng-Longmaxi shale, southern Sichuan Basin, China: Implications for pore evolution of highly overmature marine shale. *J. Nat. Gas. Sci. Eng.* **2019**, *67*, 134–146. [CrossRef]
19. He, J.W.; Xie, Y.; Hou, M.C.; Liu, J.Q.; He, L.; Lu, Y.F. Geochemical characteristics and geological significance of the Silurian Longmaxi Formation shale in Yanjin area, southwestern Sichuan Basin. *J. Palaeogeogr. Chin. Ed.* **2021**, *23*, 1174–1191.
20. Wang, Z.F.; Zhang, Y.F.; Liang, X.L.; Cheng, F.; Jing, Q.H.; Liu, W.; Zhang, H.B.; Li, H.P. Characteristics of shale lithofacies formed under different hydrodynamic conditions in the Wufeng—Longmaxi Formation, Sichuan Basin. *Acta Pet. Sin.* **2014**, *35*, 623–632.
21. Guo, W.; Liu, H.L.; Li, X.B.; Xue, H.Q. Reservoir characteristics and factors controlling gas-bearing capacity of black rocks in the northeastern Yunnan province. *Nat. Gas Ind.* **2012**, *9*, 22–27.
22. Zhang, X.; Liu, C.L.; Zhu, Y.M.; Wang, Y.; Fu, C.Q. Geological Conditions Evaluation and Favorable Areas Selection of the Shale Gas from Longmaxi Formation in the Northeast of Yunnan. *Nat. Gas Geosci.* **2015**, *26*, 1190–1199.
23. SY/T 5124-2012. Method for Determining the Reflectance of Specular Body in Sedimentary Rocks. China National Petroleum Corporation: Beijing, China, 2012.
24. GB/T 19145-2022. Determination of Total Organic Carbon in Sedimentary Rocks. China National Petroleum Corporation: Beijing, China, 2022.
25. SY/T 5163-2010. X-ray Diffraction Analysis Method of Clay Minerals and Common Non-Clay Minerals in Sedimentary Rocks. China National Petroleum Corporation: Beijing, China, 2010.
26. Cai, Y.D.; Liu, D.M.; Pan, Z.J.; Che, Y.; Liu, Z.H. Investigating the Effects of Seepage-Pores and Fractures on Coal Permeability by Fractal Analysis. *Transp. Porous Med.* **2016**, *111*, 479–497. [CrossRef]
27. GB/T 29171-2012. Determination of Rock Capillary Pressure Curve. China National Petroleum Corporation: Beijing, China, 2012.
28. Yu, Y.X.; Luo, X.R.; Wang, Z.X.; Cheng, M.; Lei, Y.H.; Zhang, L.K.; Yin, J.T. A new correction method for mercury injection capillary pressure (MICP) to characterize the pore structure of shale. *J. Nat. Gas Sci. Eng.* **2019**, *68*, 102896. [CrossRef]
29. Washburn, E.W. The Dynamics of Capillary Flow. *Phys. Rev.* **1921**, *17*, 273. [CrossRef]
30. GB/T 19587-2017. Determination of Specific Surface Area of Solids by Gas Adsorption BET Method. China National Petroleum Corporation: Beijing, China, 2017.
31. IUPAC. Reporting physisorption data for gas/solid systems with special reference to the determination of surface area and porosity. *Pure Appl. Chem* **1982**, *54*, 2201–2218. [CrossRef]
32. Barrett, E.P.; Joyner, L.G.; Halenda, P.P. The Determination of Pore Volume and Area Distributions in Porous Substances. I. Computations from Nitrogen Isotherms. *J. Am. Chem. Soc.* **1951**, *73*, 373–380. [CrossRef]
33. Jarvie, D.M.; Hill, R.J.; Ruble, T.E. Unconventional shale-gas systems: The Mississippian Barnett Shale of north-central Texas as one model for thermogenic shale-gas assessment. *AAPG Bull.* **2007**, *91*, 475–499. [CrossRef]
34. Meng, F.Z.; Wong, L.N.Y.; Zhou, H. Rock brittleness indices and their applications to different fields of rock engineering: A review. *J. Rock Mech. Geotech.* **2021**, *13*, 221–247. [CrossRef]
35. Qin, Z.Y.; Wen, X.T.; Pan, S.L.; Chen, J.Y.; Gou, Q.Y. Study on selection and direct inversion method of brittleness index for shale reservoir. *Acta Geophys.* **2023**, *71*, 1–12. [CrossRef]
36. Sui, L.L.; Yang, Y.M.; Yang, W.G.; Liu, P.; Zhang, S.C.; Han, Y.L. Comprehensive evaluation of shale fracability in Dongying subsidence zone of Shengli oil-field. *J. China Coal Soc.* **2015**, *40*, 1588–1594.
37. Kahraman, S.S.Y.C.; Altindag, R. A brittleness index to estimate fracture toughness. *Int. J. Rock Mech. Min.* **2004**, *41*, 343–348. [CrossRef]
38. Diao, H.Y. Rock mechanical properties and brittleness evaluation of shale reservoir. *Acta Petrol. Sin.* **2013**, *29*, 3300–3306.
39. Wang, Y.M.; Wang, S.F.; Dong, D.Z.; Li, X.J.; Huang, J.L.; Zhang, C.C.; Guan, Q.Z. Lithofacies characterization of Longmaxi Formation of the Lower Silurian, southern Sichuan. *Earth Sci. Front.* **2016**, *23*, 119–133.
40. Cheng, P.; Xiao, X.M. Gas content of organic-rich shales with very high maturities. *J. China Coal Soc.* **2013**, *38*, 737–741.
41. Liang, P.P.; Wang, H.Y.; Zhao, Q.; Zhang, Q.; Wang, N.; Xiao, Q.H. Accumulation condition for Wufeng-Longmaxi Formation shale gas of Yanjin-Gongxian area. *Chin. J. Eng.* **2016**, *38*, 224–231.

42. Lu, L.F.; Liu, W.X.; Wei, Z.H.; Pan, A.Y.; Zhang, Q.Z.; Teng, G.E. Diagenesis of the Silurian Shale, Sichuan Basin: Focus on pore development and preservation. *Acta Sedimentol. Sin.* **2022**, *40*, 73–87.
43. Wu, Y.J.; Wang, Y.; Li, J. Sedimentary characteristics and main reservoir control factors of deep shale in the Sichuan Basin: A case study on the Longmaxi Formation in the eastern Weiyuan area. *Nat. Gas Ind.* **2021**, *41*, 55–65.
44. Yuan, J.Z.; Ye, Y.H.; Xu, F.H.; Fang, L.Y.; Chen, H.; Ma, S.G.; Jiao, K. Lithofacies and pore characteristics of Wufeng—Longmaxi deep buried shale gas reservoirs in western Chongqing-southern Sichuan area, China. *Mineral. Petrol.* **2022**, *42*, 104–115.
45. Liu, C.; Ding, W.G.; Zhang, J.; Chen, X.; Wu, P.; Liu, X.Q.; Li, Y.B.; Ma, L.T.; Hu, W.Q.; Kong, W.; et al. Qualitative-quantitative multi scale characteristics of shale pore structure from Upper Paleozoic coal-measures in Linxing area. *Coal Geol. Explor.* **2021**, *49*, 46–57.
46. Ning, C.X.; Jiang, Z.X.; Gao, Z.Y.; Li, Z.; Zhu, R.F.; Su, S.Y.; Li, T.W.; Wang, Z.; Huang, R.Z.; Cheng, L. Quantitative evaluation of pore connectivity with nuclear magnetic resonance and high pressure mercury injection: A case study of the lower section of Es_3 in Zhanhua sag. *J. China Univ. Min. Technol.* **2017**, *46*, 578–585.
47. Zhang, D.Z. Characterization of microscopic pore structure of tight sandstone reservoirs through nitrogen adsorption experiment: Case study of Shahezi Formation in Xujiaweizi Fault Depression, Songliao Basin, China. *Nat. Gas Geosci.* **2017**, *28*, 898–908.
48. Howard, J.J. Porosimetry measurement of shale fabric and its relationship to illite/smectite diagenesis. *Clay Clay Miner.* **1991**, *39*, 355–361. [CrossRef]
49. Zhao, D.F.; Guo, Y.H.; Mao, X.X.; Lu, C.G.; Li, M.; Qian, F.C. Characteristics of macro-nanopores in anthracite coal based on mercury injection, nitrogen adsorption and FE-SEM. *J. China Coal Soc.* **2017**, *42*, 1517–1526.
50. Shi, D.S.; Xu, Q.C.; Guo, R.L.; Liu, E.R.; Zhu, D.S.; Wang, Y.H.; Wang, B.Q.; Ouyang, Z.Y. Pore structure characteristics and main controlling factors of Permian organic-richshale in Lower Yangtze Region. *Nat. Gas Geosci.* **2022**, *33*, 1911–1925.
51. Mastalerz, M.; Schimmelmann, A.; Drobniak, A. Porosity of Devonian and Mississippian New Albany Shale across a maturation gradient: Insights from organic petrology, gas adsorption, and mercury intrusion. *AAPG Bull.* **2013**, *97*, 1621–1643. [CrossRef]
52. Cao, T.T.; Deng, M.; Luo, H.Y.; Liu, H.; Liu, G.X.; Hursthouse, A.S. Characteristics of organic pores in Middle and Upper Permian shale in the Lower Yangtze region. *Pet. Geol. Exp.* **2018**, *40*, 315–322.
53. Rickman, R.; Mullen, M.J.; Petre, J.E.; Grieser, W.V.; Kundert, D. A Practical Use of Shale Petrophysics for Stimulation Design Optimization: All Shale Plays Are Not Clones of the Barnett Shale. In Proceedings of the SPE Annual Technical Conference and Exhibition, Denver, CO, USA, 21–24 September 2008.
54. Sondergeld, C.H.; Newsham, K.E.; Comisky, J.T.; Rice, M.C.; Rai, C.S. Petrophysical Considerations in Evaluating and Producing Shale Gas Resources. In Proceedings of the SPE Unconventional Gas Conference, Pittsburgh, PA, USA, 23–25 February 2010.

processes MDPI

Article

Study on the Preparation of $ZnFeO_4$ by Roasting Zinc-Containing Gossan Ore

Jinlin Yang [1], Zongyu Li [1], Xingnan Huo [1], Hangyu Li [1], Shizhen Liao [1], Shaojian Ma [1] and Hengjun Li [2,*]

1 State Key Laboratory of Featured Metal Materials and Life-Cycle Safety for Composite Structures, MOE Key Laboratory of New Processing Technology for Nonferrous Metals, Guangxi Higher School Key Laboratory of Minerals Engineering and Materials, College of Resources, Environment and Materials, Guangxi University, Nanning 530004, China; 2215394008@st.gxu.edu.cn (H.L.)
2 College of Chemistry and Chemical Engineering, Guangxi University, Nanning 530004, China
* Correspondence: 1615302001@st.gxu.edu.cn; Tel.: +86-131-5266-0958

Abstract: Gossan ore is typically abandoned after mining, which not only increases mining production costs but also wastes mineral resources, and its long-term accumulation can easily lead to environmental pollution hazards. Therefore, this paper takes zinc-containing gossan ore as the research object and, based on the high content of zinc and iron minerals in gossan ore, this study conducts a roasting experiment to prepare $ZnFeO_4$. X-ray diffraction is used to characterize and analyze the $ZnFeO_4$ sample prepared by roasting zinc-containing gossan ore. The experimental results indicate that controlling the particle size of the roasted ore sample to −0.074 mm can effectively remove impurities and facilitate the reaction. The influence of roasting temperature and time on the formation of $ZnFeO_4$ is remarkable. The conditions for roasting zinc-containing gossan ore to maximize the $ZnFeO_4$ content are as follows: −0.074 mm particle size ore sample, reaction zinc/iron molar ratio of 1:2, mechanical activation time of 120 min, roasting temperature of 1050 °C, and roasting time of 120 min. These findings provide new ideas for the utilization of gossan ore and lay a theoretical foundation for the efficient development and utilization of difficult-to-select zinc-containing gossan ore.

Keywords: gossan ore; $ZnFeO_4$; roasting; preparation

Citation: Yang, J.; Li, Z.; Huo, X.; Li, H.; Liao, S.; Ma, S.; Li, H. Study on the Preparation of $ZnFeO_4$ by Roasting Zinc-Containing Gossan Ore. *Processes* **2023**, *11*, 1991. https://doi.org/10.3390/pr11071991

Academic Editors: Ping Gao, Yidong Cai, Yingfang Zhou and Quan Gan

Received: 15 June 2023
Revised: 26 June 2023
Accepted: 27 June 2023
Published: 1 July 2023

1. Introduction

Gossan ore, as a mineral resource, has typically been overlooked by researchers due to its complex composition, low valuable metal content, and high development costs. The mineral and chemical composition of gossan ore is related to the main metal components in the primary sulfide deposit. Various types of primary sulfide deposits form gossans through surface oxidation, and their elemental combinations differ [1]. For example, gossan ore formed from sulfide ore deposits containing gold (silver) contains valuable metals such as gold (silver) and iron. The valuable metals of iron ore in sulfide copper deposits include copper and iron, whereas those in gossan ore of sulfide zinc deposits include zinc and iron. These gossan ores contain many valuable metals and have considerable potential for development and utilization. From the perspective of mineral resource sustainability, with the gradual depletion of high-quality mineral resources, resources such as gossan ore are bound to be effectively utilized in the future to compensate for the shortage of mineral resources. Therefore, gossan ore exhibits immense potential for recycling value. However, although there are several valuable metals in gossan ore, due to its oxidized nature and low metal content, conventional beneficiation and smelting techniques are difficult to effectively recover and utilize resources. For decades, gossan has been limited in its application as a guide for exploring deep sulfide ore bodies, especially those containing Cu, Pb, Zn, Mo, Ni, Au, and Pt [2,3]. As a result, studies on the application of gossan ore remain scarce, with the studies being limited to investigating the use of gossan ore in determining the type and content of ore deposits [4–6], the heavy metal composition in gossan ore and its impact on the surrounding environment [7–10], and the high value-added precious metal gold (silver) in gossan ore [11–13]. In summary, research on the development and utilization of gossan ore primarily focuses on the recovery of gold, silver, and copper, with limited research efforts being dedicated toward the development and utilization of other valuable elements in gossan ore.

$ZnFeO_4$ is an excellent material with a stable structure and does not easily decompose at high temperatures. It is nontoxic and harmless to the human body and insoluble in weak acids and alkalis. Utilizing these excellent properties of $ZnFeO_4$, high-temperature and corrosion-resistant nontoxic coatings can be prepared. $ZnFeO_4$ also exhibits excellent photocatalytic performance and can be used for the catalytic adsorption and degradation of water pollutants. Furthermore, $ZnFeO_4$ is highly sensitive to visible light, does not undergo photocorrosion, and possesses interface electron transfer characteristics, making it an excellent photoelectric conversion material [14–18]. Ebrahimi prepared $ZnFeO_4$ nanoparticles with a mixed spinel structure using the coprecipitation method at 20–80 °C [19]. Sangita et al. summarized various manufacturing methods of $ZnFeO_4$ [20]. Sun investigated the preparation of mesoporous $ZnFeO_4$ flame retardants at different scales and their performance in epoxy resins [21]. Rachna investigated the preparation, characterization, performance, and application of nano $ZnFeO_4$ [22]. These researchers used high-purity raw materials to prepare $ZnFeO_4$, which has high costs and low economic benefits.

Therefore, based on the new concepts of material processing technology for mineral resources, this study innovatively proposes a technical concept of preparing $ZnFeO_4$ from zinc-containing gossan ore. This concept is based on the high content of zinc and iron minerals in zinc-containing gossan ore. By adjusting the phase of zinc and iron minerals in the ore and applying roasting methods, $ZnFeO_4$ products are synthesized. This study lays a theoretical groundwork for the development of new processes and technologies for the preparation of $ZnFeO_4$ with industrial application value, as well as the efficient development and utilization of zinc-containing gossan ore.

2. Materials and Methods

2.1. Materials

The test sample was a zinc-containing gossan ore extracted from a certain mine, as depicted in Figure 1.

Figure 1. Raw gossan ore sample.

As shown in Figure 1, the gossan ore had an irregular block shape and an earthy yellow or brown surface with many corrosion traces. The surface of the ore had irregularly distributed small pores, consistent with the appearance characteristics of classic gossan ore. The zinc-containing gossan ore was dried naturally, crushed, and screened to obtain a −1-mm test sample, following which it was mixed well and bagged for subsequent testing. The crushing equipment was a jaw crusher, model XPC-100 × 150, and the screening equipment was a vibrating screen, model Analysette 3.

The semiquantitative multi-element analysis results of gossan ore are presented in Table 1. The equipment was an X-ray fluorescence element analyzer, model S8 TIGER.

Table 1. The multi-element semiquantitative analysis results of gossan ore (source: [23]).

Component	**Zn**	**Fe_2O_3**	**SiO_2**	**Al_2O_3**	**MgO**	**CaO**
Content (%)	8.99	68.32	10.32	5.6	1.35	0.56
Component	**Na_2O**	**K_2O**	**SO_3**	**TiO_2**	**Mn**	**Pb**
Content (%)	0.20	0.12	0.90	0.25	1.46	1.38

Table 1 shows that the chemical element composition of the sample was simple and was mainly composed of zinc, iron, silicon, oxygen, and aluminum. Furthermore, there were trace amounts of metal elements such as manganese, magnesium, calcium, and lead. The main components of the ore were zinc minerals, iron minerals, silica, and alumina, with Zn accounting for 8.99%, Fe_2O_3 for 68.32%, SiO_2 for 10.32%, and Al_2O_3 for 5.6% of the ore.

2.2. Experimental Principles and Characterization Methods

2.2.1. Test Methods and Principles

This study adopted the roasting method to prepare $ZnFeO_4$. the effects of factors such as material particle size, mechanical activation time, roasting temperature, and roasting time were studied. The specific test method was as follows.

First, a GD200 × 75 planetary grinding ball mill was used to finely grind the −1 mm-particle size gossan ore sample, and a test sample with a particle size of −0.074 mm was screened out. Second, zinc and iron assays were conducted on samples with a particle size of −0.074 mm to obtain the zinc/iron molar ratio in the gossan ore. Then, based on the zinc/iron molar ratio in the gossan ore and the zinc/iron molar ratio required by the $ZnFeO_4$ reaction theory, zinc or iron oxide was added to regulate the appropriate zinc/iron molar ratio. Then, the test sample with an adjusted zinc/iron molar ratio was placed into the GD200 × 75 planetary grinding ball mill for mechanical activation treatment. Finally, the mechanically activated test samples were taken and roasted in a GXL-15 high-temperature box furnace to obtain the roasted $ZnFeO_4$ product.

The main chemical reactions that may occur during the roasting process for preparing $ZnFeO_4$ are as follows:

$$ZnO + Fe_2O_3 = ZnFe_2O_4 \tag{1}$$

$$ZnCO_3 + Fe_2O_3 = ZnFe_2O_4 + CO_2 \tag{2}$$

2.2.2. Characterization Methods

X-ray diffraction (XRD) analysis was the main method used in the experiment for characterization and analysis, and the analysis principle is as follows.

In XRD analysis, the intensity of the diffraction rays of different substances in a mixture increases with the relative content of the substance in the sample. As a result, the diffraction peak intensity of $ZnFeO_4$ in XRD analysis can be used to calculate the relative content of $ZnFeO_4$ in the sample.

3. Results and Discussion

3.1. Effect of Ore Particle Size

Here, the research on the effect of ore particle size is primarily based on two considerations. First, one of the main impurity minerals of gossan ore is quartz, with a Mohs hardness of 7.0, whereas the Mohs hardness of smithsonite in the main zinc and iron ores is 4–4.5, that of hemimorphite is 4–4.5, and that of limonite is 1–4. The hardness of impurity mineral quartz is higher than that of zinc and iron minerals; thus, fine grinding and screening may be performed to remove some difficult-to-grind quartz. Second, the specific surface area of materials with different particle sizes differs, affecting the reaction rate. Therefore, the effect of ore particle size on the preparation of $ZnFeO_4$ by roasting was investigated. After grinding the −1 mm gossan ore sample for 1 h, the samples were sieved using 150-, 180-, 200-, and 250-mesh sieves. The corresponding particle sizes of the products under the sieves were −0.1, −0.088, −0.074, and −0.065 mm, respectively. The other experimental conditions are as follows: zinc/iron molar ratio of 1:2, mechanical activation time of 120 min, roasting temperature of 1050 °C, and roasting time of 120 min. The content of $ZnFeO_4$ in products roasted with different particle sizes is shown in Table 2.

Table 2. Content of $ZnFeO_4$ in roasted products of different particle sizes.

Particle Size (mm)	−1	−0.1	−0.088	−0.074	−0.065
Content of $ZnFeO_4$ (%)	75.3	80.1	85.3	88.6	88.4

Table 2 shows that as the particle size of zinc-containing gossan ore decreased, the amount of $ZnFeO_4$ generated continuously increased; however, there was negligible change after the −0.074 mm particle size. Among them, the $ZnFeO_4$ content in the roasted product of −1 mm particle size was 75.3%, whereas the $ZnFeO_4$ content in the roasted product of −0.074 mm particle size was 88.6%, and the $ZnFeO_4$ content was increased by 13.3%.

The multi-element semiquantitative analysis results of the −0.074 mm particle size sample are shown in Table 3.

Table 3. The multi-element semiquantitative analysis results of the −0.074 mm particle size sample.

Component	Zn	Fe_2O_3	SiO_2	Al_2O_3	MgO	CaO
Content (%)	12.32	73.56	5.81	2.6	0.78	0.68
Component	**Na_2O**	**K_2O**	**SO_3**	**TiO_2**	**Mn**	**Pb**
Content (%)	0.15	0.10	0.86	0.35	0.96	1.45

Tables 1 and 3 illustrate that after fine grinding and screening with a 200-mesh sieve, the silicon dioxide content in the −0.074 mm particle size ore sample considerably decreased by approximately 43.7%. Therefore, the relative content of zinc and iron minerals as reactants also increased to a certain extent, which was conducive to the formation of $ZnFeO_4$ and increased the $ZnFeO_4$ content in the roasted product.

XRD analysis was performed on the products after roasting the samples with particle sizes of −1, −0.1, −0.088, −0.074, and −0.065 mm. The XRD diffraction patterns are displayed in Figure 2.

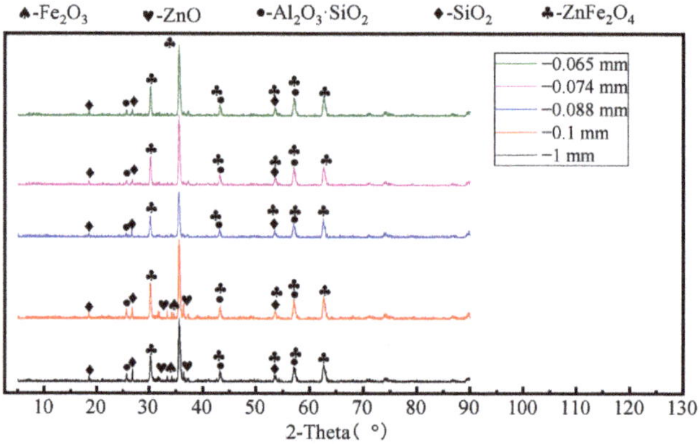

Figure 2. XRD patterns of the roasted products at different particle sizes.

Figure 2 shows that among the five roasted ore products, the diffraction peaks of silicon and aluminum compounds in the roasted products with −1 and −0.1 mm particle sizes exhibited almost no change in intensity, indicating that the use of a 150-mesh sieve did not substantially remove silicon and aluminum minerals. After sieving with a 180-mesh or finer sieve, silicon and aluminum minerals can be effectively removed. A comparison of the roasted products with particle sizes of −0.074 and −0.065 mm revealed that the diffraction peaks of silicon and aluminum compounds were the smallest. However, the difference between the two was insignificant. Therefore, the subsequent experimental ore samples were selected as follows: the original −0.1 mm particle size gossan ore was finely ground for 1 h and screened with a 200-mesh sieve, and the −0.074 mm particle size ore sample was used as the material to prepare $ZnFeO_4$.

3.2. *Effect of Roasting Temperature*

The roasting temperature test conditions were −0.074 mm particle size ore sample; zinc/iron molar ratio of 1:2; mechanical activation time of 120 min; roasting time of 120 min; and roasting temperatures of 700 °C, 800 °C, 900 °C, 1050 °C, and 1200 °C, respectively. The XRD patterns of the roasted products at different roasting temperatures are depicted in Figure 3.

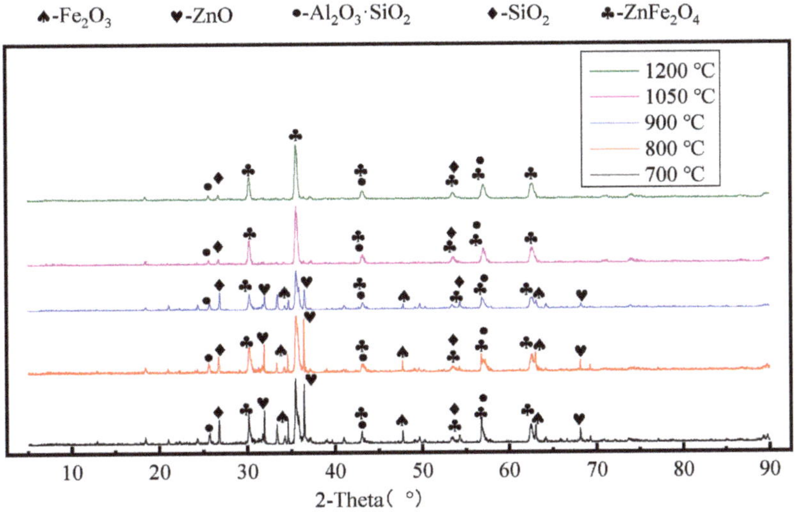

Figure 3. XRD patterns of the roasted products at different roasting temperatures.

Figure 3 shows that at roasting temperatures of 700 °C and 800 °C, the diffraction peaks of zinc and iron oxide minerals strengthened, indicating that several reactants had yet to undergo the reaction. When the roasting temperature reached 900 °C, the diffraction peaks of zinc and iron oxide minerals weakened, indicating that at this roasting temperature, the reactant content in the sample decreased and the reaction proceeded more fully. When the roasting temperature increased from 700 °C to 900 °C, the diffraction peak intensity represented by zinc oxide minerals in the XRD pattern exhibited the largest decrease, and the diffraction peak of iron oxide minerals also weakened to a certain extent. This indicates that a considerable amount of zinc and iron oxide minerals, such as smithsonite and limonite, participated in the reaction when the roasting temperature increased from 700 °C to 900 °C. When the roasting temperature was 1050 °C, the diffraction peak of $ZnFeO_4$ in the roasted product was evident, whereas those of zinc and iron oxide minerals disappeared, indicating that the reaction was complete at this time. When the roasting temperature was above 1050 °C, the diffraction curve of the sample remained almost unchanged, and the diffraction peaks of $ZnFeO_4$ were also visible, whereas the diffraction peaks of the zinc and iron oxide minerals disappeared. According to XRD quantitative analysis, the $ZnFeO_4$ content in the roasted product reached 75.3% at a roasting temperature of 800 °C. At a roasting temperature of 1050 °C, the $ZnFeO_4$ content in the roasted product was 88.6%.

To investigate the changes in $ZnFeO_4$ content in roasted products at different roasting temperatures, the following experiments were conducted. The experimental conditions were as follows: −0.074 mm particle size ore sample; zinc/iron molar ratio of 1:2; mechanical activation time of 120 min; roasting time of 120 min; and roasting temperatures of 600 °C, 700 °C, 800 °C, 900 °C, 950 °C, 1000 °C, 1050 °C, 1100 °C, and 1200 °C. The variation in $ZnFeO_4$ content among the roasted products with roasting temperature is depicted in Figure 4.

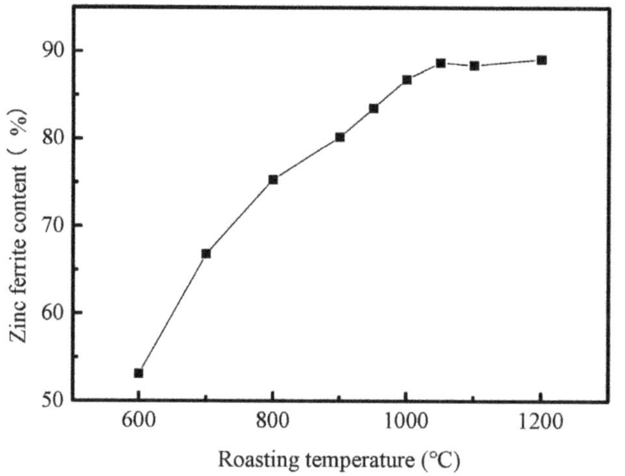

Figure 4. $ZnFeO_4$ content of the roasted products at different temperatures.

Figure 4 reveals that prior to 1050 °C, the increase in roasting temperature had a remarkable impact on the formation of $ZnFeO_4$, resulting in a rapid increase in $ZnFeO_4$ content. After 1050 °C, there was little change in the $ZnFeO_4$ content. Therefore, a roasting temperature of approximately 1050 °C is the optimal temperature in terms of maximizing $ZnFeO_4$ content. However, a high temperature of 1050 °C consumes a substantial amount of energy, and the higher the temperature, the more time and energy required to raise the temperature. Therefore, from the perspective of saving energy consumption and based on the XRD spectrum analysis in Figure 3, the diffraction peaks of zinc and iron oxide minerals weaken when the roasting temperature reaches 800 °C. At this roasting temperature, the content of reactants in the sample considerably decreases and the reaction proceeds more fully. Therefore, the roasting temperature is a crucial factor affecting the formation of $ZnFeO_4$.

3.3. *Effect of Mechanical Activation Time*

The mechanical activation time test conditions were −0.074 mm particle size ore sample; zinc/iron molar ratio of 1:2; roasting temperatures of 800 °C, 900 °C, 950 °C, 1000 °C, and 1050 °C; roasting time of 120 min; and mechanical activation times of 0, 60, 120, and 200 min. The XRD patterns of the roasted products at different roasting temperatures and mechanical activation times are depicted in Figure 5.

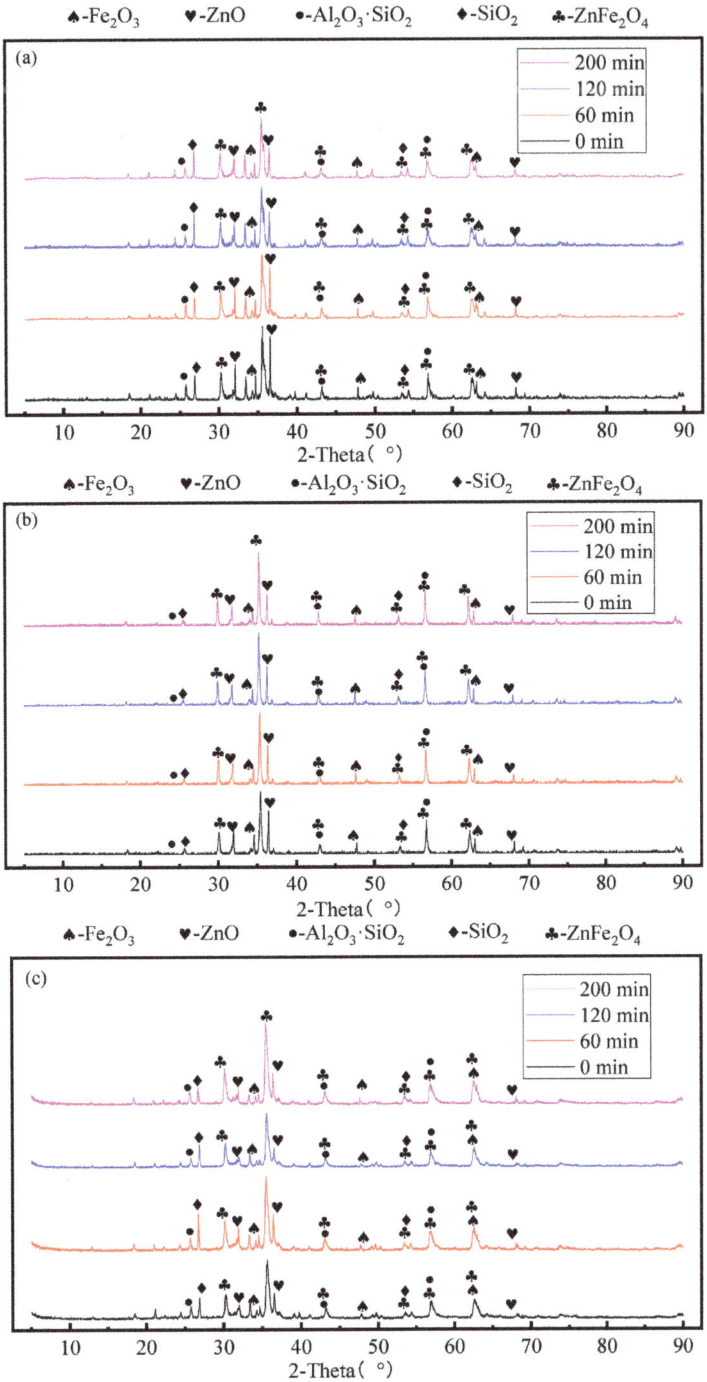

Figure 5. *Cont.*

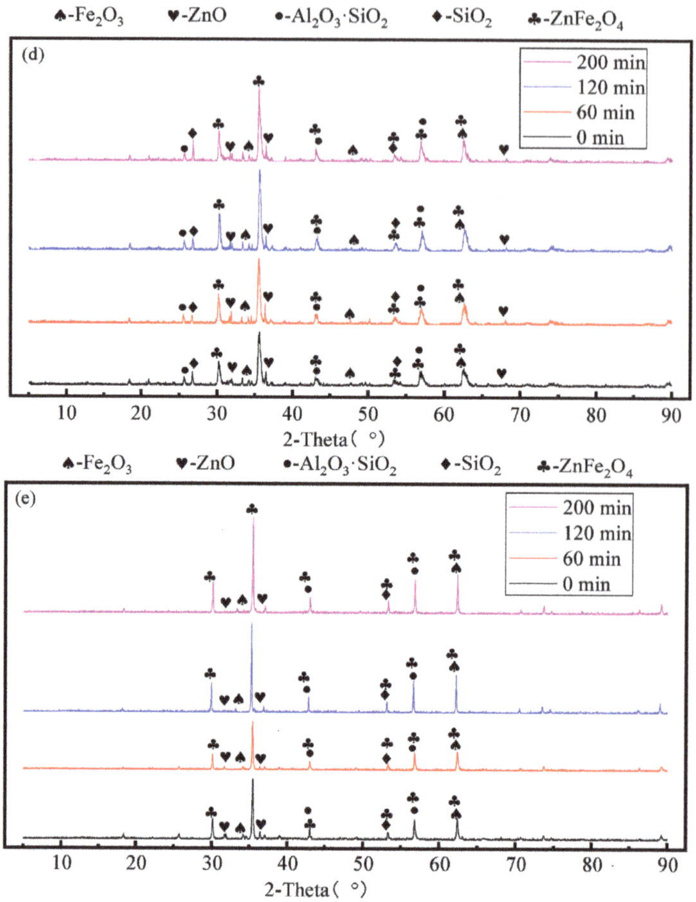

Figure 5. XRD patterns of the roasted products at different roasting temperatures and mechanical activation times: (**a**): 800 °C; (**b**): 900 °C; (**c**): 950 °C; (**d**): 1000 °C; (**e**): 1050 °C.

Figure 5a illustrates that at a roasting temperature of 800 °C, with prolonging mechanical activation time, the diffraction peak of $ZnFeO_4$ in the roasted product strengthened, indicating an increase in the $ZnFeO_4$ content. XRD quantitative analysis revealed that the content of $ZnFeO_4$ in the roasted product increased from 68.4% under 0 min mechanical activation to 75.3% under 120 min mechanical activation. This may be attributed to the fact that during mechanical activation, the ore produces several lattice dislocations at a microscopic level. The aggregation of many lattice dislocations can generate microcracks. The rapid propagation of these microcracks results in a temperature exceeding 1300 K at the top of the crack. On a microscopic scale, the high-temperature and high-pressure state at the top of the crack corresponds to the high-speed movement of atoms, promoting the migration and diffusion between atoms, which is conducive to the progress of the reaction. Furthermore, some atoms are excited, and electrons are excited in extremely small regions to form a plasma region, further accelerating the reaction. In addition, mechanical activation plays a role in the ability to completely crush and grind the material, increasing the specific surface area of the sample and the possibility of direct contact between the reaction materials, thereby effectively improving the reaction rate. A comparison of Figure 5a–e reveals that the increase in roasting temperature led to a considerable enhancement in the diffraction peak of $ZnFeO_4$ in the XRD spectrum of the sample. However, when the roasting temperature exceeded 1000 °C, the XRD spectrum of the roasted sample exhibited little change under different mechanical activation times, indicating that the promotion effect of mechanical activation on the preparation of $ZnFeO_4$ by roasting was insignificant at this temperature.

To investigate the effect of mechanical activation time on the preparation of $ZnFeO_4$ at different roasting temperatures, the following experiments were conducted. The experimental conditions were −0.074 mm particle size ore sample; zinc/iron molar ratio of 1:2; roasting temperatures of 800 °C, 900 °C, 950 °C, 1000 °C, and 1050 °C; roasting time of 120 min; and mechanical activation times of 0, 30, 60, 90, 120, 150, and 200 min. The effect of mechanical activation time on the content of $ZnFeO_4$ in the roasted products at different roasting temperatures is depicted in Figure 6.

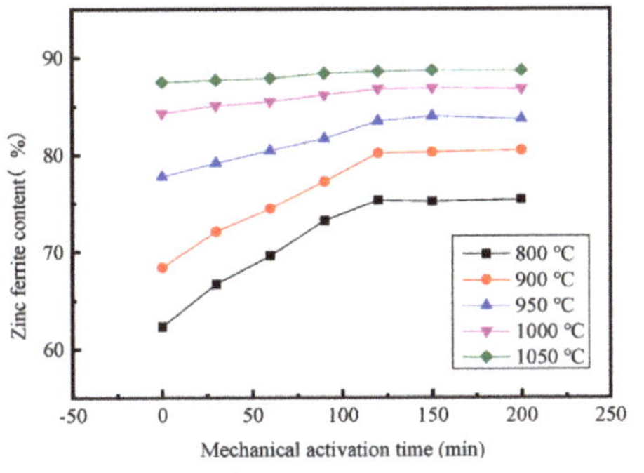

Figure 6. Effect of mechanical activation time on $ZnFeO_4$ content of the roasted products at different roasting temperatures.

Figure 6 shows that at a roasting temperature of 800 °C, the mechanical activation time had a remarkable impact on the $ZnFeO_4$ content in the roasted product. According to the analysis results, the $ZnFeO_4$ content was 62.3% in the absence of mechanical activation. After 120 min of mechanical activation, the $ZnFeO_4$ content in the roasted product reached 75.3%, demonstrating an increase of 13% compared with the case of $ZnFeO_4$ content without mechanical activation. At a roasting temperature of 900 °C, the content of $ZnFeO_4$ is 68.4% without mechanical activation. After 120 min of mechanical activation, the content of $ZnFeO_4$ in the roasted product reached 80.2%. Consequently, the content of $ZnFeO_4$ increased by 11.8%, which was lower than the increase in $ZnFeO_4$ content at 800 °C. At higher roasting temperatures of 1000 °C and 1050 °C, the effect of mechanical activation time on the preparation of $ZnFeO_4$ by roasting was insignificant. According to the analysis results, within the range of the above mechanical activation times, the highest $ZnFeO_4$ content in the sample was approximately 86.8% at a roasting temperature of 1000 °C, whereas the highest $ZnFeO_4$ content in the sample was approximately 88.6% at a roasting temperature of 1050 °C. This could be attributed to the roasting temperature being sufficiently high for the −0.074 mm particle size sample and the particles being capable of completely diffusing. At this time, mechanical activation has marginal utility. Because the reactants could fully react under a short mechanical activation time, extending the mechanical activation time has no considerable impact on the formation of $ZnFeO_4$. This further elucidates the point that roasting temperature is a crucial factor affecting the formation of $ZnFeO_4$.

3.4. *Effect of Roasting Time*

The roasting time test conditions were −0.074 mm particle size ore sample; zinc/iron molar ratio of 1:2; mechanical activation time of 120 min; roasting temperatures of 800 °C, 900 °C, 950 °C, 1000 °C, and 1050 °C; and roasting times of 30, 60, 90, 120, and 150 min. The XRD patterns of the roasted products at different roasting temperatures and times are depicted in Figure 7.

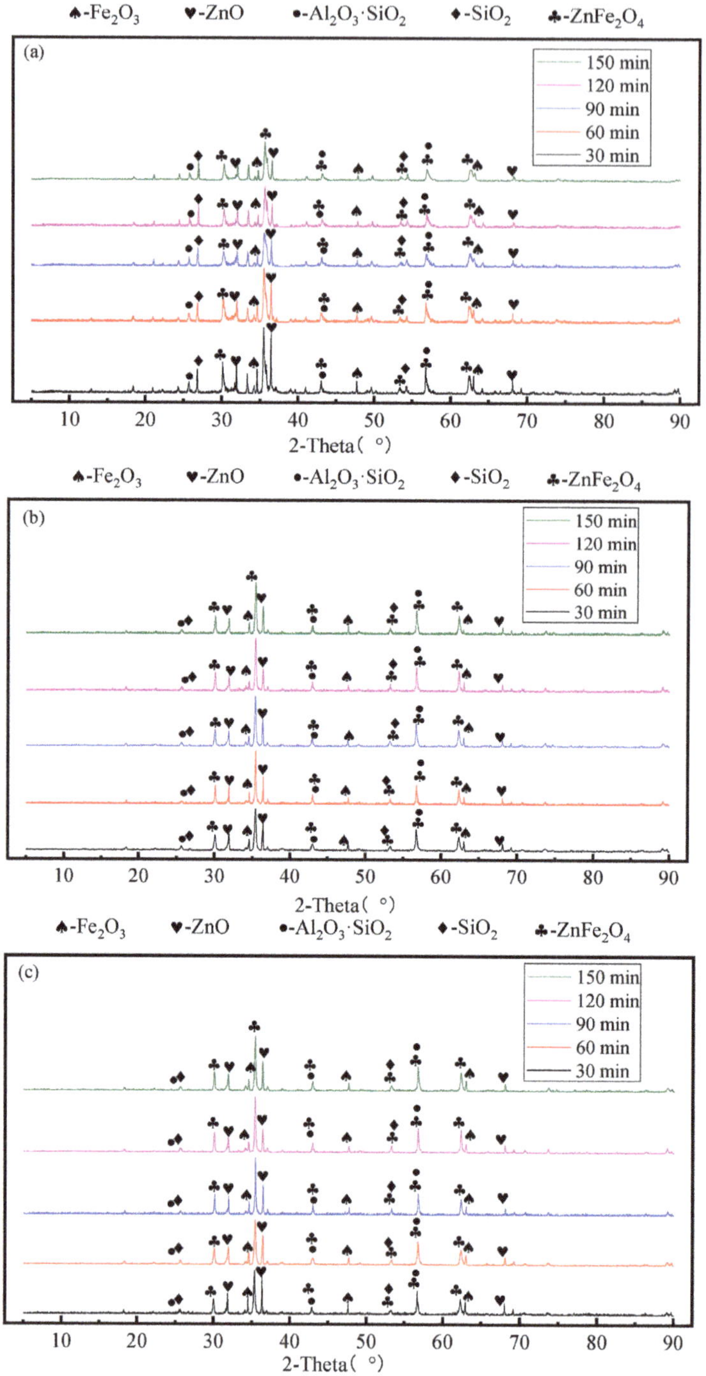

Figure 7. *Cont.*

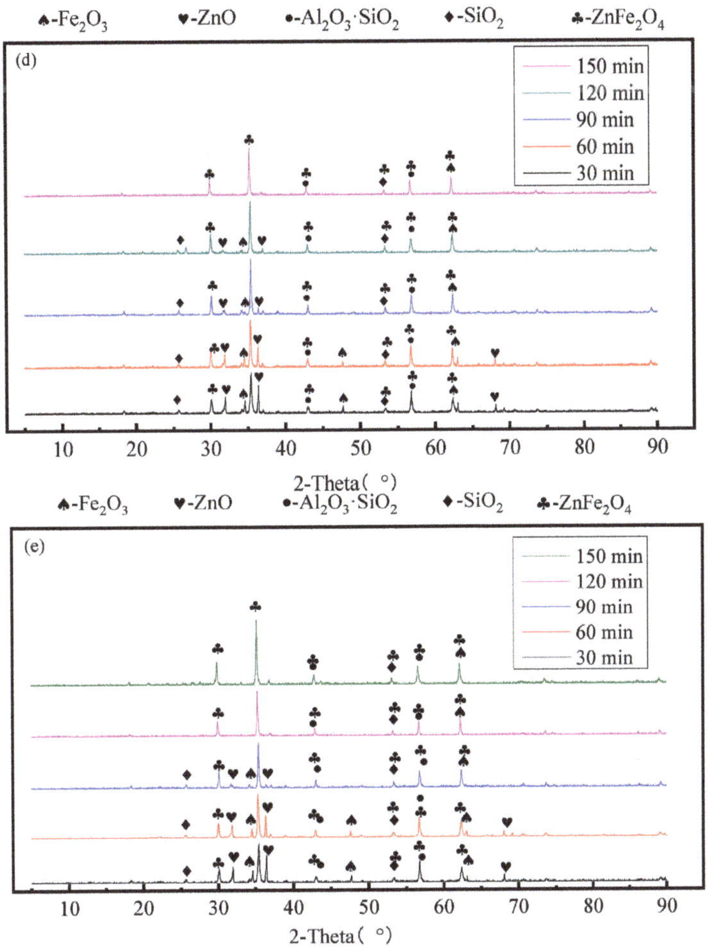

Figure 7. XRD patterns of the roasted products at different roasting times and temperatures: (**a**): 800 °C; (**b**): 900 °C; (**c**): 950 °C; (**d**): 1000 °C; (**e**): 1050 °C.

Figure 7 shows that for roasting times of 30 and 60 min, within the experimental roasting temperature range, the diffraction peaks of iron and zinc oxides in all roasted products were evident to a larger degree, indicating a considerable amount of unreacted reactants in the sample. After the roasting time reached 90 min, the diffraction peaks of zinc and iron oxides in the roasted products at 1000 °C and 1050 °C considerably weakened, indicating a complete reaction. At roasting temperatures above 1000 °C, the roasting time was extended, and the $ZnFeO_4$ content was substantially increased within the 30–90 min roasting time range. When the roasting time was 120 min, the diffraction peaks of iron and zinc oxides were weak, whereas those of $ZnFeO_4$ were strong. When the roasting time reached 120 min, the reaction was complete. When the roasting time was 150 min, in a manner similar to that when the roasting time was 120 min, there were almost no diffraction peaks of zinc and iron oxides, but the diffraction peaks of $ZnFeO_4$ were very strong, indicating that the reaction was sufficient at this time. Note that the XRD-pattern waveforms of the samples for roasting times of 120 and 150 min are very similar. Because the diffraction peak intensity of a substance represents the relative content of the substance, the peaks of the spectra of the samples are similar, indicating that the reaction was complete at this time. Moreover, the peak diffraction intensity of $ZnFeO_4$ was high, and the difference was insignificant, indicating that the $ZnFeO_4$ content in the samples was high. At roasting temperatures of 800 °C and 900 °C and roasting time of 90 min or more, the diffraction peak of zinc/iron impurities in the sample was evident. This indicates that the reaction of zinc-containing gossan ore was not fully complete at this temperature and the roasted product contained many impurities.

To investigate the effect of roasting time at different roasting temperatures on the formation of $ZnFeO_4$, the following experiments were conducted. The experimental conditions were as follows: −0.074 mm particle size ore sample; zinc/iron molar ratio of 1:2; mechanical activation time of 120 min; roasting temperatures of 800 °C, 900 °C, 950 °C, 1000 °C, and 1050 °C; and roasting times of 30, 45, 60, 75, 90, 105, 120, 135, 150, 165, and 180 min. The effect of roasting time at different roasting temperatures on the preparation of $ZnFeO_4$ is depicted in Figure 8.

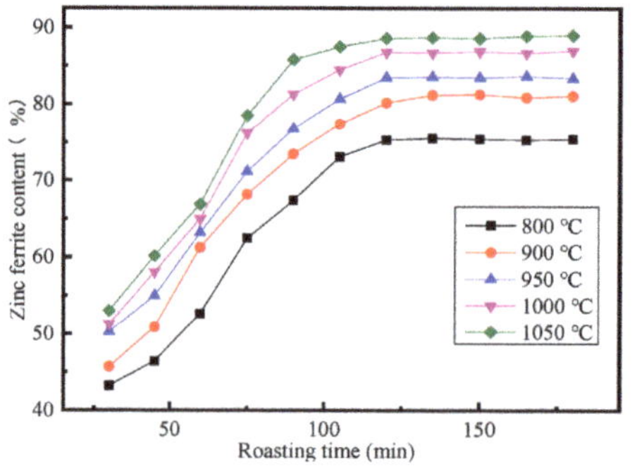

Figure 8. Effect of roasting time on $ZnFeO_4$ content of the roasted products at different roasting temperatures.

Figure 8 shows that at low roasting temperatures (800 °C and 900 °C) and a roasting time of less than 75 min, the $ZnFeO_4$ content in the roasted product increased rapidly, exhibiting a near-linear trend. When the roasting time was 75–105 min, the growth rate of the zinc ferrite content slowed. When the roasting time was above 120 min, the $ZnFeO_4$ content in the roasted product remained almost unchanged. At high roasting temperatures (1000 °C and 1050 °C), the pattern exhibited was similar to that at lower roasting temperatures, except for the faster growth rate of $ZnFeO_4$ in the early stage. When the roasting time reached 120 min, the $ZnFeO_4$ content was almost maximum. If the roasting time was extended further, the $ZnFeO_4$ content in the product would not change considerably, indicating that the sample had fully reacted at this time. Figure 8 also shows that at a roasting time above 120 min, the $ZnFeO_4$ content in the roasted products at different roasting temperatures reached equilibrium. Moreover, as the roasting temperature increased, the slope of the curve also slightly increased, indicating that higher roasting temperatures lead to a faster increase in $ZnFeO_4$ content.

4. Conclusions

(1) Before roasting, fine grinding and screening of the −1 mm gossan ore sample can effectively remove impurities and facilitate the reaction by controlling the particle size of the ore to −0.074 mm. Compared with nonscreening after fine grinding of the ore, the $ZnFeO_4$ content in the roasted product increased by 13.3% after roasting for a particle size of −0.074 mm.

(2) When the roasting temperature was low (800 °C), the mechanical activation time had a remarkable impact on the preparation of $ZnFeO_4$ by roasting. Following 120 min mechanical activation, the $ZnFeO_4$ content in the roasted product reached 75.3%, which was 13% higher than that without mechanical activation. When the roasting temperature

was high (1050 °C or above), the effect of mechanical activation time on the formation of $ZnFeO_4$ during roasting was insignificant.

(3) Roasting temperature and roasting time considerably influenced the formation of $ZnFeO_4$. As the roasting temperature and roasting time increased, the $ZnFeO_4$ content in the roasted products gradually increased. When the roasting temperature reached 1050 °C or above and the roasting time reached 120 min, the reaction was complete.

(4) Considering the maximization of $ZnFeO_4$ content, the suitable conditions for preparing $ZnFeO_4$ from zinc-containing gossan ore by roasting are as follows: −0.074 mm-particle size ore sample, reaction zinc/iron molar ratio of 1:2, mechanical activation time of 120 min, roasting temperature of 1050 °C, and roasting time of 120 min. Under these conditions, the $ZnFeO_4$ content in the roasted product reached 88.6%.

(5) From the perspective of saving energy consumption and simplifying the phase composition of reaction products, the suitable conditions for preparing $ZnFeO_4$ from zinc-containing gossan ore by roasting are as follows: −0.074 mm particle size ore sample, reaction zinc/iron molar ratio of 1:2, mechanical activation time of 120 min, roasting temperature of 800 °C, and roasting time of 120 min. Under these conditions, the $ZnFeO_4$ content in the roasted product reached 75.3%.

Author Contributions: Conceptualization, J.Y. and Z.L.; data curation, Z.L. and H.L. (Hengjun Li); formal analysis, X.H. and S.L.; funding acquisition, J.Y.; investigation, S.L. and H.L. (Hangyu Li); methodology, J.Y. and S.M.; project administration, J.Y. and H.L. (Hengjun Li); validation, H.L. (Hengjun Li) and Z.L.; writing—original draft, H.L. (Hengjun Li) and Z.L.; writing—review and editing, J.Y. and S.M. All authors have read and agreed to the published version of the manuscript.

Funding: This research was funded by the National Natural Science Foundation of China (No. 52264020, No. 51774099).

Data Availability Statement: Not applicable.

Conflicts of Interest: The authors declare no conflict of interest.

References

1. Pabllo, H.C.S.; Marcondes, L.C. Mineralogical and textural evolution of the Alvo 118 copper-bearing gossan: Implications for supergene metallogenesis in Carajás Mineral Province. *Brazil. J. S. Am. Earth Sci.* **2023**, *121*, 104108.
2. Andreu, E.; Torró, L.; Proenza, J.A.; Domenech, C.; García-Casco, A.; Villanova de Benavent, C.; Chavez, C.; Espaillat, J.; Lewis, J.F. Weathering profile of the Cerro de Maimón VMS deposit (Dominican Republic): Textures, mineralogy, gossan evolution and mobility of gold and silver. *Ore Geol. Rev.* **2015**, *65*, 165–179. [CrossRef]
3. Pires, G.L.C.; Renac, C.; Bongiolo, E.M.; Neumann, R. Gossan mineralogy, textures, and gold enrichment over the Au (As, Bi, Ag) deposit in the Buracão Area (Brasília Fold Belt, Brazil): Implications for gold prospecting in weathering profiles. *J. Geochem. Explor.* **2020**, *218*, 106615. [CrossRef]
4. Křìbek, B.; Zachariáš, J.; Knésl, I.; Míková, J.; Mihaljevič, M.; Veselovský, F.; Bamba, O. Geochemistry, mineralogy, and isotope composition of Pb, Zn, and Cu in primary ores, gossan and barren ferruginous crust from the Perkoa base metal deposit, Burkina Faso. *J. Geochem. Explor.* **2016**, *168*, 49–64. [CrossRef]
5. Baggio, S.B.; Hartmann, L.A.; Massonne, H.J.; Theye, T.; Antunes, L.M. Silica gossan as a prospective guide for amethyst geode deposits in the Ametista do Sul mining district, Paraná volcanic province, southern Brazil. *J. Geochem. Explor.* **2015**, *159*, 213–226. [CrossRef]
6. Yesares, L.; Sáez, R.; Ruiz De Almodóvar, G.R.; Nieto, J.M.; Gómez, C.; Ovejero, G. Mineralogical evolution of the Las Cruces gossan cap (Iberian Pyrite Belt): From subaerial to underground conditions. *Ore Geol. Rev.* **2017**, *80*, 377–405. [CrossRef]
7. Laakso, K.; Rivard, B.; Rogge, D. Enhanced detection of gossans using hyperspectral data: Example from the Cape Smith Belt of northern Quebec, Canada. *ISPRS J. Photogramm.* **2016**, *114*, 137–150. [CrossRef]
8. Velasco, F.; Herrero, J.M.; Suárez, S.; Yusta, I.; Alvaro, A.; Tornos, F. Supergene features and evolution of gossans capping massive sulphide deposits in the Iberian pyrite Belt. *Ore Geol. Rev.* **2013**, *53*, 181–203. [CrossRef]
9. Costa, M.L.; Angélica, R.S.; Costa, N.C. The geochemical association Au–As–B-(Cu)−Sn−W in latosol, colluvium, lateritic iron crust, and gossan in Carajás, Brazil: Importance for primary ore identification. *J. Geochem. Explor.* **1999**, *67*, 33–49. [CrossRef]
10. Assawincharoenkij, T.; Hauzenberger, C.; Ettinger, K.; Sutthirat, C. Mineralogical and geochemical characterization of waste rocks from a gold mine in northeastern Thailand: Application for environmental impact protection. *Environ. Sci. Pollut. Res. Int.* **2018**, *25*, 3488–3500. [CrossRef]
11. Li, J.M. Experimental study on cyanide leaching of a gossan type gold and silver ore in Qinghai. *Gold Sci. Technol.* **2015**, *23*, 88–93. (In Chinese)

12. Celep, O.; Serbest, V. Characterization of an iron oxy/hydroxide (gossan type) bearing refractory gold and silver ore by diagnostic leaching. *Trans. Nonferrous Met. Soc. China* **2015**, *25*, 1286–1297. [CrossRef]
13. Zhang, R.; Lu, X.C.; Liu, H. Mineral phase transition and heavy metal release during the reduction of iron caps by Shewanella oneidensis MR-1. *Bull. Mineral. Rock Geochem.* **2015**, *34*, 316–322. (In Chinese)
14. Hankare, P.P.; Patil, R.P.; Jadhav, A.V.; Garadkar, K.M.; Sasikala, R. Enhanced photocatalytic degradation of methyl red and thymol blue using titania-alumina-zinc ferrite nanocomposite. *Appl. Catal. B* **2011**, *107*, 333–339. [CrossRef]
15. Jadhav, S.V.; Jinka, K.M.; Bajaj, H.C. Nanosized sulfated zinc ferrite as catalyst for the synthesis of nopol and other fine chemicals. *Catal. Today* **2012**, *198*, 98–105. [CrossRef]
16. Matinise, N.; Kaviyarasu, K.; Mongwaketsi, N.; Khamlich, S.; Kotsedi, L.; Mayedwa, N.; Maaza, M. Green synthesis of novel zinc iron oxide($ZnFe_2O_4$) nanocomposite via Moringa oleifera natural extract for electrochemical applications. *Appl. Surf. Sci.* **2018**, *446*, 66–73. [CrossRef]
17. Raut, S.D.; Awasarmol, V.V.; Ghule, B.G.; Shaikh, S.F.; Gore, S.K.; Sharma, R.P.; Pawar, P.P.; Mane, R.S. γ-irradiation induced zinc ferrites and their enhanced room-temperature ammonia gas sensing properties. *Mater. Res. Express* **2018**, *5*, 035702. [CrossRef]
18. Sai, R.; Kulkarni, S.D.; Yamaguchi, M.; Bhat, N.; Shivashankar, S.A. Integrated X-band inductor with a nanoferrite film core. *IEEE Magn. Lett.* **2017**, *8*, 3703904. [CrossRef]
19. Ebrahimi, M.; Raeisi Shahraki, R.; Seyyed Ebrahimi, S.A.; Masoudpanah, S.M. Magnetic properties of zinc ferrite nanoparticles synthesized by coprecipitation method. *J. Supercond. Nov. Magn.* **2014**, *27*, 1587–1592. [CrossRef]
20. Sangita, N.P.; Pratik, A.N.; Arvind, V.N.; Shankar, R.T.; Arun, V.B. Preparation techniques for zinc ferrites and their applications: A review. *Mater. Today Proc.* **2022**, *60*, 2194–2208.
21. Sun, X.M.; Li, Z.W.; Li, X.H.; Zhang, Z.J. Preparation of mesoporous zinc ferrite flame retardant with different scales and its performance in epoxy resin. *Polym. Test.* **2022**, *110*, 107549. [CrossRef]
22. Rachna, S.; Singh, N.B.; Agarwal, A. Preparation, characterization, properties, and applications of nano zinc ferrite. *Mater. Today Proc.* **2018**, *5*, 9148–9155. [CrossRef]
23. Yang, J.; Zhu, P.; Ma, S. Study on properties of gossan ore from zinc sulfide deposit. *J. Phys. Conf. Ser.* **2021**, *2044*, 012081. [CrossRef]

processes

MDPI

Article

Evaluation of Reconstruction Potential for Low-Production Vertical Wells of CBM in the Southern Qinshui Basin

Kaihong Xue, Beilei Sun * and Chao Liu

Department of Earth Science and Engineering, College of Mining Engineering, Taiyuan University of Technology, Taiyuan 030024, China; xuekaihong0110@link.tyut.edu.cn (K.X.); tyutliuchao@163.com (C.L.)
* Correspondence: sunbeilei@tyut.edu.cn

Abstract: Production practice has shown that not all low-production coalbed methane (CBM) wells can be reconstructed into high-production wells through secondary stimulation, so it is necessary and timely to establish an evaluation index system, form an evaluation method, and evaluate the reconstruction potential of low-production wells. Based on the development practice of CBM in the southern Qinshui Basin, this paper analyzes the influencing factors of low production in vertical wells from the aspects of coal and rock reservoir conditions, drilling and completion engineering, and drainage engineering. It is proposed that the evaluation of the reconstruction potential of low-production wells should focus on the quality of CBM resources, the difficulty of CBM desorption and diffusion, and the degree of damage to coal reservoirs caused by the initial reservoir stimulation. Twelve parameters, including gas content, gas saturation, reservoir pressure gradient, critical desorption–reservoir pressure ratio, and permeability, were systematically selected as evaluation indicators, and the grading reference values for each evaluation indicator were comparatively given. Then, a multi-factor comprehensive evaluation method for the reconstruction potential of low-production wells based on gray correlation analysis method was established. The reconstruction potential of low-production wells was divided into three levels: high, medium, and low. When reconstructing low-production wells, it is recommended to prioritize the low-production wells with high reconstruction potential, followed by those with medium reconstruction potential, while low-production wells with low reconstruction potential are not recommended for reconstruction. Finally, the evaluation method was used to evaluate the reconstruction potential of five low-production wells in a CBM block, and suggestions for the reconstruction order and reconstruction potential levels for each well were given.

Keywords: CBM vertical well; low production; Qinshui Basin; reconstruction potential evaluation

Citation: Xue, K.; Sun, B.; Liu, C. Evaluation of Reconstruction Potential for Low-Production Vertical Wells of CBM in the Southern Qinshui Basin. *Processes* **2023**, *11*, 1741. https://doi.org/10.3390/pr11061741

Academic Editors: Yidong Cai, Ping Gao, Yingfang Zhou and Quan Gan

Received: 10 April 2023
Revised: 30 May 2023
Accepted: 2 June 2023
Published: 7 June 2023

1. Introduction

The growing global demand for energy requires sustainable energy supplies. Coalbed methane (CBM) is a kind of energy which is efficient, clean, and pollution-free [1]. It can alleviate the energy crisis and the greenhouse effect [2–5]. Moreover, coalbed methane extraction can effectively reduce the occurrence of coal mine fire and explosion accidents [6]. Safe and efficient production of coalbed methane is of strategic significance to global sustainable development [4]. Coal and coalbed methane coupling coordinated exploitation is a key technology for the safe exploitation of both resources [7]. China has huge coalbed methane reserves, but the exploitation and utilization of coalbed methane resources fall behind the United States, Canada, Australia, and other countries [1,8]. The exploitation of coalbed methane in China is hampered by the "four lows": low effective producing ratio, low remaining producible reserves, low single-well gas production rate, and low exploitation profit [9]. These factors are seriously restricting China's sustainable CBM development. In the southern Qinshui Basin, there are many low-production wells that need to be reconstructed to increase individual well production. However, not every low-production well can achieve a significant increase in production through secondary stimulation. Therefore, it is very necessary and urgent to select the right wells for reconstruction.

Qinshui Basin, located in Shanxi province (Figure 1), exhibits desirable characteristics for a CBM field, including high permeability and gas content, making it especially suitable for exploration and commercial development, particularly in the southern basin where stable gas production has already been achieved via large-scale investment and development [10]. However, even areas that were properly developed, including the Fanzhuang block, present considerable proportions (32%) of low-production (<500 m^3/d) vertical wells due, in part, to the geological complexity of the coal reservoirs and other factors, including the compatibility of reservoir fracturing and drainage processes with geological conditions [11]. Other blocks that present more complex geological conditions, including Zhengzhuang, Gujiao, and Shizhuang, exhibit even higher proportions of low-production wells, while the proportion is only 10% in the Black Warrior Basin in the United States [5,12–15] (Figure 2). Such high proportions of low-production wells significantly hinder the development and efficient production of CBM reservoirs and drive away potential investors. To remedy this situation, researchers have conducted in-depth research on the underlying causes of low production and the stimulation operations that can be utilized and technological reconstruction of low-production wells of late; and comprehensive conclusions on the general causes of low productivity have been drawn by researchers by analyzing the geological conditions, engineering, drainage of CBM reservoirs, and hydraulic fracturing characteristics [11,16,17]. Furthermore, the following suggestions for targeted improvements have been made, including nitrogen-foam unblocking, refracturing with active water, acidization, pulse fracturing, and intelligent drainage. Significantly enhanced productivity has been exhibited by some enhanced wells [18–23]; however, the lack of considerations for reconstruction potential leads to the blind utilization of stimulation which, in turn, induced a major waste of manpower and resources as stimulation was applied on many unsuitable wells. Thus, this study contends that a method of selecting low-production wells with the most stimulation potential for successful enhancement is direly needed, which can enhance the efficiency of reconstructing low-production CBM wells, reduce the waste of human and material resources, and promote sustainable social and economic development.

Therefore, the objective of this study is to find a method to select low-production wells with the highest enhancement potential in order to achieve successful production increase. This method aims to improve the efficiency of wellbore reconstruction for low- production coalbed methane wells in the study area, reduce resource waste, and promote sustainable socio-economic development. However, it should be noted that in practical operations, there are various limitations to the on-site verification of the reconstruction potential of low-production wells, such as limitations related to funding, company planning, and negotiation and cooperation. Therefore, currently, on-site verification of the research results is not feasible.

Additionally, due to significant variations in reservoir characteristics, geological conditions, drilling, and fracturing techniques among different basins and even different blocks within the same basin, the selected evaluation indicators and their criteria in this thesis have certain limitations and are applicable only to the study area. Nonetheless, considering the practical significance of the research and its potential for future development, we are considering integrating the research model with software development to facilitate the evaluation of reconstruction potential. By developing a computer software, we can utilize the evaluation indicators and methods derived from the research to provide a convenient and efficient tool for calculating and outputting the evaluation results of reconstruction potential.

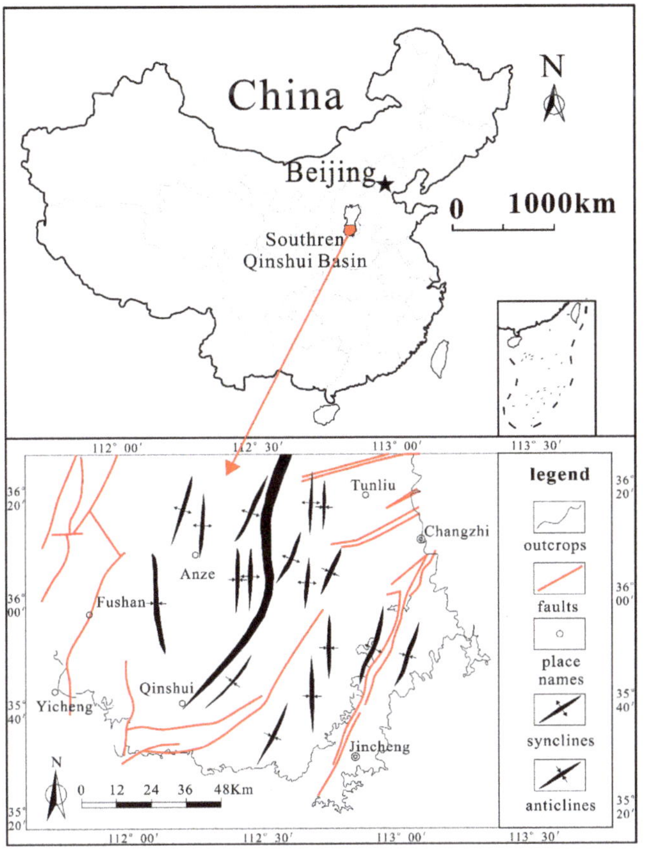

Figure 1. Location and structure outline of the southern Qinshui Basin.

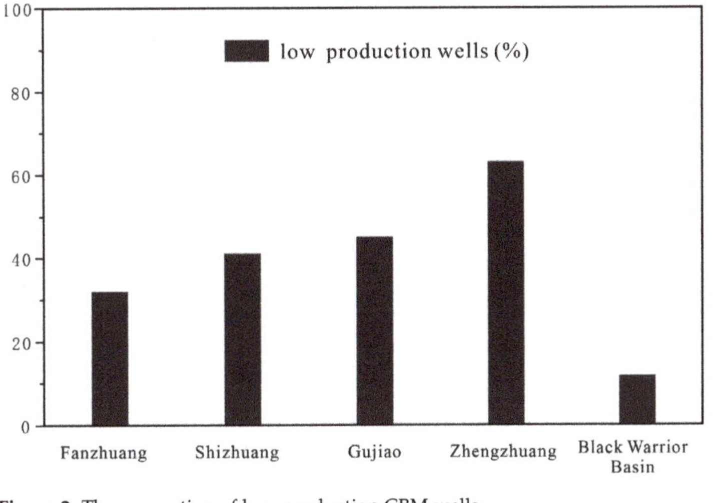

Figure 2. The proportion of low-production CBM wells.

2. Research Method

Reconstruction of low-production vertical CBM wells requires consideration of multiple factors, including reservoir characteristics, drilling and initial stimulation-induced damage, and drainage technology. The impact of these factors on the reconstruction process is complex. To establish an evaluation index system for the potential reconstruction of low-production CBM wells, statistical analysis was performed to identify the factors that affect CBM production. The classification standard of each evaluation index was then determined. Due to the varying degrees of impact on the potential for reconstruction, the weights of each evaluation index were determined through the application of the AHP model. Finally, the gray correlation analysis method is used for target evaluation, which completes the reconstruction potential evaluation through quantitative and qualitative analyses.

2.1. AHP Model for Determination of Weights of Each Evaluation Index

The Analytic Hierarchy Process (AHP) is a widely recognized multicriteria decision-making method commonly used to address complex decision problems [24]. Many scholars have applied AHP to evaluate oil and gas reservoirs and productivity [25–29]. The basic idea is to decompose the problem into multiple levels, which aids in the understanding of the essence of the problem more clearly and considers the influence of multiple factors, thus obtaining the relative weights of each influencing parameter. The hierarchical analysis method consists of several steps, as depicted in Figure 3, which are further detailed in Section 4.2.

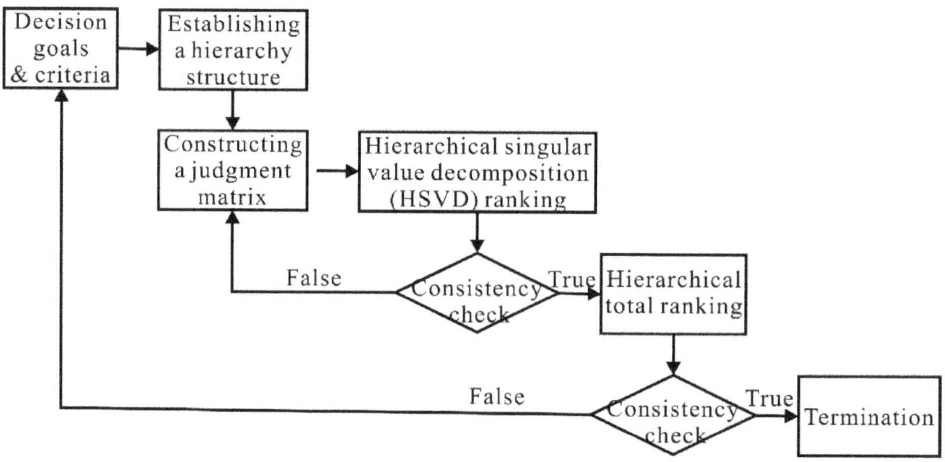

Figure 3. AHP flow chart.

2.2. The Gray Correlation Analysis Method

The gray correlation analysis (GRA) method, a quantitative analysis technique, is utilized to assess the interrelationships among different factors within a system. It has gained extensive application across various scientific domains due to its modeling, control, prediction, and decision-making capabilities [11,30–32]. In fact, the assessment of the reconstruction potential of low-production vertical CBM wells involves complex multi-objective and multi-factor approaches. Therefore, it is appropriate to view the evaluation system as a gray system and utilize the GRA methodology to evaluate the reconstruction potential of each individual well.

(1) Determine the comparison sequence and reference sequence

The comparative sequence is composed of the values of the evaluation parameters for the reconstruction potential of the well to be evaluated, denoted as: $X_i(k)$

$(k = 1, 2, \cdots, n;\ i > 1)$; the reference sequence is composed of the standards of the evaluation parameters for the reconstruction potential, denoted as: $Y_j(k)$ $(k = 1, 2, \cdots, m;\ j > 1)$.

(2) Normalization of data

As the dimensions of each parameter vary and there are magnitudes of differences between each parameter value, the initialization method is used to normalize the data to the interval of [0, 1] to eliminate the influences of dimensionality when comparing the parameters.

When $X_i(k)$ and $Y_j(k)$ are negatively correlated, the reference sequence is normalized according to the following Equation:

$$Y_j'(k) = \frac{Y_m(k) - Y_j(k)}{Y_m(k) - Y_1(k)}(j = 1, 2, \cdots, m; k = 1, 2, \cdots n) \tag{1}$$

The comparison sequence is normalized according to the following Equation:

$$X_i'(k) = \begin{cases} 1 & X_i(k) \le Y_1(k) \\ \frac{Y_m(k) - X_i(k)}{Y_m(k) - Y_1(k)} & Y_m(k) > X_i(k) > Y_1(k) \\ 0 & X_i(k) \ge Y_m(k) \end{cases} \tag{2}$$

When $X_i(k)$ and $Y_j(k)$ are positively correlated, the reference sequence is normalized according to the following Equation:

$$Y_j'(k) = \frac{Y_j(k) - Y_m(k)}{Y_1(k) - Y_m(k)}(j = 1, 2, \cdots, m; k = 1, 2, \cdots n) \tag{3}$$

The comparison sequence is normalized according to the following Equation:

$$X_i'(k) = \begin{cases} 1 & X_i(k) \ge Y_1(k) \\ \frac{X_j(k) - Y_m(k)}{Y_1(k) - Y_m(k)} & Y_m(k) < X_i(k) < Y_1(k) \\ 0 & X_i(k) \le Y_m(k) \end{cases} \tag{4}$$

(3) Calculation of correlation coefficient

A sequence matrix is formed by taking any sample vector from the comparison sequence $X_i(k)$ and comparing it with the reference sequence $Y_j(k)$. The absolute difference of the corresponding index factor between each vector in the sample and the comparison sequence is calculated as follows:

$$\Delta_{ij}(k) = \left|Y_j'(k) - X_i'(k)\right| \tag{5}$$

The matrix composed of absolute values $\Delta_{ij}(k)$ is analyzed to determine the $\underset{i}{max}\underset{j}{max}\left|Y_j'(k) - X_i'(k)\right|$ and $\underset{i}{min}\underset{j}{min}\left|Y_j'(k) - X_i'(k)\right|$ I in the matrix, which are denoted as Δ_1 and Δ_2, respectively. The calculation Equation for correlation coefficient $\xi_{ij}(k)$ is as follows:

$$\xi_{ij}(k) = \frac{\Delta_2 + \rho\Delta_1}{\Delta_{ij}(k) + \rho\Delta_1} \tag{6}$$

where ρ is the discrimination coefficient, which is generally taken as 0.5.

(4) Calculation of correlation degree

Based on the obtained correlation coefficients and the weight of each parameter, the degree of correlation between each vector in the reference and comparison sequences can be calculated according to the following Equation:

$$r_{ij} = \frac{1}{n}\sum\nolimits_{k=1}^{n} \omega_i \xi_{ij}(k) \tag{7}$$

where, ω_i is the weight of i parameter; r_{ij} is the correlation degree between i sequence in the comparison sequence and the j- level sequence in the reference sequence.

The maximum value of r_{ij} indicates the best correlation between comparison sequence i and the j- level sequence in the reference sequence. Based on this, the evaluation results can be used to determine whether a candidate well has high, medium, or low potential for reconstruction. When carrying out reconstruction, priority should be given to the low-production wells with high reconstruction potential, followed by those with medium potential, while wells with low potential are not recommended for reconstruction.

3. The Factors Controlling the Production of CBM

CBM is originally in a state of adsorption in reservoirs, and the majority of CBM wells will produce gas after water drainage and pressure drops occur, which indicates that CBM wells mostly lack the ability to produce free gas [3,33,34]. As such, the alteration of CBM state from adsorption to free via pressure reduction desorption, which is a physical desorption process in which adsorbed CBM molecules become increasingly active with decreases in "external pressure," and van der Waals forces are reduced enough for the change of state, is a major goal of enhancement operations [3]. The pressure at which the adsorbed gas on micropores begins to desorb is called the critical desorption pressure [33]. After CBM is desorbed, it flows from high- to low-pressure areas near the bottom of the well and into the wellbore [35]. When the bottom flow pressure drops to the minimum reservoir pressure (i.e., the abandonment pressure of the CBM well), the coalbed methane well stops producing gas. The factors that affect the development results of CBM wells mainly include coal reservoir characteristics, engineering, and drainage techniques [11,13,36–41].

3.1. Coal Reservoir Characteristics

(1) Gas Content

High gas content is a prerequisite for high-production CBM wells [42]. The coalbed gas content is between 10 and 22 m^3/t in the developed blocks in the south of Qinshui Basin. The coal seam in the Fanzhuang block has a gas content of less than 12 m^3/t, and the average daily gas production of vertical wells in this seam is less than 500 m^3/d. In contrast, the coal seams in Puchi and Guxian areas generally have a gas content of more than 20 m^3/t, and the average daily gas production per well in these areas exceeds 1500 m^3/d, with a maximum of 5000~6000 m^3/d [11].

(2) Gas Saturation

Coalbed gas saturation, which refers to the ratio of actual gas content to theoretical gas content, is related to pressure reduction and gas desorption [43]. The southern Qinshui Basin exhibits overall coalbed gas saturation in a wide range of 8.2–90% [36]. The Fanzhuang block has the lowest saturation, ranging from 8.2% to 43.8% and averaging only 24.8%, while that of the Guxian block is between 50% and 90%. There is a positive correlation between the gas production of CBM wells and CBM saturation. At higher gas saturations, free gas enters the wellbore earlier, which induces higher total gas production and productivity of CBM wells [44,45].

(3) Critical desorption–reservoir pressure ratio and recoverable coefficient

Critical desorption–reservoir pressure ratio reflects the difficulty of reducing the reservoir pressure to the critical desorption pressure. The greater the ratio, the easier it is to desorb and extract CBM. There is a good positive correlation between critical desorption–reservoir ratio and the daily gas production of CBM wells. The higher the ratio, the higher the proportion of high-production wells. For the high-rank coals, the CBM wells in the southern Qinshui Basin generally exhibit low productivity when gas saturation is less than 60% and critical desorption–reservoir pressure ratio is less than 0.55 [36].

The recoverable coefficient represents the proportion of gas that can be extracted from the reservoir when pressure drops from the critical desorption pressure to the abandonment pressure [5]. When other parameters remain unchanged, a higher recoverable coefficient indicates a greater amount of CBM that can be extracted per unit volume of coal. CBM reservoirs with higher gas saturations also have higher critical desorption pressures, which results in an overall higher gas production. Conversely, when the coal reservoir is an undersaturated gas reservoir, its critical desorption pressure is relatively low, resulting in a relatively low gas well production. For CBM reservoir potential evaluation, both the reservoir pressure and the critical desorption pressure must be high to be considered as a high-quality reservoir [46].

(4) Coal seam thickness

Coal seam thickness plays a significant role in CBM production, wherein with increased effective coal seam thickness, more CBM gathers at the wellbore, which consequently increases total CBM gas production. The above stated trend has been proven to hold true for the southern Qinshui Basin via statistical analyses of CBM wells of varying thicknesses. Additional numerical simulation calculations have also shown that, for a coal seam with a thickness of 10 m, peak daily and cumulative gas production over 10 years are 2.5 times those of a 4 m thick coal seam when other parameters remain unchanged [44].

(5) Permeability

Permeability, the ability of fluids to spread and diffuse through pore spaces within the coal matrix [21], is typically expressed in millidarcies (mD). Permeability is a key factor that affects CBM gas production and is primarily controlled by the modern tectonic stress field. Xiao et al. [47] established a new apparent permeability model, revealing the evolution of permeability under the combined action of effective stress and slippage in the full pore pressure range. The permeability decreases as the water content increases in wet coal under non-equilibrium state (Xiao et al., 2023). Based on comparative statistical analyses of CBM wells in the Zhengzhuang, Fanzhuang, Chengzhuang, and Zhengcun blocks, there is a power index relationship between CBM permeability and daily gas production [41,48].

(6) Geological Structure

Due to early inadequacies in the understanding of structural geology, some CBM wells were drilled near tensional normal faults or collapsed columns and presented with high water production and low gas production, which are due to inherently low gas content (less than 14 m^3/t) and adsorption saturation (desorption pressure less than 1.5 MPa) [17]. CBM wells that are too close to faults generally produce significant amounts of water and little to no gas.

The Zhengzhuang M well started production in 2012 [39]. Due to the ineffective sealing of the coal reservoir caused by the nearby tensional normal fault (50 m away), the reservoir has only produced 726 m^3 of gas, 3564 m^3 of water, and has a casing pressure of 0.05 MPa after more than 4 years of production [49]. Similar to the previous case, the N well was drilled near a collapsed column, resulting in a discontinuous coal reservoir with low gas content, producing only 763.1 m^3 of water after more than 4 years of production, with no gas being produced [49].

(7) In situ Stress

In situ stress is closely related to CBM permeability as well as the efficacy of the enhancement of CBM wells. In areas with low in situ stress, permeability is higher and gas production is high; whereas, in areas with high in situ stress, permeability is low, which leads to difficulties in gas production [50]. In deep burial depth (>1000 m) of coal seams, such as those of the Zhengzhuang block, the gas production from vertical wells is between 500 and 1000 m^3/d. In contrast, in low to medium burial depth (300–600 m) of coal seams, such as those in the Panzhuang block, gas production from vertical wells is comparatively high, with an average daily production of 2000 m^3/d.

(8) Reservoir Pressure

Reservoir pressure, also known as formation pressure or fluid pressure in reservoir fractures, can be determined via well testing that pushes the water and gas from coal fractures to the wellbore. In the Zhengzhuang block, under-pressured formations result in low-production wells, whereas over-pressured reservoirs result in high gas production [49].

(9) Coal texture

In coal seams with original texture, hydraulic fracturing creates cracks that initiate and propagate along the direction of maximum principal stress, which avoids the problem of fracture turning and allows for the formation of a simple, long, and straight crack [51]. However, in areas with fragmented or granulated coal, hydraulic fracturing cracks extend along preexisting fractures within the coal matrix, which results in short fracture lengths due to the high fracturing resistance and difficulty in crack extension [51]. The fracturing construction curve in the fragmented coal zone shows an abnormal increase or significant fluctuation of construction pressure, and the characteristics of gas production are high desorption pressure, low gas production, and the inability to sustain gas production [19]. A comparison of original-textured and fractured coal shows that during the adsorption of methane in coal, the total change in surface free energy of mylonitized coal is the largest, and its system energy is the most stable. Methane molecules adsorbed on mylonitized coal surface require greater potential energy to desorb [52]; therefore, the larger the proportion of mylonitized coal in the coal seam, the more unfavorable it is for the stimulation of low-production gas wells in the coal seam. Studies involving animals or humans, and other studies that require ethical approval, must list the authority that provided approval and the corresponding ethical approval code.

3.2. Engineering and Drainage Techniques

(1) Drilling engineering

In drilling operations, the improper usage of drilling fluid, poor cementing quality, and excessive wellbore enlargement rate can all limit the eventual productivity of vertical wells [39]. Adding a small amount of polymer to the drilling fluid can prevent wellbore collapse and improve carrying capacity; however, the adsorption and hydration effects of polymer can cause clay flocculation and swelling plugging, which can decrease coal reservoir permeability, making drainage and pressure reduction difficult and resulting in low gas production. Poor cementing quality can cause excessive or continuous water production, leading to long drainage times and difficulty in CBM production. Excessive wellbore enlargement rate can result in a thicker cement ring, affecting perforation quality, making fracturing construction difficult, and ultimately reducing CBM productivity.

(2) Fracturing engineering

Both the selection of fracturing fluid and method of fracturing can affect the productivity of vertical wells. The differences between the properties of fracturing and formation fluid are the main causes of permeability damage to the fractured reservoir matrix, wherein the invasion of fracturing fluid and physical–chemical changes of fracturing fluid in formation pores and throats can cause permeability damage of the fractured reservoir matrix. Therefore, the improper selection of fracturing fluid can result in significant permeability damage to the reservoir matrix, which directly affects gas production [53].

(3) Drainage technology

The reduction of pressure through drainage is a crucial step in the production of CBM wells, and, as such, the design of drainage systems has become an important factor affecting CBM production. The reasonable and effective control of casing pressure is a crucial aspect of the CBM drainage process. Both too rapid or too slow drainage modes will fail to establish an effective pressure relief range, thus preventing stable and high-yield production. Unreasonable casing pressure control and frequent system adjustments can cause liquid level oscillations, which can damage the reservoir, increase coal powder output, increase pump inspection frequency, and reduce drainage efficiency, resulting in low CBM production [16].

(4) Water production

Reservoir pressure drops will result from the inflow of free water into the wellbore, and afterwards, once the CBM desorption pressure is reached, the CBM wells will begin to produce gas. According to the currently available production data of CBM wells, drainage volume is relatively low and gas production is high when the CBM reservoir is in an area without free water. When the CBM well is in an area with a source of free water, the depressurization cone in the reservoir near the wellbore expands slowly, and a desorption pressure zone cannot be formed in a relatively large range, resulting in low gas production of the coalbed methane well [54]. In the field, a considerable proportion of gas wells exhibit little to no gas production and significant water drainage, which is caused by an overabundant water recharge. Consequently, effective depressurization cone cannot be established close to the wellbore. The statistical results of the water production and gas production of 20 trial wells in the Fanzhuang block show that wells have low production efficiency and do not produce gas when water production exceeds 1 and 10 m^3/d, respectively [11].

4. Results and Discussion

Experience obtained in the field indicates that only some low-production wells have the capacity to be reconstructed into high-production wells through secondary stimulation. Therefore, an evaluation index system and evaluation method must be established to evaluate the potential of low-production wells for further stimulation.

4.1. Evaluation Index

When selecting evaluation indexes, two primary aspects are taken into consideration: (1) The reservoir possesses high production potential. Considering that geological factors are crucial in determining the enrichment and production of coalbed methane and are intrinsic factors influencing well productivity, this study primarily focuses on evaluating geological factors near low-production wells in terms of their impact on reconstruction potential. The geological conditions of the reservoir are given significant consideration, particularly gas content, gas saturation, and recoverability; (2) Whether the primary stimulation has caused any damage to the reservoir and the extent of such damage.

Therefore, resource and geological structure (U_1), coal texture and gas saturation (U_2), in situ stress and permeability (U_3), and reservoir damage and water production (U_4) are selected to be the primary indicators for the evaluation of reconstruction potential of low-production well in the southern Qinshui Basin. To establish the grading standard for each evaluation index, Figure 4 presents scatter plots of gas production and various factors in the southern Qinshui Basin to facilitate analysis. Figure 4 depicts the classification criteria of gas content, permeability, critical desorption–reservoir pressure ratio, gas saturation, and distance to fault (date source: [11,13,36,55]). The remaining parameters were derived from field experience. The evaluation parameters and their classification standards for the reconstruction potential of low-production wells are shown in Table 1.

4.2. Determination of Weights for Each Evaluation Index (AHP Model)

(1) Establishing a hierarchical structure via extensive analyses: it is believed that resource and geological structure (U_1), coal texture and gas saturation (U_2), in situ stress and permeability (U_3), and reservoir damage and water production (U_4) are the primary indicators for the evaluation of the potential of low-production well reconstruction in the southern Qinshui Basin. Each primary indicator is further categorized into multiple secondary parameters, and the hierarchical structure of the evaluation index system is illustrated in Figure 5.

(2) Constructing the judgment matrix: to quantify each evaluation parameter, it is necessary to construct judgment matrices for the relevant evaluation indicators. The judgment matrix represents the relative importance of all parameters in this layer compared to a certain parameter in the previous layer. The element of judgment matrix is defined

as X_{ij}, namely $X_{ij} = U_i : U_j$. The judgment matrices are constructed based on 1–9 scale method (Table 2) to rank the importance of the evaluation indicators [24]. It should be noted that if the indexes are very close, the value of X_{ij} can be 1.1 to 1.9 [24]. Firstly, a judgement matrix is constructed for the primary evaluation parameters, U_1, U_2, U_3, and U_4, under the overall goal (parameters affecting the reconstruction potential); subsequently, judgement matrices for the secondary evaluation parameters are constructed with respect to the primary evaluation parameters U_1, U_2, U_3, and U_4, as shown in Table 3.

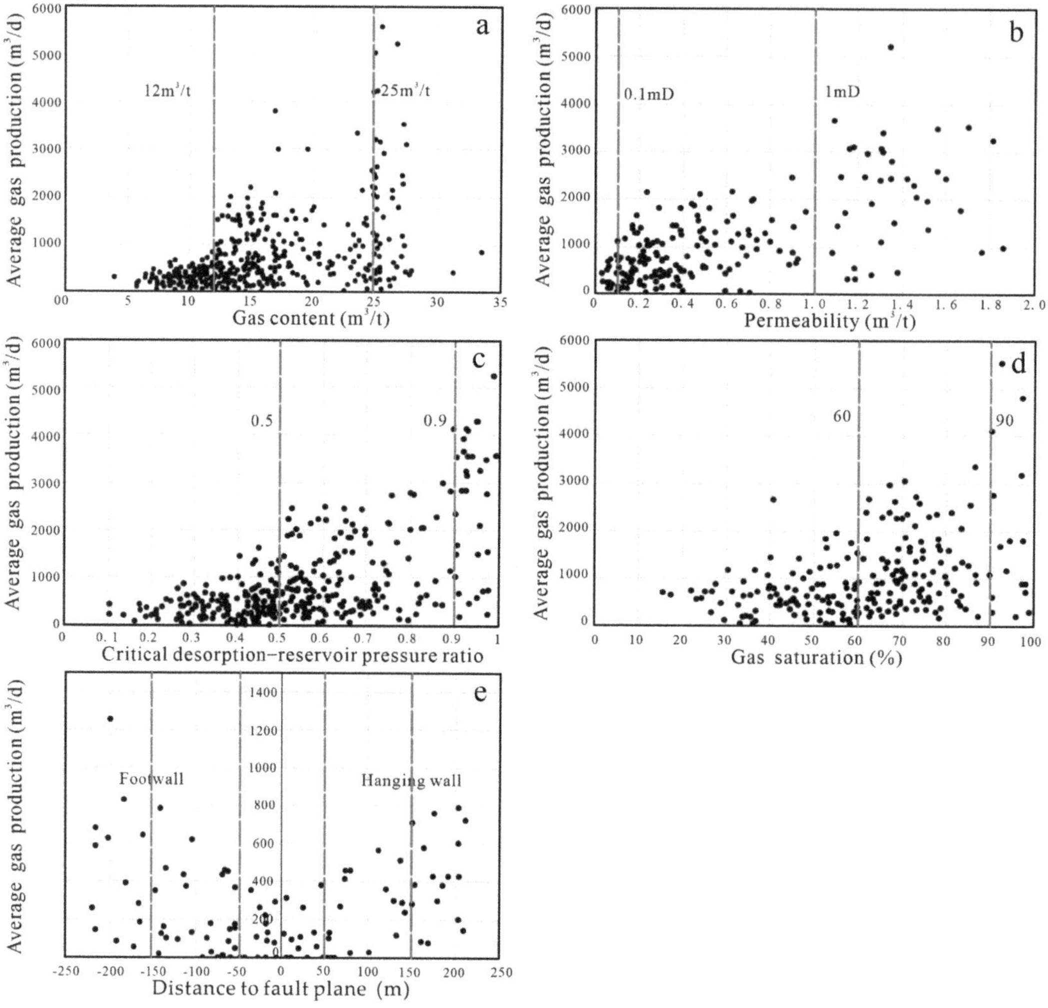

Figure 4. Scatter plots of gas production and various factors for the southern Qinshui Basin (**a**, gas content; **b**, permeability; **c**, critical desorption−reservoir pressure; **d**, gas saturation; **e**, distance to fault plane).

Table 1. Evaluation indexes and classification of reconstruction potential of low-production wells.

Primary Evaluation Factors	Secondary Evaluation Factors	Classification Standards Low	Medium	High
Resources and geological structure (U_1)	(U_{11}) Gas content (m^3/t)	≤12	12–25	≥25
	(U_{12}) Coal seam thickness (m)	≤3	3–6	≥6
	(U_{13}) Recoverable coefficient (%)	≤50	50–80	≥80
	(U_{14}) Distance to structure (m)	≤50	50–150	≥150
Coal texture and gas saturation (U_2)	(U_{21}) Critical desorption–reservoir pressure ratio (%)	≤0.5	0.5–0.9	≥0.9
	(U_{22}) Gas saturation (%)	≤60	60–90	≥90
	(U_{23}) * Coal texture (%)	≥50	20–50	≤20
In situ stress and permeability (U_3)	(U_{31}) Permeability (mD)	≤0.1	0.1–1	≥1
	(U_{32}) Pressure Gradient (kPa/m)	≤9.50	9.5–10	≥10
	(U_{33}) Burial depth (m)	≥1000	500–1000	≤500
Reservoir damage and water production (U_4)	(U_{41}) Water production rate (m^3/d)	≥5	0.5–5	≤0.5
	(U_{42}) Reservoir damage ratio (%)	≥40	20–40	≤20

* Coal texture is described as the proportion of mylonitized coal to the total coal thickness.

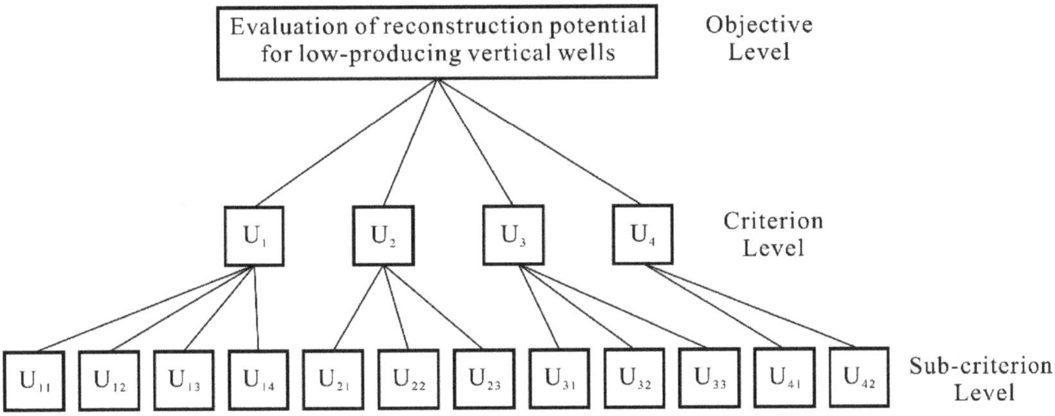

Figure 5. Hierarchy for the evaluation of reconstruction potential of low-production wells.

Table 2. The fundamental scale of absolute numbers.

Intensity of Importance	Definition	Explanation
1	Equal Importance	Two indexes contribute equally to the objective
2	Weak or slight	
3	Moderate importance	Experience and judgement slightly favor one index over another
4	Moderate plus	
5	Strong importance	Experience and judgement strongly favor one index over another
6	Strong plus	
7	Very strong	An index is favored very strongly over another; its dominance demonstrated in practice
8	Very, very strong	
9	Extreme importance	The evidence favoring one index over another is of the highest possible order of affirmation
1.1–1.9	If the indexes are very close	May be difficult to assign the best value, but when compared with other contrasting indexes, the size of the small numbers would not be too noticeable, yet they can still indicate the relative importance of the indexes

Table 3. Each parameter weight coefficient and random consistency ratio.

Evaluation Indicators and Matrices						Eigenvector W	Maximum Eigenvalue λ_{MAX}	Random Consistency Ratio (CR)
		U_1	U_2	U_3	U_4			
U	U_1	1	2	1	3	0.35	4.01	0.01
	U_2	0.5	1	2	2	0.28		
	U_3	1	0.5	1	3	0.26		
	U_4	0.33	0.5	0.33	1	0.11		
		U_{11}	U_{12}	U_{13}	U_{14}			
U_1	U_{11}	1	2	5	2	0.44	4.24	0.09
	U_{12}	0.5	1	3	2	0.28		
	U_{13}	0.2	0.33	1	0.33	0.08		
	U_{14}	0.5	0.5	3	1	0.20		
		U_{21}	U_{22}	U_{23}				
U_2	U_{21}	1	0.9	1		0.32	3.00	0
	U_{22}	1.1	1	1.1		0.36		
	U_{23}	1	0.9	1		0.32		
		U_{31}	U_{32}	U_{33}				
U_3	U_{31}	1	0.83	0.83		0.29	3.00	0
	U_{32}	1.2	1	1		0.36		
	U_{33}	1.2	1	1		0.36		
		U_{41}	U_{42}					
U_4	U_{41}	1	1			0.5	2.00	0
	U_{42}	1	1			0.5		

(3) Calculation of single-layer weight: after constructing the judgment matrix, the maximum eigenvalue and eigenvector of the matrix are obtained to determine the relative importance weight of the parameters in this layer with respect to a certain index in the previous layer. The calculation results are shown in Table 4.

Table 4. Evaluation index weights results.

Objective Level	Criterion Level	Weight	Sub-Criterion Level	Weight
Evaluation of transformation potential	Resources and geological structure (U_1)	0.35	U_{11}	0.15
			U_{12}	0.10
			U_{13}	0.03
			U_{14}	0.07
	Coal texture and gas saturation (U_2)	0.28	U_{21}	0.09
			U_{22}	0.10
			U_{23}	0.09
	In situ stress and permeability (U_3)	0.26	U_{31}	0.08
			U_{32}	0.09
			U_{33}	0.09
	Reservoir damage and water production (U_4)	0.11	U_{41}	0.06
			U_{42}	0.06

(4) Consistency check: the insurance of credibility and accuracy of calculation results necessitates the performance of a consistency check on the matrix, wherein, in this study, the random consistency ratio (*CR*) proposed by T.L. Saaty [24] is used to judge the consistency of the matrix. The Equation for calculating *CR* is:

$$CR = CI/RI \tag{8}$$

where CI is the consistency index and its value is $(\lambda_{max} - n)/(n-1)$, λ_{max} is the maximum eigenvalue of the matrix, and n is the order of the matrix. RI is the random consistency index of the same order, and its value can be obtained from Table 5 [24].

Table 5. The look-up table of RI.

Matrix Order	1	2	3	4	5	6	7	8	9
RI	0.00	0.00	0.58	0.9	1.12	1.24	1.32	1.41	1.42

If $CR < 10\%$, the judgment matrix has been deemed to have an acceptable consistency, whereas if $CR > 10\%$, the values and calculations need to be adjusted and revised until consistency is achieved. The results of the consistency check are shown in Table 2.

(5) Calculation of overall weights: The analytic hierarchy process requires layer-by-layer calculation of weights. The weights of the first level parameters in the table are relative to the target layer, while the weights of the second level parameters are relative to the first level parameters. The product of the two yields the total ranking of the second level factors relative to the target layer. The results show that the weight of gas content is the largest, and recoverable coefficient is the lowest in the evaluation of the reconstruction potential of low-production wells (Table 4).

4.3. Gray Correlation Degree

Several vertical wells have been stimulated by fracturing in the early-stage development of a certain coalbed methane development block, but some low-production wells still require repeated productivity enhancement. To this end, a productivity improvement potential assessment and optimization were conducted on five vertical wells in the block. Relevant data for the candidate wells were collected, and the results are shown in Table 6. The reconstruction potential evaluation criteria used are shown in Table 1.

Table 6. Table of relevant evaluation parameters of candidate low-production wells.

Well	U_{11} (m^3/t)	U_{12} (m)	U_{13} (%)	U_{14} (m)	U_{21}	U_{22} (%)	U_{23} (%)	U_{31} (mD)	U_{32} (kPa/m)	U_{33} (m)	U_{41} (m^3/d)	U_{42} (%)
1	28	5.5	60	200	0.9	90	10	1.5	10	500	1	20
2	30	6.5	65	170	0.8	70	10	0.5	9	800	0.8	15
3	14	5.5	50	70	0.6	70	5	0.5	10	1000	6	20
4	28	5.5	75	45	0.9	80	5	0.5	5	800	0.5	40
5	27	5.5	85	210	0.8	90	0	1.5	18	600	0.5	25

Among the selected 12 evaluation parameters, U_{23}, U_{33}, U_{41}, and U_{42} are negatively correlated with the potential for low-production well improvement, and Equations (1) and (2) are used to normalize these evaluation parameters in Tables 6 and 7. U_{11}, U_{12}, U_{13}, U_{14}, U_{21}, U_{22}, U_{31}, and U_{32} are positively correlated with the potential for low-production well reconstruction, and Equations (3) and (4) are used to normalize the relevant evaluation parameters in Tables 6 and 7, obtaining the following matrix sequences: $X_{5\times12}$, $Y_{3\times12}$.

$$X_{5\times12} = \begin{bmatrix} 0 & 0.1667 & 0.6667 & 0 & 0 & 0 & 0 & 0 & 0 & 0 & 0.3333 & 0 \\ 0 & 0 & 0.5 & 0 & 0.25 & 0.6667 & 0 & 0.5556 & 1 & 0.6 & 0.2 & 0 \\ 0.8462 & 0.1667 & 1 & 0.8 & 0.75 & 0.6667 & 0 & 0.5556 & 0 & 1 & 1 & 0 \\ 0 & 0.1667 & 0.1667 & 1 & 0 & 0.3333 & 0 & 0.5556 & 1 & 0.6 & 0 & 1 \\ 0 & 0.1667 & 0 & 0 & 0.25 & 0 & 0 & 0 & 0 & 0.2 & 0 & 0.25 \end{bmatrix} \quad (9)$$

$$Y_{3\times12} = \begin{bmatrix} 1 & 1 & 1 & 1 & 1 & 1 & 0 & 1 & 1 & 0 & 0 & 0 \\ 0.5 & 0.5 & 0.5 & 0.5 & 0.5 & 0.5 & 0.5 & 0.5 & 0.5 & 0.5 & 0.5 & 0.5 \\ 0 & 0 & 0 & 0 & 0 & 0 & 1 & 0 & 0 & 1 & 1 & 1 \end{bmatrix} \quad (10)$$

Table 7. Calculation value and evaluation results of gray correlation degree.

Candidate Well	Correlation Coefficient			Results
	Low	Medium	High	
1	0.5248	0.5375	0.7735	High
2	0.5837	0.6300	0.6117	Medium
3	0.6245	0.6093	0.5743	Low
4	0.5848	0.6050	0.6487	High
5	0.5011	0.5512	0.7636	High

Calculation of correlation coefficient and degree of association for No.1 candidate well is presented as an example. Firstly, the following evaluation sequence is established:

$$Y_{4\times 12} = \begin{bmatrix} 0 & 0.1667 & 0.1667 & 0 & 0 & 0 & 0 & 0 & 0 & 0 & 0.3333 & 0 \\ 1 & 1 & 1 & 1 & 1 & 1 & 0 & 1 & 1 & 0 & 0 & 0 \\ 0.5 & 0.5 & 0.5 & 0.5 & 0.5 & 0.5 & 0.5 & 0.5 & 0.5 & 0.5 & 0.5 & 0.5 \\ 0 & 0 & 0 & 0 & 0 & 0 & 1 & 0 & 0 & 1 & 1 & 1 \end{bmatrix} \quad (11)$$

Using Equation (5), the calculation was performed on the matrix (11), and the result is:

$$\Delta_{ij} = \begin{bmatrix} 1 & 0.8333 & 0.3333 & 1 & 1 & 1 & 0 & 1 & 1 & 0 & 0.3333 & 0 \\ 0.5 & 0.3333 & 0.1667 & 0.5 & 0.5 & 0.5 & 0.5 & 0.5 & 0.5 & 0.5 & 0.1667 & 0.5 \\ 0 & 0.1667 & 0.6667 & 0 & 0 & 0 & 1 & 0 & 0 & 1 & 0.6667 & 1 \end{bmatrix} \quad (12)$$

Using Equations (6) and (7), the correlation degrees between the evaluation factors of the first candidate well and the low-production well reconstruction potential evaluation criteria can be obtained as follows: $r_1 = (0.5251,\ 0.5290,\ 0.7651)$. Following the above principles and steps, the correlation degrees for the other four candidate wells can be obtained. The results are shown in Table 7.

According to the data in Table 7, candidate well 1 exhibits gray correlation coefficients of 0.5251, 0.5290, and 0.7651 with respect to low, medium, and high reconstruction potential, respectively. The evaluation of reconstruction potential for candidate well 1 shows the closest correlation with high reconstruction potential, indicating that candidate well 1 possesses a significant potential for reconstruction. Similarly, we can determine the reconstruction potential levels for candidate wells 4 and 5 are also high, while well 2 is medium and well 3 is low.

5. Conclusions

(1) Through extensive data analysis and field practical experience, the evaluation of reconstruction potential for low-production wells in the southern Qinshui Basin focuses on geological conditions and the degree of damage caused by initial fracturing to the coal reservoir. For this purpose, a comprehensive set of 12 indicators and their corresponding grading standards have been established to evaluate the reconstruction potential. These indicators encompass crucial factors, such as gas content, coal seam thickness, recoverable coefficient, distance to structure, critical desorption–reservoir pressure ratio, gas saturation, coal texture, permeability, pressure gradient, burial depth, water production rate, and reservoir damage ratio.

(2) The weights for each evaluation indicator were obtained using the Analytic Hierarchy Process (AHP). The results indicate that the gas content has the highest weight, with a value of 0.15. On the other hand, the recovery coefficient has the lowest weight, with a value of 0.03. The weights for the remaining indicators fall between these two values, reflecting their relative importance in the evaluation process.

(3) The reconstruction potential of five wells was evaluated using the gray correlation analysis method. The results indicate that candidate wells 1, 4, and 5 have high

reconstruction potential, candidate well 2 has a moderate reconstruction potential, and candidate well 3 has a low reconstruction potential.

(4) The developed evaluation method for reconstruction potential is primarily applicable to the Qinshui Basin. Due to significant differences in geological characteristics and coal reservoir conditions in other regions, the applicability of this evaluation method in other areas requires further research and validation.

Author Contributions: Conceptualization, B.S. and C.L.; methodology, B.S.; validation, K.X., B.S. and C.L.; formal analysis, K.X.; investigation, C.L.; resources, B.S.; data curation, B.S.; writing—original draft preparation, K.X.; writing—review and editing, K.X.; supervision, C.L.; project administration, B.S.; funding acquisition, B.S. All authors have read and agreed to the published version of the manuscript.

Funding: This research was funded by Shanxi Province Science and Technology Major Project, grant number 20191102001.

Institutional Review Board Statement: Not applicable.

Informed Consent Statement: Not applicable.

Data Availability Statement: All relevant data are within the paper.

Conflicts of Interest: The authors declare no conflict of interest.

References

1. Mastalerz, M.; Drobniak, A. 5–Coalbed Methane: Reserves, Production, and Future Outlook. In *Future Energy*, 3rd ed.; Elsevier: Amsterdam, The Netherlands, 2020; pp. 97–109. [CrossRef]
2. Wen, S.; Zhou, K.; Lu, Q. A discussion on CBM development strategies in China: A case study of PetroChina Coalbed Methane Co., Ltd. *Nat. Gas Ind. B* **2019**, *6*, 610–618. [CrossRef]
3. Gilman, A.; Beckie, R. Flow of Coal-Bed Methane to a Gallery. *Transp. Porous Media* **2000**, *41*, 1–16. [CrossRef]
4. Altowilib, A.; AlSaihati, A.; Alhamood, H.; Alafnan, S.; Alarifi, S. Reserves Estimation for Coalbed Methane Reservoirs: A Review. *Sustainability* **2020**, *12*, 10621. [CrossRef]
5. Mu, F.; Zhong, W.; Zhao, X.; Che, C.; Chen, Y.; Zhu, J.; Wang, B. Strategies for the development of CBM gas industry in China. *Nat. Gas Ind. B* **2015**, *2*, 383–389. [CrossRef]
6. Bosikov, I.I.; Martyushev, N.V.; Klyuev, R.V.; Savchenko, I.A.; Kukartsev, V.V.; Kukartsev, V.A.; Tynchenko, Y.A. Modeling and Complex Analysis of the Topology Parameters of Ventilation Networks When Ensuring Fire Safety While Developing Coal and Gas Deposits. *Fire* **2023**, *6*, 95. [CrossRef]
7. Zou, Q.; Chen, Z.; Cheng, Z.; Liang, Y.; Xu, W.; Wen, P.; Zhang, B.; Liu, H.; Kong, F. Evaluation and intelligent deployment of coal and coalbed methane coupling coordinated exploitation based on Bayesian network and cuckoo search. *Int. J. Min. Sci. Technol.* **2022**, *32*, 1315–1328. [CrossRef]
8. Pashin, J.C. Variable gas saturation in coalbed methane reservoirs of the Black Warrior Basin: Implications for exploration and production. *Int. J. Coal Geol.* **2010**, *82*, 135–146. [CrossRef]
9. Liang, W.; Yan, J.; Zhang, B.; Hou, D. Review on Coal Bed Methane Recovery Theory and Technology: Recent Progress and Perspectives. *Energy Fuels* **2021**, *35*, 4633–4643. [CrossRef]
10. Zhu, Q.; Zuo, Y.; Yang, Y. How to solve the technical problems in CBM development: A case study of a CBM gas reservoir in the southern Qinshui Basin. *Nat. Gas Ind. B* **2015**, *2*, 277–281. [CrossRef]
11. Tao, S.; Tang, D.; Xu, H.; Gao, L.; Fang, Y. Factors controlling high-yield coalbed methane vertical wells in the Fanzhuang Block, Southern Qinshui Basin. *Int. J. Coal Geol.* **2014**, *134–135*, 38–45. [CrossRef]
12. Karacan, C.Ö. Production history matching to determine reservoir properties of important coal groups in the Upper Pottsville formation, Brookwood and Oak Grove fields, Black Warrior Basin, Alabama. *J. Nat. Gas Sci. Eng.* **2013**, *10*, 51–67. [CrossRef]
13. Peng, C.; Zou, C.; Zhou, T.; Li, K.; Yang, Y.; Zhang, G.; Wang, W. Factors affecting coalbed methane (CBM) well productivity in the Shizhuangnan block of southern Qinshui basin, North China: Investigation by geophysical log, experiment and production data. *Fuel* **2017**, *191*, 427–441. [CrossRef]
14. Li, Z.; Liu, D.; Wang, Y.; Si, G.; Cai, Y.; Wang, Y. Evaluation of multistage characteristics for coalbed methane desorption-diffusion and their geological controls: A case study of the northern Gujiao Block of Qinshui Basin, China. *J. Pet. Sci. Eng.* **2021**, *204*, 108704. [CrossRef]
15. Zhang, P.; Meng, Z.; Jiang, S.; Chen, X. Characteristics of in-situ stress distribution in Zhengzhuang Region, Southern Qinshui Basin, China and its stress path during depletion. *Eng. Geol.* **2020**, *264*, 105413. [CrossRef]

16. Yang, G.; Tang, S.; Hu, W.; Song, Z.; Zhang, S.; Xi, Z.; Wang, K.; Yan, X. Analysis of abnormally high water production in coalbed methane vertical wells: A case study of the Shizhuangnan block in the southern Qinshui Basin, China. *J. Pet. Sci. Eng.* **2020**, *190*, 107100. [CrossRef]
17. Wang, Q.; Su, X.; Su, L.; Zhou, F. CBM geological characteristics and exploration potential in the Sunan Syncline block, southern north China basin. *J. Pet. Sci. Eng.* **2020**, *186*, 106713. [CrossRef]
18. Zhao, L.; Chen, X.; Zou, H.; Liu, P.; Liang, C.; Zhang, N.; Li, N.; Luo, Z.; Du, J. A review of diverting agents for reservoir stimulation. *J. Pet. Sci. Eng.* **2020**, *187*, 106734. [CrossRef]
19. Huang, Q.; Liu, S.; Wang, G.; Wu, B.; Zhang, Y. Coalbed methane reservoir stimulation using guar-based fracturing fluid: A review. *J. Nat. Gas Sci. Eng.* **2019**, *66*, 107–125. [CrossRef]
20. Keshavarz, A.; Badalyan, A.; Johnson, R.; Bedrikovetski, P. Improving the efficiency of hydraulic fracturing treatment in CBM reservoirs by stimulating the surrounding natural fracture system. *APPEA J.* **2015**, *55*, 351–360. [CrossRef]
21. Keshavarz, A.; Badalyan, A.; Carageorgos, T.; Bedrikovetsky, P.; Johnson, R., Jr. Stimulation of coal seam permeability by micro-sized graded proppant placement using selective fluid properties. *Fuel* **2015**, *144*, 228–236. [CrossRef]
22. Sampath, K.H.S.M.; Perera, M.S.A.; Ranjith, P.G.; Matthai, S.K.; Rathnaweera, T.; Zhang, G.; Tao, X. CH_4–CO_2 gas exchange and supercritical CO_2 based hydraulic fracturing as CBM production-accelerating techniques: A review. *J. CO_2 Util.* **2017**, *22*, 212–230. [CrossRef]
23. Danesh, N.N.; Zhao, Y.; Teng, T.; Masoudian, M.S. Prediction of interactive effects of CBM production, faulting stress regime, and fault in coal reservoir: Numerical simulation. *J. Nat. Gas Sci. Eng.* **2022**, *99*, 104419. [CrossRef]
24. Basak, I.; Saaty, T. Group decision making using the analytic hierarchy process. *Math. Comput. Model.* **1993**, *17*, 101–109. [CrossRef]
25. Ni, X.; Tan, X.; Wang, B.; Fu, X. An evaluation method for types of low-production coalbed methane reservoirs and its application. *Energy Rep.* **2021**, *7*, 5305–5315. [CrossRef]
26. Yao, Y.; Liu, D.; Tang, D.; Huang, W.; Tang, S.; Chen, Y. A Comprehensive Model for Evaluating Coalbed Methane Reservoirs in China. *Acta Geol. Sin. Engl. Ed.* **2010**, *82*, 1253–1270. [CrossRef]
27. Yao, Y.; Liu, D.; Tang, D.; Tang, S.; Che, Y.; Huang, W. Preliminary evaluation of the coalbed me-thane production potential and its geological controls in the Weibei Coalfield, southeastern Ordos Basin, China. *Int. J. Coal Geol.* **2009**, *78*, 1–15. [CrossRef]
28. Cai, Y.; Liu, D.; Yao, Y.; Li, J.; Qiu, Y. Geological controls on prediction of coalbed methane of No. 3 coal seam in Southern Qinshui Basin, North China. *Int. J. Coal Geol.* **2011**, *88*, 101–112. [CrossRef]
29. Tapia, J.F.D.; Promentilla, M.A.B.; Tseng, M.-L.; Tan, R.R. Screening of carbon dioxide utilization options using hybrid Analytic Hierarchy Process-Data Envelopment Analysis method. *J. Clean. Prod.* **2017**, *165*, 1361–1370. [CrossRef]
30. Lin, S.J.; Lu, I.J.; Lewis, C. Grey relation performance correlations among economics, energy use and carbon dioxide emission in Taiwan. *Energy Policy* **2017**, *35*, 1948–1955. [CrossRef]
31. Wang, J.J.; Jing, Y.Y.; Zhang, C.F.; Zhang, X.T.; Shi, G.-H. Integrated evaluation of distributed triple-generation systems using improved grey incidence approach. *Energy* **2008**, *33*, 1427–1437. [CrossRef]
32. ASLAN, N. Use of the grey analysis to determine optimal oil agglomeration with multiple performance characteristics. *Fuel* **2013**, *109*, 109373–109378. [CrossRef]
33. Kiani, A.; Sakurovs, R.; Grigore, M.; Sokolova, A. Gas sorption capacity, gas sorption rates and nanoporosity in coals. *Int. J. Coal Geol.* **2018**, *200*, 77–86. [CrossRef]
34. Salmachi, A.; Yarmohammadtooski, Z. Production data analysis of coalbed methane wells to estimate the time required to reach to peak of gas production. *Int. J. Coal Geol.* **2015**, *141–142*, 33–41. [CrossRef]
35. Pillalamarry, M.; Harpalani, S.; Liu, S. Gas diffusion behavior of coal and its impact on production from coalbed methane reservoirs. *Int. J. Coal Geol.* **2011**, *86*, 342–348. [CrossRef]
36. Li, P.; Zhang, X.; Li, J.; Zhao, J.; Huang, J.; Zhang, S.; Zhou, S. Analysis of the Key Factors Affecting the Productivity of Coalbed Methane Wells: A Case Study of a High-Rank Coal Reservoir in the Central and Southern Qinshui Basin, China. *ACS Omega* **2020**, *5*, 28012–28026. [CrossRef]
37. Chen, S.; Tao, S.; Tian, W.; Tang, D.; Zhang, B.; Liu, P. Hydrogeological control on the accumulation and production of coalbed methane in the Anze Block, southern Qinshui Basin, China. *J. Pet. Sci. Eng.* **2021**, *198*, 108138. [CrossRef]
38. Sun, F.; Wang, B.; Li, M.; Liang, H. Major geological factors controlling the enrichment and high yield of coalbed methane in the southern Qinshui Basin. *Acta Pet. Sin.* **2014**, *35*, 1070–1079. [CrossRef]
39. Tao, S.; Pan, Z.; Tang, S.; Chen, S. Current status and geological conditions for the applicability of CBM drilling technologies in China: A review. *Int. J. Coal Geol.* **2018**, *202*, 95–108. [CrossRef]
40. Wang, H.; Yao, Y.; Liu, D.; Cai, Y.; Yang, Y.; Zhou, S. Determination of the degree of coal deformation and its effects on gas production in the southern Qinshui Basin, North China. *J. Pet. Sci. Eng.* **2022**, *216*, 110746. [CrossRef]
41. Sang, S.; Liu, H.; Li, Y.; Li, M.; Li, L. Geological controls over coal-bed methane well production in southern Qinshui basin. *Procedia Earth Planet. Sci.* **2009**, *1*, 917–922. [CrossRef]
42. Karthikeyan, G.; Chand, J.; Chatterjee, R. Impact of geomechanics in coal bed methane development and production, Barakar coals in central India. *J. Pet. Sci. Eng.* **2020**, *194*, 194107515. [CrossRef]
43. Perera, M.S.A.; Ranjith, P.G.; Peter, M. Effects of saturation medium and pressure on strength parameters of Latrobe Valley brown coal: Carbon dioxide, water and nitrogen saturations. *Energy* **2011**, *36*, 6941–6947. [CrossRef]

44. Zhou, R.; Zuo, Y.; Wang, Y.; Zhang, X.; Shuai, Z. Study on Optimization and evaluation indexes of coalbed methane development units in south of Qinshui Basin. In Proceedings of the 5th National Symposium on Efficient Development Technology of Natural Gas Reser, Nanchang, China, 1 March 2014.
45. Zheng, A.; Wang, X.; Wang, X.; Wu, M.; Yuan, Y. The Effect of Coal Seam Gas saturation on CBM Well Productivity—A Case Study of Central Region of Hedong Area. *Procedia Eng.* **2011**, *26*, 1205–1213. [CrossRef]
46. Han, S.-J.; Sang, S.-X.; Duan, P.-P.; Zhang, J.-C.; Xiang, W.-X.; Xu, A. The effect of the density difference between supercritical CO_2 and supercritical CH_4 on their adsorption capacities: An experimental study on anthracite in the Qinshui Basin. *Pet. Sci.* **2022**, *19*, 1516–1526. [CrossRef]
47. Xiao, Z.; Wang, C.; Wang, G.; Jiang, Y.; Yu, J. An improved apparent permeability model considering full pore pressure range, variable intrinsic permeability and slippage coefficient. *Int. J. Min. Sci. Technol.* **2022**, *32*, 1233–1244. [CrossRef]
48. Zhu, H.; Tang, X.; Liu, Q.; Liu, S.; Zhang, B.; Jiang, S.; McLennan, J.D. Permeability stress-sensitivity in 4D flow-geomechanical coupling of Shouyang CBM reservoir, Qinshui Basin, China. *Fuel* **2018**, *232*, 817–832. [CrossRef]
49. Wupeng, Z.H.A.O.; Chunchun, L.I.U.; Xingwei, S.H.E.N.; He, P.E.N.G.; Hongtao, H.U.O. Study on the stimulation technologies for low-yield CBM wells in Zhengzhuang block. *Oil Drill. Prod. Technol.* **2017**, *39*, 491–494. [CrossRef]
50. Meng, Z.; Zhang, J.; Wang, R. In-situ stress, pore pressure and stress-dependent permeability in the Southern Qinshui Basin. *Int. J. Rock Mech. Min. Sci.* **2011**, *48*, 122–131. [CrossRef]
51. McMillan, D.W.; Palanyk, V.S. CBM: Fracture Stimulation—An Australian Experience. In Proceedings of the SPE Annual Technical Conference and Exhibition, Anaheim, CA, USA, 11–14 November 2007. [CrossRef]
52. Jia, H.; Hu, Q.; Qi, K.; Liu, C.; Fan, B.; He, J. Reasons of low yield and stimulation measures for vertical CBM wells in high-rank coal. Coal Geology & Exploration. *Coal Geol. Explor.* **2019**, *47*, 104–110. [CrossRef]
53. Zhang, L.; Shan, B.; Zhao, Y. Production performance laws of vertical wells by volume fracturing in CBM reservoirs. *Nat. Gas Ind. B* **2017**, *4*, 189–196. [CrossRef]
54. Jamshidi, M.; Jessen, K. Water production in enhanced coalbed methane operations. *J. Pet. Sci. Eng.* **2012**, *92–93*, 56–64. [CrossRef]
55. Ni, X.; Wang, Y.; Han, W.; Liu, D.; Li, Y.; Tao, C.; Gao, X.; Zhao, S. Division and application of development geological units for coalbed methane. *J. China Coal Soc.* **2020**, *45*, 2562–2574. [CrossRef]

processes

Article

Characteristics and Controlling Role in Hydrocarbon Accumulation of Strike-Slip Faults in the Maigaiti Slope

Bin Wang, Changchao Chen *, Jiangwei Shang, Ming Lei, Wenhui Zhu, Yang Qu, Di Sun, Chunyan Sun and Li Li

Exploration and Development Research Institute, PetroChina Tarim Oilfield Company, Korla 841000, China
* Correspondence: chencc-tlm@petrochina.com.cn

Abstract: The Maigaiti Slope is a significant oil-gas-bearing field in the Tarim Basin. Based on 3D and 2D seismic data, systematic interpretation, stage sorting and genetic analysis of strike-slip faults in the Maigaiti Slope were carried out for the first time. The relationship between strike-slip faults and hydrocarbon accumulation was studied in combination with the fine dissection of hydrocarbon reservoirs. The study suggested that: (1) Staging and segmentation characteristics of strike-slip faults are evidently presented in the Maigaiti Slope. According to active periods, strike-slip faults can be divided into early Caledonian period, late Caledonian period, Hercynian period, and Himalayan period. According to plane distribution characteristics, strike-slip faults can be divided into the west Maigaiti Segment, mid-Maigaiti Segment, Madong Segment, and Bachu Segment. The main active periods and plane distribution of strike-slip faults in different sections are remarkably different. This analysis suggests that it is the response to multi-period and multi-directional tectonic movements, which are primarily dominated by the migration and evolution of the Hetian paleo-uplifts. (2) The coupling relationship between the active period of strike-slip faults and the trap forming period is the key to hydrocarbon accumulation in the Carboniferous–Ordovician, which determines the petroleum properties and enrichment horizon of the Cambrian post-salt system; medium-heavy oil is in the Caledonian period, light oil in the Indosinian period, and dry gas in the Himalayan period.

Keywords: strike-slip fault; active period; hydrocarbon accumulation; Maigaiti Slope; Tarim Basin

Citation: Wang, B.; Chen, C.; Shang, J.; Lei, M.; Zhu, W.; Qu, Y.; Sun, D.; Sun, C.; Li, L. Characteristics and Controlling Role in Hydrocarbon Accumulation of Strike-Slip Faults in the Maigaiti Slope. *Processes* **2023**, *11*, 1049. https://doi.org/10.3390/pr11041049

Academic Editors: Ping Gao, Yidong Cai, Yingfang Zhou, Quan Gan and Qingbang Meng

Received: 5 January 2023
Revised: 13 February 2023
Accepted: 27 March 2023
Published: 31 March 2023

1. Introduction

The Maigaiti Slope has always been considered as one of the important areas of exploration for petroleum in the Tarim Basin. So far, the Bashituopu oil field, Yubei 1-Madong 3 oil field, Niaoshan gas reservoir, Hetianhe gas field, Luosi 2 gas reservoir, and the Yasongdi gas reservoir have been discovered in the Carboniferous, Ordovician and other strata, and it shows superior geological conditions for petroleum formation. However, the resources of the discovered oil and gas reservoirs are small in scale and distributed in scattered spots on the plane. Oil and gas exploration has not been carried out on a large scale. The Maigaiti Slope is in the early exploration stage, at present. The hydrocarbon accumulation pattern and controlling factors are unclear, which are the main reasons restricting the discovery of oil and gas at scale in this area.

In recent years, many scholars have continuously explored the petroleum entrapment, hydrocarbon accumulation periods, paleo-uplift evolution, reservoir evolution, and fault activity characteristics of the Maigaiti Slope. It is believed that the oil and gas sources of the discovered reservoirs are mainly from the lower Cambrian–Precambrian marine source rocks [1–9]. The Maigaiti Slope mainly experienced oil charging in the late Hercynian period and natural gas charging in the Xishan period, which has the characteristics of early oil and late gas accumulation [10–15]. The evolution and fault activity of paleo-uplift show certain characteristics of controlling the circle, reservoir, and accumulation [16–21]. These studies have effectively promoted the understanding of oil and gas geological conditions in this area. However, due to limited and scattered drilling and seismic data in the study area, the previous research on petroleum geology in the slope area mainly focus on comprehensive and general understanding, and lack in-depth research on a single factor. In particular, few systematic studies have been carried out on the oil and gas drainage system, which leads to a lack of in-depth understanding of the laws of oil and gas accumulation in the Maigaiti Slope, and, to a certain extent, restricts the selection of oil and gas exploration zones and strata in this area. Therefore, combined with previous research results, this paper makes full use of drilling and seismic data from recent years, takes drilling and 3D continuous seismic data in the three zones of Maixi, Maizhong, and Yubei as the entry point, and combines 2D seismic data with whole-layer detailed seismic interpretation; 3D seismic coherence attribute analysis; regional tectonic evolution analysis; and fracture formation time study etc. The strike-slip fault system in the study area was systematically studied for the first time, and its formation mechanism was discussed. It is clear that the coupling relationship between active stages of strike-slip faults and trap formation stages is the key controlling factor for hydrocarbon accumulation in the Ordovician and Carboniferous, and determines the hydrocarbon properties and enrichment strata of the upper salt exploration prospects of the Cambrian. These studies have greatly enriched understanding of the petroleum geological laws of the Maigaiti Slope, and will provide new ideas for the selection of petroleum exploration zones and targets.

2. The Geological Outline

The Maigaiti Slope is located in the front of the Southwest Depression of the Tarim Basin, adjacent to the Xikeer structural belt in the north, connected to the Bachu Uplift by the Selibuya mazatake fault in the northeast, and bounded by the Kashi Depression in the west and Yecheng Depression in the southwest. It is a slope structure with rapid northeast tilting in the late Neogene. Its exploration area is about 40,000 square kilometers (Figure 1). The Maigaiti Slope strikes from northwest to southeast. Influenced by multi-stage tectonic activities, the Maigaiti Slope has east–west segmentation in plane. From west to east, several fault structural belts formed, such as the Kuoshilake structural belt, Qunkuqiake structural belt, Selibuya structural belt, Luositage structural belt, Luonan-Niaoshan structural belt, and the Mazhatage structural belt.

The Maigaiti Slope inclines to the southwest, where deposited strata from PreCambrian to Neogene without Mesozoic, and Silurian and Devonian developed locally only. According to seismic and drilling data, the strata are thicker in the west than in the east. The thickness of the lower Paleozoic is about 4000 to 5200 m, and the thickness of the upper Paleozoic is about 900 to 2500 m. The Middle and Upper Ordovician, Silurian and Devonian were overlaid on the older weathering crust, and peaked from north to south under the control of paleo-uplift. For example, in the western Maxi area, the Silurian deposited on the weathering crust of the Ordovician Yingshan Formation, and in the eastern Manan area, the lower Carboniferous mudstone directly covered on the weathering crust of the Yingshan Formation. The Mesozoic strata were almost denudated due to the overall uplift in the Mesozoic. In the Cenozoic, overall tectonic subsidence and inversion developed, causing the tectonic tendency changes from northwest to southeast. Therefore, the Cenozoic was distributed along the whole slope area, but its thickness increased from north to south, varying from 2000 to 9000 m.

The history of oil and gas exploration on the Maigaiti Slope dates back to the 1960s. Through the petroleum exploration of several generations of geologists, a basic understanding has formed, such as multiple types of reservoirs, multiple series of strata containing oil and gas, and multi-layer vertical accumulation models. The Maigaiti Slope can be divided into over-salt and sub-salt exploration fields by the boundary of thick paste salt of the Middle Cambrian. The reservoir and cap assemblages of the Middle and Lower Cambrian developed in the sub-salt, and the favorable reservoirs of the Ordovician, Devonian, Carboniferous and Permian developed in the over-salt. The main source rocks of the Lower Cambrian—PreCambrian developed, and multi-stage strike faults communicated source rocks and reservoirs. This is an important channel for hydrocarbon migration, as well as adjustment.

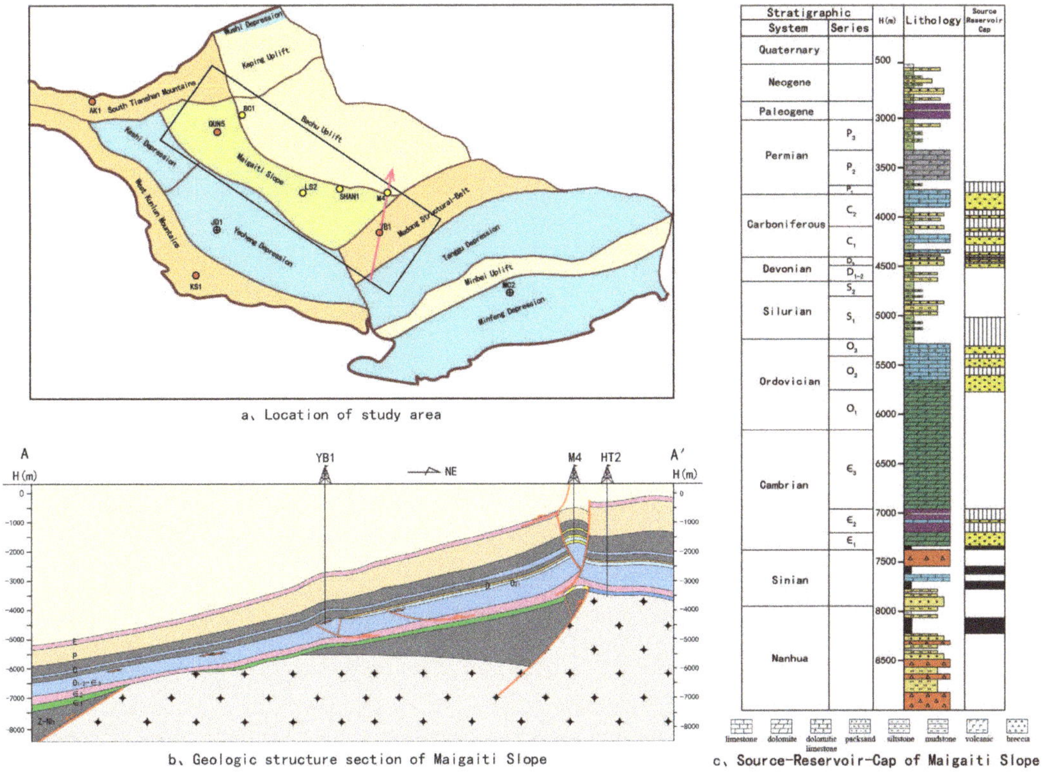

Figure 1. Geological profile map of the study area.

3. Staged and Segmental Characteristics of Maigaiti Slope Strike-Slip Faults

3.1. *Staging Characteristics of Strike-Slip Faults*

The stage characteristics of strike-slip faults of the Maigaiti Slope were studied by detailed seismic interpretation of the whole layer, and the tectonic evolution of the basin [22–24]. The Maigaiti Slope strike-slip faults can be roughly divided into four stages: early Caledonian, late Caledonian, Hercynian, and Himalayan. Early Caledonian strike-slip fault: the longitudinal fault disappeared in the Middle Cambrian. It can be seen from the three-dimensional earthquake that this fault has a certain control on the deposition of the Middle Cambrian, and the Middle Cambrian shows certain synsedimentary characteristics (Figure 2a); Late Caledonian strike-slip fault: the fault extends vertically to the top of Ordovician limestone, seismic facies phenomena such as "beading" and "flake reflection" of Ordovician carbonate rocks related to the strike-slip fault can be seen in the seismic section, which is direct evidence of the development of the strike-slip fault in this period (Figure 2b); Hercynian strike-slip fault: the longitudinal upper fault to Carboniferous to Lower Permian disappeared, the seismic profile shows the characteristics of "flower structure", Hercynian is Maigaiti Slope strike-slip fault mainly formed stage (Figure 2c); Strike-slip faults in Himalayan period: this fault traverses the Paleogene system vertically, mostly to the Neogene, and partially to the Quaternary system (Figure 2d).

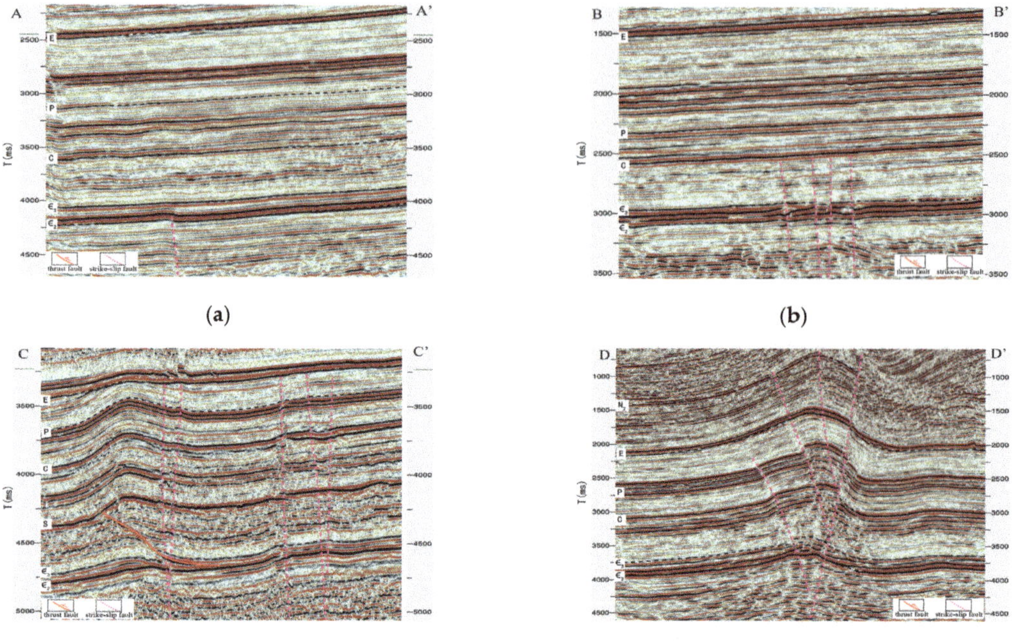

Figure 2. Characteristics of strike-slip fault profiles in different periods, Maigaiti Slope. (**a**) Characteristics of strike-slip fault profiles—early Caledonian. (**b**) Characteristics of strike-slip fault profiles—late Caledonian. (**c**) Characteristics of strike-slip fault profiles—Hercynian. (**d**) Characteristics of strike-slip fault profiles—Himalayan.

3.2. Segmentation Characteristics of Strike-Slip Faults

The development characteristics of thrust and strike-slip faults in the longitudinal section were described through full-layer seismic interpretation, the distribution characteristics of plane faults were described by 3D seismic coherence attribute analysis, and the fault characteristics of the Maigaiti Slope were understood in 2–3D seismic combination, and 3D vertical and horizontal. According to the research findings, two groups of strike-slip fault systems are developed in the main part of the Maigaiti Slope, which are NE and NW-trending. However, there are certain differences in fracture-formation period, fault strike, and tectonic style in different regions. Therefore, the study area can be divided into four zones: Maixi Member, Maidong Member, Madong Member, and Bachu Member (Figure 3).

(1) The Maixi segment: From seismic coherence properties and seismic profile characteristics, it can be seen that two groups of strike-slip faults are mainly developed in this area: NW trending and NW trending, the strike-slip fault groups are longitudinally faulted upward to the Permian system, which are formed in the late Hercynian period. The characteristics of "flower structure" of these strike-slip faults can be seen clearly in the seismic section. Each group of strike-slip faults is spread in parallel. The two groups of faults intersect at a low Angle, and show an "X" shear relationship. The strike-slip fault system weakens from NW to SE, and exhibits dextral strike-slip characteristics, with a plane length of 600 km and disappearing southward in the area of Well BT7. In addition, NE and NW-trending strike-slip faults are locally developed in the southeast end. The strike-slip faults disappear in the Middle Cambrian paste rock profile; they were formed in the Middle Caledonian period. The strike-slip faults are short in plane length and have obvious features of being cut into wrong segments by late strike-slip faults (Figure 4).

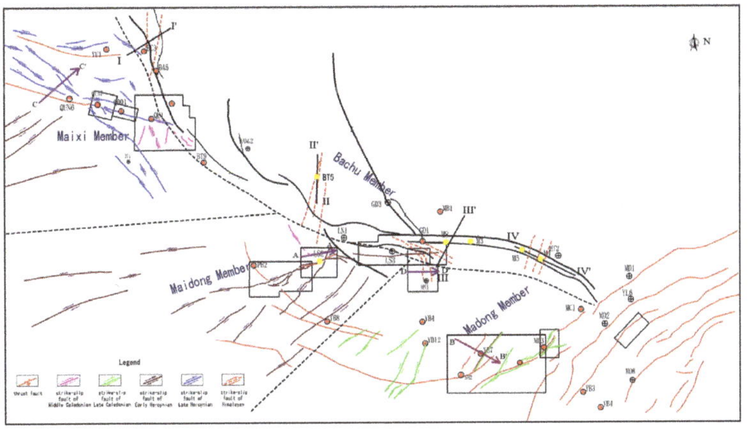

Figure 3. Fault pattern of the Maigaiti Slope.

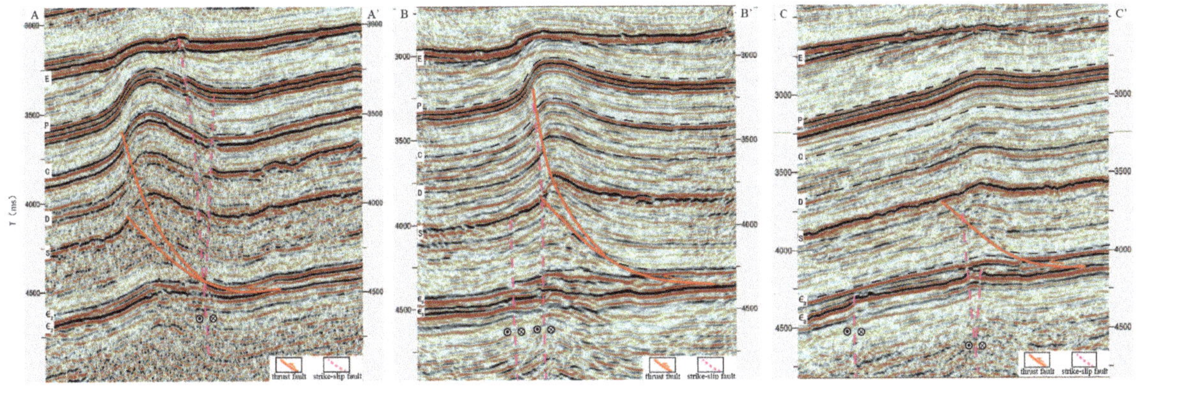

(a)

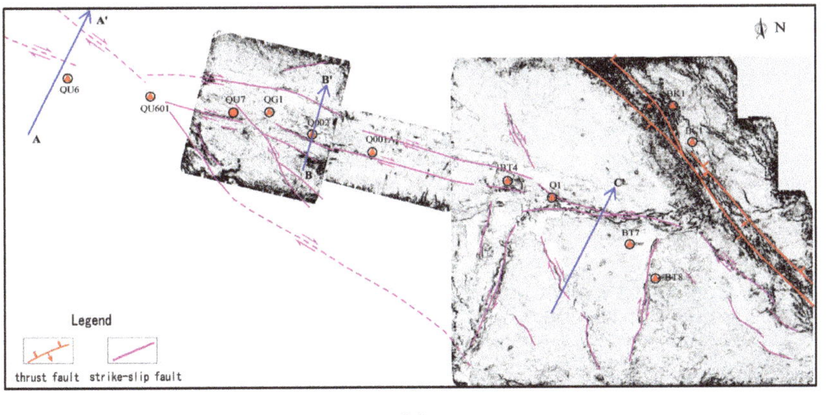

(b)

Figure 4. Characteristics of strike-slip fault in western part of Maigaiti Slope. (**a**) Characteristics of strike-slip fault profiles in western part of Maigaiti Slope. (**b**) Middle Cambrian bottom coherence of strike-slip fault about 3D seismic data in western part of Maigaiti Slope.

(2) The Maidong Segment: It can be clearly identified by the 3D seismic coherence attribute that this segment mainly develops NE-trending and a NE-trending early Hercynian dextral strike-slip fault system. The fault zone extends about 250 km in plane length and converges from west to east, showing the characteristics of "wide in the west and narrow in the east, strong in the west and weak in the east". From the seismic profile, it can be seen that the strike-slip fault in the west reaches to the Permian igneous rock segment, the fault zone is 4–7 km wide, and the branch faults are developed. In the periphery of the LS2 Well area in the middle part, the strike-slip fault breaks to the top of the carboniferous system, and the fault zone is narrow—about 2–4 km—showing continuous linear distribution in plane. In the periphery of the eastern LS3 Well area, strike-slip faults extend to the Cambrian system; most of them disappear into the Middle Cambrian gypsum rock layer, showing discontinuous linear distribution characteristics in plane (Figure 5).

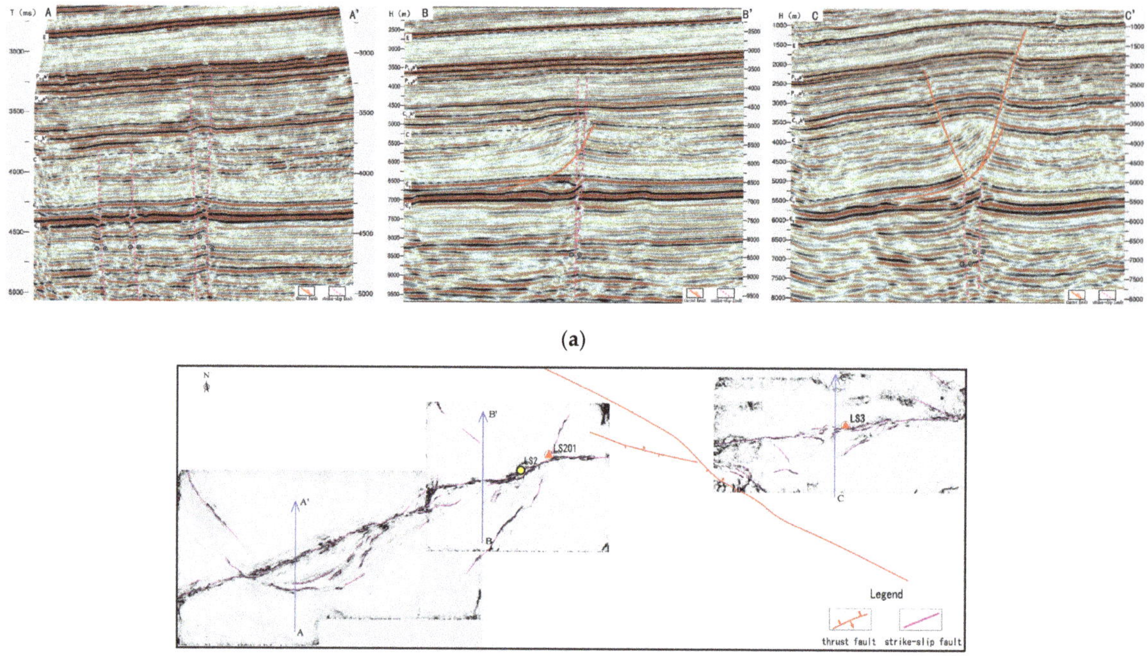

(**a**)

(**b**)

Figure 5. Characteristics of strike-slip fault in middle of Maigaiti Slope. (**a**) Characteristics of strike-slip fault profiles in middle of Maigaiti Slope. (**b**) Middle Cambrian bottom coherence of strike-slip fault about 3D seismic data in middle of Maigaiti Slope.

(3) The Madong Segment: The late Caledonian salt-slip thrust fault is developed as a whole, and the NE-trending late Caledonian strike-slip fault is only developed in the southwest end. The fault zone has a short extension length of about 50 km and limited distribution, which is associated with the late Caledonian thrust fault (Figure 3).

(4) The Bachu Segment: Himalayan strike-slip faults are mainly developed, distributed on the southern slope of circum-Bachu Uplift. The Selibuya fault-Tuohetage-Kangtakumu fault zone, Haimiluositage-Gudongshan fault zone, and Mazhatage fault zone are successively developed in the southern slope of Bachu Uplift from north to south [25,26]. The strike-slip faults are mainly developed between the fault zones, which are the adjustment faults between the fault zones. With the characteristics of dextral strike-slip

faults, four concentrated development zones of strike-slip faults can be roughly identified (Figures 3 and 6).

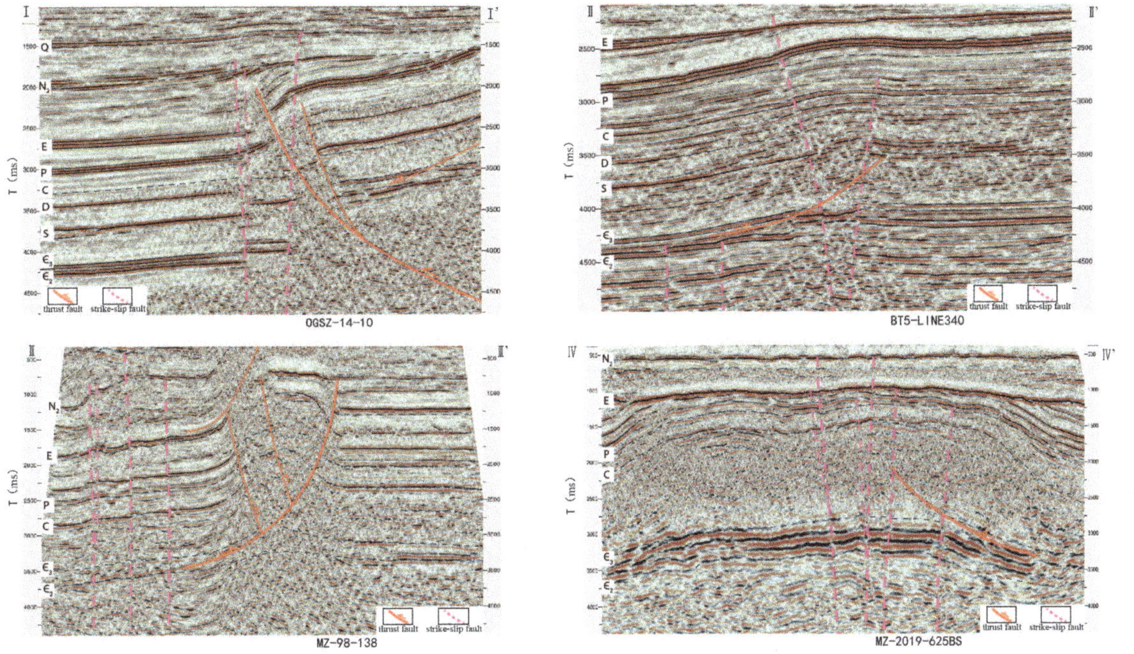

Figure 6. Characteristics of strike-slip fault profiles in Bachu Lift (The profile position is shown in Figure 3).

3.3. Genetic Mechanism of Strike-Slip Faults

Due to the interaction of India, northern Tibet, and the Eurasia plate, the Maigaiti Slope in the Tarim Basin is held by several orogenic systems, including the South Tianshan Mountains, West Kunlun Mountains and Aerjin Mountains. Influenced by different period orogeny, the region's tectonic stress field changes drastically, and the formation of the strike-slip fracture system is a comprehensive response to the different periods and the direction of tectonic movement. It is closely related to the formation and migration of the Hotan paleo-uplift (Figure 7).

During the late Caledonian, under the influence of the compression stress transfer from south to north of the Aerjin orogenic belt, the Tanggu depression in the front of the orogenic belt formed a NE-trending multi-row thrust structure. Restricted by the Hetian paleo-uplift at the front, a NE-trending compression strike-slip component was formed between the Hetian paleo-uplift and the Tanggu Depression (Figure 7a).

During the early Hercynian to Indosinian period, with the continuous uplift of the Hetian paleo-uplift, the distribution range of the paleo-uplift converged to the south and migrated to the northeast, forming the strike-slip fault system in the active margin of the north flank of the Hetian paleo-uplift. During the late Hercynian period, the distribution range and scale of the Hetian paleo-uplift remained basically unchanged. Under the action of the south Tianshan orogenic belt from north to south, Maixi paleo-uplift was formed, and the strike-slip fault system was formed in the active margin of the southwest wing of Maixi paleo-uplift (Figure 7b,c).

The Himalayan Hetian paleo-uplift migrated north with strong uplift, and the Maixi paleo-uplift merged to form the early prototype of the Bachu uplift. As a result of differences in uplift, the Bachu uplift southern boundary faults show obvious differences in deformation [27,28], forming multiple fault zones (such as Selibuya-Tuohetage-Kangtakumu fault zone, Haimiluositage-Gudongshan fault zone, Mazhatage fault zone), between the fault zones, forming a series of NE-trending strike-slip fault systems of structure adjustment (Figure 7d).

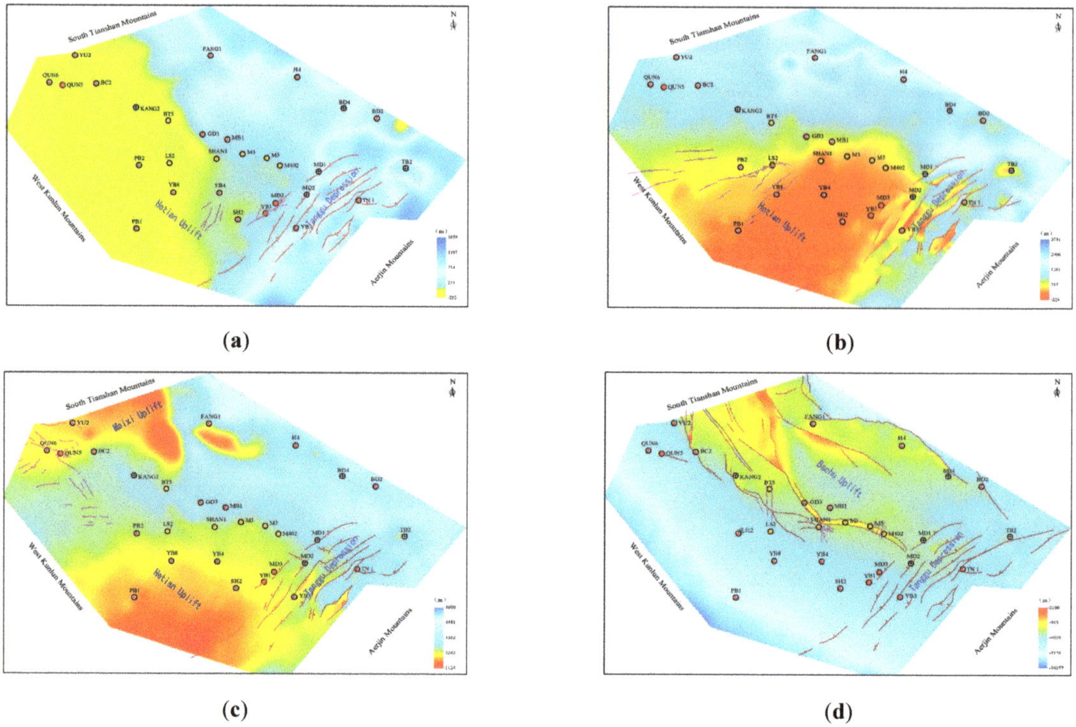

Figure 7. Structure of Ordovician top and strike-slip fault in different periods, Bachu-Maigaiti area. (**a**) Paleotectonic map of top surface of Ordovician–late Caledonian. (**b**) Paleotectonic map of top surface of Ordovician–early Hercynian. (**c**) Paleotectonic map of top surface of Ordovician–late Hercynian. (**d**) Tectonic map of top surface of Ordovician–present day.

4. Control of Strike-Slip Faults on Hydrocarbon Accumulation

Strike-slip fault, as an important channel for hydrocarbon migration and accumulation, is the key factor controlling hydrocarbon reservoirs forming in the Maigaiti Slope. There are four strike-slip faults in the early Caledonian, late Caledonian, Hercynian, and Himalayan stages in the Maigaiti Slope. The spatial-temporal matching relationship between the active stages of strike-slip faults and the trap forming stage is the key to hydrocarbon accumulation, and determines the enrichment layers and hydrocarbon properties of the reservoirs in the Maigaiti Slope (Figure 8).

The early Caledonian strike-slip faults did not break through the Middle Cambrian gypsolith and salt, and the oil and gas were preserved under the Cambrian gypsolith and salt. The strike-slip faults in the early Caledonian became the necessary conditions for the continuous accumulation of oil and gas in the dolomite reservoirs of the Lower Cambrian as a pathway system close to the source rocks.

The late Caledonian strike-slip fault is well matched with the Ordovician thrust buried-hill trap formation stage, which is the key for hydrocarbon accumulation in the Ordovician carbonate fracture-cavity reservoir. In this period, the deep source Cambrian rocks are of lower maturity and buried less deeply. In addition, the buried depth of the Middle

Cambrian gypsum rock (about 2300 m) is brittle [29]; the oil and gas easily break through the thick Middle Cambrian gypsum rock and accumulate in the Ordovician carbonate reservoir. The crude oil properties are of medium-heavy oil, such as in Yubei 1-Madong 3 reservoir, where the average crude oil density is 0.918–0.937 g/cm^3. There is a small amount of mature associated gas (Figure 8a).

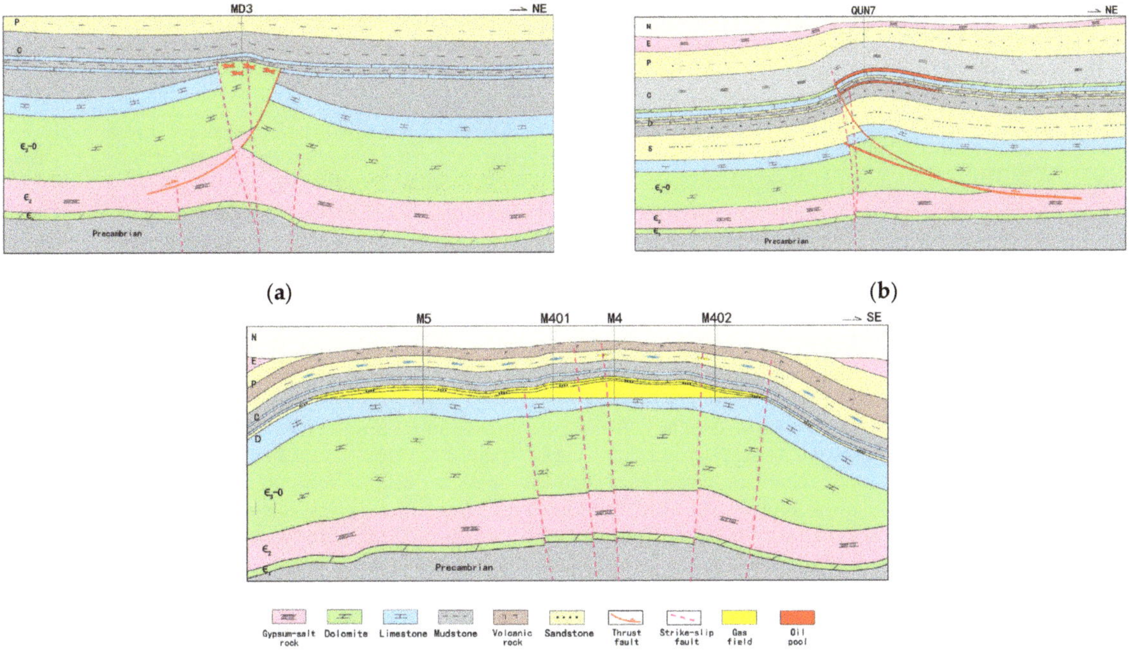

(a) (b) (c)

Figure 8. Control of hydrocarbon accumulation about strike-slip fault in different periods, Maigaiti Slope. (**a**) Reservoir profile of Madong 3 oil pool—late Caledonian. (**b**) Reservoir profile of Bashituopu oil pool—late Hercynian. (**c**) Reservoir profile of Hetianhe gas field—late Hercynian.

Hercynian strike-slip faults and structural traps formed in this period constituted a superior trap-migration-preservation spatial-temporal matching relationship, forming the characteristics of Ordovician–Carboniferous multi-purpose strata containing oil and gas, and the strike-slip faults directly determine the layers of oil-gas accumulation. In the Cambrian system, during the period where hydrocarbon source rocks were buried deep, the highly mature overall stage of oil and gas, by strike-slip fault in the Ordovician, Devonian, and Carboniferous systems gathered in many sets of reservoirs, and presents multi compound hydrocarbon accumulation characteristics; the crude oil is characterized by light oil, such as the Bashituopu oil pool, oil density of about 0.797 to 0.820 g/cm^3, accompanied by a certain amount of gas (Figure 8b).

The strike-slip faults of the Himalayan period determine the characteristics of multi-purpose gas reservoirs in this study area. During this period, the buried depth of the deep source rocks of the Cambrian was more than 6500 m, and it was in the over-mature stage of gas generation. Natural gas accumulated in the structural traps of the Ordovician, Devonian, Carboniferous and Permian through strike-slip faults (Figure 8c), and showed the characteristics of dry gas, forming the Hetianhe and Yasongdi gas fields.

5. Discussion

The exploration of oil and gas in the Maigaiti Slope started some time ago and has taken a long time. It was considered by the older geologists as one of the four most important favorable exploration fields in the Tarim Basin at the stage of regional oil check. Up to now, nine small and medium-sized oil and gas reservoirs have been discovered in the Maigaiti Slope, mainly distributed on the edge of the paleo-uplift. However, exploration work has not been carried out on a large scale. The oil and gas in the Maigaiti Slope is all derived from Cambrian source rocks; therefore, resources are abundant, but hardly proved. As a result, the Maigaiti Slope is still one of the most important exploration and research areas in the Tarim Basin.

According to current understanding of tectonic stratification and the allocation of hydrocarbon accumulation factors, three oil and gas accumulation assemblages developed in the Maigaiti Slope: Carboniferous–Devonian, Ordovician, and Cambrian sub-salt. The multi-layer vertical accumulation model also developed, with strike-slip and thrust faults forming at different times as the key factors for hydrocarbon accumulation.

Donghe sandstone, Carboniferous bioclastic limestone, and Carboniferous high-energy beach carbonate rocks developed in the Upper Carboniferous and Devonian. The Upper combination is mainly explored for stratigraphic lithologic oil and gas reservoirs, but its single layer is thin, and the reservoir varies rapidly laterally. The existing seismic data are mainly two-dimensional seismic, with a limited area of three-dimensional seismic. It is, therefore, difficult to carry out effective reservoir characterization. Since the Donghe sandstone of Well No. 7 obtained industrial oil flow in 2007, it is still in the preparatory research stage, without new exploration in the Upper combination over the past 15 years.

There are three types of reservoirs in the Middle combination: Ordovician carbonate buried-hill, fault-controlled karst body, and karst slope weathering crust. On the basis of the current 2D and 3D seismic data, it is difficult to make new progress in the search for and implementation of buried-hill traps. Furthermore, it is also very difficult to depict the palaeogeomorphology because of the limited accuracy of weathering crust traps. Therefore, current exploration research's focus on fault-controlled karst reservoirs; fully absorbing the successful exploration experience of fault-controlled karst reservoirs in the Fuman area in the past five years; depict the Tongyuan strike-slip fault with the idea that "there must be faults". To carry out the fault-controlled karst reservoir prediction on the Maigaiti Slope, we firstly depict source rocks-connected strike-slip faults based on the understanding that where there is a string of beads, there are faults.

The Lower combination, which is Cambrian sub-salt, is located in the favorable facies zone of the subtidal grain beach, developing favorable reservoir space such as intergranular pore and intergranular solution pore of dolomite, which form a good reservoir-cap assemblage with the gypsum salt of the Middle Cambrian. However, drilling explorations in recent years all failed because strong reflection of Permian igneous rock and Cambrian gypsum rock produces multiple waves in the deep area, which can greatly mislead the interpretation of the Cambrian bottom boundary and the characterization of source rocks. Therefore, at present, the Lower combination is at a key stage of processing seismic data to eliminate multiple waves.

At present, it is considered that the dark mudstone of the Lower Cambrian Yuertusi Formation is the most important source rock in the Maigaiti Slope. Based on the latest seismic and drilling data, the Cambrian source rocks in the Maigaiti Slope are re-recognized. According to the analysis of paleostructure and paleofault, the source rocks in the Maigaiti Slope are mainly developed in the lower slope, and the source rocks in the bastion zone are thinning or missing. According to the existing data, three sub-pools developed in the Maigaiti Slope at the time when Yuertusi Formation source rock was deposited. The source rocks are about 0 to 40 m thick in the Maixi and Xikeer sub-pools, and about 0 to 20 m thick in the Manan sub-pool, covering a whole area of 26,000 square kilometers.

In conclusion, the characterization of source rocks, and the study of hydrocarbon accumulation rules are the two keys to exploration of, and research in, the Maigaiti Slope.

The oil and gas accumulation in the Maigaiti Slope can be characterized by "multi-layer accumulation vertically, but short-distance extension horizontally". Strike-slip faults with multiple stages, different grades and various tendencies connecting source rocks and favorable traps are the most important secondary migration channels for oil and gas. Strike-slip faults control the distribution of oil and gas in plane, and the accumulation layer of oil and gas in vertical. In the next stage of oil and gas exploration, we should take the characterization of the strike-slip fault as the breakthrough point, strengthen the processing and interpretation of seismic data, use the new seismic and new drilling data to continuously promote the characterization of strike-slip faults, and explore the favorable exploration target of the effective allocation of hydrocarbon, reservoir, fault, and cap rock in the slope area.

6. Conclusions

(1) The fault-formation period, fault strike, and tectonic style are different in different areas of the Maigaiti Slope, which can be divided into four sections: Maixi Segment, Maidong Segment, Madong Segment, and Bachu Segment. The Maixi Segment mainly develops the NW-trending late Hercynian strike-slip fault system, the Maidong Segment mainly develops the NW-trending early Hercynian strike-slip fault system, the Maidong Segment mainly develops the NW-trending late Caledonian strike-slip fault system, and the Bachu Segment mainly develops the Himalayan strike-slip fault system.

(2) The tectonic stress field of the Maigaiti Slope varies greatly under the influence of orogenic movement in different periods. The formation of a strike-slip fault system in the Maigaiti Slope is a comprehensive response to multi-period and multi-direction tectonic movement, which is closely related to the formation and migration of Hetian paleo-uplift and other tectonic events.

(3) The coupling relationship between the active period of strike-slip faults and the trap-forming period is the key to hydrocarbon accumulation in the Carboniferous–Ordovician, which determines the petroleum properties and enrichment horizon of the Cambrian post-salt system. These are: medium-heavy oil in the Caledonian period, light oil in the Indosinian period, and dry gas in the Himalayan period.

Author Contributions: Conceptualization, B.W. and C.C.; methodology, J.S. and M.L.; formal analysis, W.Z. and Y.Q.; investigation, D.S. and C.S.; resources, L.L.; data curation, W.Z.; writing—original draft preparation, B.W. and C.C.; writing—review and editing, C.C.; visualization, B.W. and C.C. All authors have read and agreed to the published version of the manuscript.

Funding: This research was funded by a major science and technology project of China National Petroleum Co., Ltd. and grant number [2022KT0201].

Data Availability Statement: Not applicable.

Acknowledgments: Support from the major science and technology project of China National Petroleum Co., Ltd. is highly acknowledged.

Conflicts of Interest: The authors declare no conflict of interest.

References

1. Yan, L.; Zhu, G.; Chen, Y.; Han, C.; Yang, M.; Du, D.; Zhu, W. Distribution of Lower Cambrian source rocks in the Tarim Basin. *Nat. Gas Geosci.* **2019**, *30*, 1569–1578.
2. Chen, Y.; Yan, W.; Han, C.; Yan, L.; Ran, Z.; Kang, Q.; He, H.; Ma, Y. Strucural and sedimentary basin transformation at the Cambrian/Neoproterozoic interval in Tarim Basin: Implication to subsalt dolistone exploration. *Nat. Gas Geosci.* **2019**, *30*, 730–739.
3. Zhu, G.; Du, D.; Chen, W.; Sun, Q.; Li, T.; Zhang, Z.; Chen, Z. The discovery and exploration significance of the old thick black mudstones in the southwest margin of Tarim Basin. *Acta Petro. Sin.* **2017**, *38*, 1335–1343.
4. Cui, H.; Liu, J.; Tian, L.; Liu, J.; Zhang, N. Palaeotectonic pattern at the end of Sinian and its hydrocarbon significance in the southwest depression of Tarim Basin. *Chin. Petro. Expl.* **2018**, *23*, 67–75.
5. Zhu, X.; Chen, J.; He, L.; Wang, Y.; Zhang, W.; Zhang, B.; Zhang, K. Geochemical characteristics and source correlation of hydrocarbons in the Well Luosi 2 of Maigaiti Slope, Tarim Basin, China. *Nat. Gas Geosci.* **2017**, *28*, 565–574.

6. Yang, X.; Li, H.; Yue, Y.; Liu, S.; Li, J.; Xiong, P. The strata and palaeo-geomorphology framework at the end of Neoproterozoic and development mode of source rocks at the beginning of Cambrian. *Nat. Gas Geosci.* **2017**, *28*, 189–198. [CrossRef]
7. Jiang, H.; Chu, C.; Yang, X.; Chen, Q.; Yue, Y.; Shao, Z. Lithofacies Paleogeography of Early-Middle Cambrian in Southwestern Tarim Basin. *Marine Origin Petrol. Geol.* **2017**, *22*, 32–38.
8. Pan, W.; Chen, Y.; Xiong, Y.; Li, B.; Xiong, R. Sedimentary Facies Research and Implications to Advantaged Exploration Regions on Lower Cambrian Source Rocks, Tarim Basin. *Nat. Gas Geosci.* **2015**, *26*, 1224–1232.
9. Yang, X.; Xu, X.; Chen, Q.; Qian, Y.; Chen, Y.; Chu, C. Palaeotectonics Pattern in Pre-Cambrian and Its Control on the Deposition of the Lower Cambrian Source Rocks in Tarim Basin, NW China. *Nat. Gas Geosci.* **2014**, *25*, 1164–1171.
10. Yue, Y.; Luo, S. Structural Characteristics and Their Control over Ordovician Hydrocarbon Migration Pathway System in Yubei Area, Tarim Basin. *Geol. Sci. Tech. Infor.* **2019**, *38*, 20–30.
11. Ni, B.; Tang, L.; Guo, Y.; Yu, T.; Yue, Y. Analysis of Burial History and Thermal History in Yubei Area, Tarim Basin. *Geoscience* **2017**, *31*, 357–366.
12. Yao, Y.; Xiao, Z.; Wang, Y.; Zhang, B.; Gao, S. Fluid characteristics and accumulating patterns of adjusting reservoirs in the Tarim Basin: Taking Bashituo hydrocarbon reservoir as an example. *Nat. Gas Geosci.* **2016**, *27*, 1003–1013.
13. Si, S.; Chen, H.; Tan, X.; Li, C.; Wu, Y.; Li, N. Hydrocarbon Accumulation Period and Its Carrier Systems in Ordovician Reservoir of Yubei Area, Markit Slope Tarim Basin. *Earth Sci.* **2013**, *38*, 1271–1280.
14. Li, H.; Dong, Q.; Lou, R.; Zhang, K.; Zhang, Y.; Sun, L. Regulation of Silurian–Devonian hydrocarbon accumulation in Bachu–Maigaiti region of Tarim Basin. *G. Geo.* **2012**, *31*, 730–739.
15. Cui, J.; Wang, T.; Hu, J.; Li, M. Maturity of light oil and its significance in indicating oil source in Hetianhe gas field, Tarim Basin. *Oil Gas Geol.* **2013**, *34*, 27–36.
16. Wang, B.; Wang, W.; Zhu, L.; Yin, H.; Qu, Y.; Jia, D.; Li, C. Fault-related Fold and Its Application to Madong Fold and Thrust Belt. *Geol. J. Chin. Univ.* **2019**, *25*, 268–275.
17. Wang, B.; Zeng, C.; Fu, X.; Wang, X.; Qu, Y.; Ren, P. Reservoir characteristics and hydrocarbon accumulation model of carbonate buried-hill in Luosi-Madong area, Tarim Basin. *Marine Ogn. Petrol. Geol.* **2019**, *24*, 65–72.
18. Yue, Y.; Tian, J.; Zhao, Y.; Wang, M.; Xie, P. Control of Hetian Pala-Uplift on Hydrocarbon Accumulation of Ordovician, Tarim Basin. *Earth Sci.* **2018**, *43*, 4215–4225.
19. Ding, W.; Qi, L.; Yun, L.; Yu, T.; Wu, L.; Cao, Z.; You, S. The tectonic evolution and its controlling effects on the development of Ordovician reservoir in Bachu-Markit Tarim basin. *Acta Petrol. Sin.* **2012**, *28*, 2542–2556.
20. Chen, G.; Tang, L.; Yu, T.; Guo, Y.; Yue, Y.; Li, J.; Zhang, X.; Long, Y.; Xie, X. Implications of Precambrian Unconformity to Basement Paleo-uplift and Its Tectonic Evolution of Bachu-Markit Area, Tarim Basin. *Geoscience* **2015**, *29*, 2542–2556.
21. Guo, Y.; Tang, L.; Yu, T.; Li, J.; Yue, Y. Structural Characteristics and Pertoleum Geological Significances of Mazhatage Fault Belt in Tarim Basin. *J. Earth Sci. Environ.* **2016**, *38*, 104–114.
22. He, G.; He, Z.; Zhang, H.; Lin, L.; Chen, Q.; Qian, Y.; Zhu, G. Paleozoic structural deformation of Bachu Uplift, Tarim Basin of Northwest China: Implications for plate drifting. *J. Earth Sci.* **2009**, *20*, 755–762. [CrossRef]
23. Tong, D.; Zhang, J.; Yang, H.; Hu, D.; Ren, J. Fault system, deformation style and development mechanism of the Bachu Uplift, Tarim Basin. *J. Earth Sci.* **2012**, *23*, 529–541. [CrossRef]
24. Gao, Z.Q.; Fan, T.L. Extensional tectonics and sedimentary response of the Early-Middle Cambrian passive continental margin, Tarim Basin, Northwest China. *Geosci. Fro.* **2012**, *3*, 661–668. [CrossRef]
25. Lin, C.S.; Yang, H.J.; Liu, J.Y. Paleostructural geomorphology of the Paleozoic central uplift belt and its constraint on the devel-opment of depositional facies in the Tarim Basin. *Sci. China Ser. D* **2009**, *52*, 823–834. [CrossRef]
26. Burtman, V.S. Cenozoic crustal shortening between the Pamir and Tien Shan and a reconstruction of the Pamir-Tien Shan transition zone for the Cretaceous and Palaeogene. *Tectonophysics* **2000**, *319*, 69–92. [CrossRef]
27. Tang, D.; Chen, H.; Yun, L.; Zhang, H.; Li, N.; Deng, Z. The differential activity characteristics and its evolution of the faults in the Bachu Uplift, Tarim Basin. *Chin. J. Geol.* **2014**, *49*, 49–68.
28. Yang, Y.; Tang, L.; Jiang, H.; Chen, G.; Xie, D.; Li, M.; Cao, Z. Characteristics and deformation mechanism of staging differential fault activities in Bachu Uplift, Tarim Basin. *Petro. Geol. Exp.* **2014**, *36*, 275–284.
29. Zhuo, Q.; Zhao, M.; Li, Y.; Wang, Y. Dynamic sealing evolution and hydrocarbon accumulation of evaporate cap rocks: An example from Kuqa foreland basin thrust belt. *Acta Petro. Sin.* **2014**, *35*, 847–856.

processes

MDPI

Article

Genesis Types and Migration of Middle and Lower Assemblages of Natural Gas in the Eastern Belt around the Penyijingxi Sag of the Junggar Basin, NW China

Keshun Liu [1], Jiangxiu Qu [1,*], Ming Zha [1], Hailei Liu [2], Xiujian Ding [1], Minghui Zhou [3] and Tianze Gao [1]

1 School of Geosciences, China University of Petroleum (East China), Qingdao 266580, China
2 Research Institute of Exploration and Development, PetroChina Xinjiang Oilfield Company, Karamay 834000, China
3 Baikouquan Oil Production Plant, PetroChina Xinjiang Oilfield Company, Karamay 834000, China
* Correspondence: 20030040@upc.edu.cn

Abstract: This study analyzes the geochemical characteristics of natural gas composition, carbon isotope, and light hydrocarbon in the eastern belt around the Penyijingxi sag of the Junggar Basin. The result shows the that natural gas content is dominated by alkane gas, with low contents of heavy hydrocarbon and non-hydrocarbon components. The overall carbon isotopic composition of the alkanes shows a trend as δ^{13}methane (C_1) < δ^{13}ethane (C_2) < δ^{13}propane (C_3) < δ^{13}butane (C_4), and all $\delta^{13}C_1$ values are <−30‰, which are typical of gases of organic origin. The natural gas is mainly coal-derived gas from the Lower Urho formation, mixed with a small amount of oil-associated gas from the Fengcheng formation. The vertical migration of natural gas resulted in the mixing of oil-associated gas and coal-derived gas and the mixing of alkane gas at different stages of the same origin, which should be the origin of carbon isotope inversion. The diffusion migration of carboniferous oil and gas reservoirs has led to differences in gas geochemical characteristics among gas wells. These migration characteristics of natural gas may indicate that the shallow layers are a favorable stratum for the next step of oil and gas exploration in the eastern belt around the Penyijingxi sag.

Keywords: Junggar Basin; natural gas genesis; migration characteristics; carbon isotopes; light hydrocarbons

Citation: Liu, K.; Qu, J.; Zha, M.; Liu, H.; Ding, X.; Zhou, M.; Gao, T. Genesis Types and Migration of Middle and Lower Assemblages of Natural Gas in the Eastern Belt around the Penyijingxi Sag of the Junggar Basin, NW China. *Processes* **2023**, *11*, 689. https://doi.org/10.3390/pr11030689

Academic Editors: Ping Gao, Yidong Cai, Yingfang Zhou and Quan Gan

Received: 3 February 2023
Revised: 20 February 2023
Accepted: 22 February 2023
Published: 24 February 2023

1. Introduction

There are two types of natural gas in sedimentary basins, inorganic and organic [1], and organic gas is further divided into oil-associated and coal-derived gas [2–4]. Inorganic gas is potentially associated with magmatic and deep-sea hydrothermal activity [5,6], with methane isotopes ($\delta^{13}C_1$) within the range of −50‰ (generally −30‰) to 10‰ [7,8], while organic gas is derived from the pyrolysis of kerogen in sedimentary rocks and the secondary cracking gas of crude oil, with methane isotopes in the range of −75‰ to −30‰ [7]. The inorganic alkane gas polymerized step-by-step to form long-chain alkanes through C-C bonding and the lower bond energy of ^{12}C-^{12}C caused ^{12}C to join the polymerization reaction first, showing a negative carbon isotope series of δ^{13}methane (C_1) > δ^{13}ethane (C_2) > δ^{13}propane (C_3) > δ^{13}butane (C_4) [9]. When alkane gas was generated from the degradation of kerogen, ^{12}C-^{12}C with lower bond energy broke preferentially than ^{13}C-^{13}C, leading to the gradual enrichment of $\delta^{13}C$ in organic alkane gas with the increase of the carbon atom number, thus forming a positive carbon isotope series of $\delta^{13}C_1 < \delta^{13}C_2 < \delta^{13}C_3 < \delta^{13}C_4$ [10]. In cases of mixing alkane gases of different genesis or sources and oxidation by microorganisms (propane bacteria), the arrangement of $\delta^{13}C$ may be confused [10–12]. The sedimentary environment controls the original carbon isotope composition of kerogen, and the carbon isotope of humic kerogen is greater than that of sapropelic kerogen [13]. Ethane has well inherited the difference of the original parent material, so that $\delta^{13}C_2$ is used as an important indicator to identify the genetic type of natural gas [14,15]. The $\delta^{13}C_2$ of alkane gas generated from sapropelic kerogen is generally lower than −29‰, and $\delta^{13}C_2$ of alkane gas generated by humic kerogen is generally higher than −28‰ [11].

Compared with crude oil, natural gas has greater molecular activity, and its migration process and migration phase state are more complex and changeable [16,17]. Natural gas can migrate not only laterally along sand bodies and nonconforming surfaces, but also vertically through faults, fractures, and pores [16,18,19]. Under formation conditions, natural gas may successively appear in one or more phases: the water-soluble phase, the oil-soluble phase, the free phase, and the diffusion phase [19–21]. Geochemical parameters, such as CH_4 content, C_1/C_2 value, stable carbon isotope, iC_4/nC_4 value, nitrogen-containing compounds and isotopes, and noble gas isotopes, are widely used in the research of natural gas migration [22–24]. During the migration of natural gas, the heavier hydrocarbon components like methane and isoalkanes will migrate preferentially over normal alkanes. Therefore, with the increase in migration distance, natural gas will have the trend of "methanation" and "isomerization" [20,25]. At the same time, isotope fractionation will also occur due to the "mass fractionation effect" and the "dissolution fractionation effect" [26].

Natural gas exploration in the Junggar Basin began in the early 1980s and made no significant breakthroughs until the discovery of the Mahe and Kelameili gas fields in the 21st century. The Basin's total proven reserve has reached 2000×10^8 m^3 [27]. According to China's 4th fourth assessment of oil and gas resources, the proportion of proven gas reserves in the lower and middle assemblages of the Junggar Basin is about 12% and 5.1% [28], respectively, which is significantly lower than other petroliferous basins [29]. In the past few years, the carbon isotopic composition and source of alkane gas in the Junggar Basin have been studied [27,30]. However, there are some problems: (i) analyzing the genesis and source of natural gas from the perspective of the whole basin will lead to some work that is not deep enough; for instance, determining from which source rocks the natural gas comes. (ii) Regional research is mainly concentrated in the basin's eastern, northwestern, and southern margins, with little research on the basin's central portion. The early proven small gas reservoirs in the eastern belt around the Penyijingxi sag of the Junggar Basin, such as the Pen 5, Mobei 2, and Mobei 5 gas reservoirs, are all secondary hydrocarbon reservoirs formed by the re-accumulation of primary oil and gas reservoirs after damage and adjustment [31,32]. The phenomenon of damage and adjustment of such primary oil and gas reservoirs also occurs in the Tarim Basin, the Sichuan Basin, the Georgina Basin, the Lower Indus Basin, and other structurally active basins [33–36].

Therefore, by analyzing the geochemical characteristics of natural gas in the eastern ring of the Penyijingxi sag, we hope to explain the genesis and source of natural gas as well as clarify the migration characteristics of natural gas after reservoir formation. This study can not only provide some important information for hydrocarbon exploration in the central part of the Junggar Basin, but also provide some ideas for natural gas research in other similar basins.

2. Geological Setting

Located between the Siberian plate, the Kazakhstan plate, and the Tarim plate, the Junggar Basin is an important part of the Central Asian orogenic belt [37]. The Junggar Basin has experienced multiple tectonic movements, such as Hercynian, Indosinian, Yanshan, and Himalayan, and has formed the current tectonic framework [38]. It can be divided into six primary tectonic units: two depressions (Ulungu and Central Depressions), three uplifts (Luliang, Western, and Eastern Uplifts), and one piedmont thrust belt (Northern Tianshan Piedmont Thrust Belt) (Figure 1a). These six primary tectonic units can be further divided into 44 secondary tectonic units. In the basin, there are three reservoir-caprock assemblages (upper, middle, and lower ones), bounded by two regional mudstone caprocks in the Lower Cretaceous Tugulu Group (including the Qingshuihe formation, Hutubihe formation, Shengjinkou formation, and Lianqinmu formation) and the Upper Triassic

Baijiantan formation [39]. Specifically, the lower assemblage mainly consists of the Permian and Carboniferous and the middle assemblage mainly consists of the Jurassic strata. In this study, the eastern belt around the Penyijingxi sag consists of the southern part of the Shixi bulge, the Mobei bulge, the western member of the Mosuowan bulge, and the eastern part of the Penyijingxi sag (Mobei Slope) (Figure 1b). The sedimentary sequence of the study area is Carboniferous, Permian, Triassic, Jurassic, Cretaceous, Paleogene, Neogene, and Quaternary (Figure 1c, the shallow stratum of the Qingshuihe formation, is not listed). The Carboniferous, Jiamuhe formation, Fengcheng formation, Lower Urho formation, Badaowan formation, and Xishanyao formation source rocks are deposited (Figure 1c) [40]. In the past few years, many commercial gas wells have been found in the study area (Figure 1b). Affected by the drilling depth, the proved natural gas in the north of the study area is mainly distributed in the Carboniferous and Jurassic; the proved natural gas in the south is mainly distributed in the Jurassic.

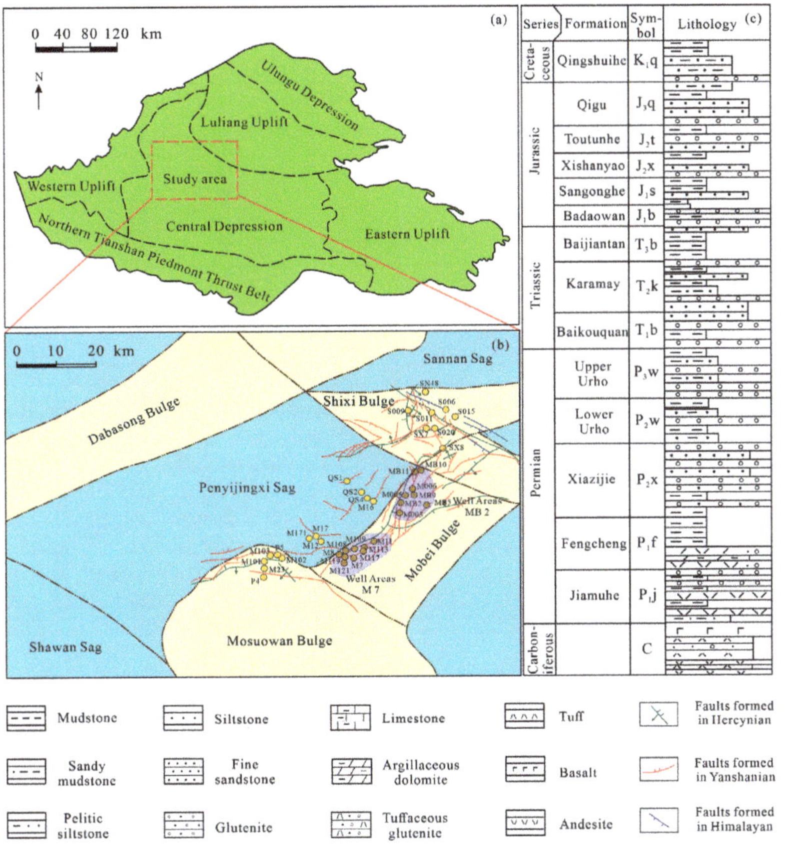

Figure 1. Geological overview of the eastern belt around the Penyijingxi sag in the Junggar Basin, NW China (modified from refs. [32,37]). (**a**) Division of tectonic units in the Junggar Basin, (**b**) geological overview of the eastern belt around Penyijingxi sag, and (**c**) stratigraphic column of the Penyijingxi sag.

3. Analytical Methods

In this study, 54 gas samples were collected from 36 wells in the eastern belt around the Penyijingxi sag for analysis of natural gas components and carbon isotopes of alkanes and light hydrocarbons (Table 1). The gas phase samples were directly collected from the wellhead using stainless steel cylinders with a diameter of 25 cm, and the gas pressure was about 2~3 MPa. After sample collection, the cylinders were placed in water to check the airtightness. The analysis and testing of the samples were completed by the Experimental Testing Research Institute of PetroChina's Xinjiang Oilfield Branch.

Table 1. Depth and source of natural gas samples.

No.	Well	Formation	Depth/m	No.	Well	Formation	Depth/m	No.	Well	Formation	Depth/m
1	M121	J1s	4222	19	QS2	J1s	3981	37	MB5	J1s	3726
2	M121	J1s	4255	20	QS4	J1s	4014	38	M108	J1s	4179
3	M109	J1s	4158	21	M101	J1s	4204	39	M109	J1s	4185
4	M113	J1s	4205	22	S006	J1s	3577	40	M11	J1s	4139
5	M115	J1s	4204	23	S006	C	4373	41	M11	J1s	4177
6	M116	J1s	4195	24	S007	C	4409	42	M7	J1s	4228
7	M117	J1s	4238	25	SX1	C	4438	43	M7	J1s	4260
8	M119	J1s	4258	26	SX1	C	4473	44	M8	J1s	4233
9	M119	J1s	4236	27	S015	J1s	nd	45	M8	J1s	4266
10	M003	J1s	3915	28	SX8	J1s	nd	46	M16	J1s	4041
11	M003	J1s	3975	29	SX14	J1s	nd	47	M171	J1s	4473
12	MB2	J1s	3921	30	MB11	J1s	3711	48	M17	J1s	4162
13	MB2	J1s	3921	31	MB2	J1s	3907	49	M12	J1s	4235
14	MB5	J1s	3726	32	M003	J1s	3972	50	M17	J1s	4192
15	MB10	J1s	3666	33	MB9	J1s	3761	51	M101	J1s	4209
16	M16	J1s	4047	34	MB9	J1s	3778	52	M102	J1s	4251
17	QS1	J1s	3945	35	M005	J1s	3890	53	M103	J1s	4251
18	QS1	J1s	3945	36	M006	J1s	3759	54	P5	J1s	4250

Notes: J1s means Jurassic Sangonghe formation, C means Carboniferous, nd means no data, M means Mo, MB means Mobei, QS means Qianshao, SX means Shixi, P means Pen.

3.1. Components of the Natural Gas

The composition (methane-pentane) of natural gas was analyzed by an Agilent 7890A gas chromatograph. The sample pretreatment and test process refer to the standard of natural gas composition analysis of the People's Republic of China (GB/T 13610-2020). Before each experiment, we carried out two or more consecutive standard gas injection checks to control the difference between the response values of each component within 1%. Therefore, the experimental results are reliable. The instrument was equipped with two thermal conductivity detectors and one flame ionization detector. In the experiment, a constant-temperature heating furnace was used to keep the sample temperature around 75 °C and make the sample composition uniform. High-purity (99.999%) helium was used as the carrier gas, with a flow rate of 2 mL/min. The outlet pressure of the cylinder was controlled at 0.2 MPa and the air flow rate at 80 mL/min. The split ratio was controlled at 150:1. DB-1 chromatographic columns were used in the experiment. The initial temperature of the chromatographic column box was 40 °C (for 2 min), and then the temperature rose to 90 °C at a rate of 10 °C/min, and then to 200 °C at a rate of 5 °C/min (for 5 min). Finally, the composition of the test sample was determined by the retention time of the standard gas.

3.2. Natural Gas Light Hydrocarbon

The Agilent 6890B gas chromatograph was used for light hydrocarbon (pentane-octane) analysis of natural gas. The sample pretreatment and experimental process refer to the oil and natural gas industry standard of the People's Republic of China for stable light component analysis (SY/T 0542-2008). The instrument was also equipped with a thermal conductivity detector and a flame ionization detector. The constant temperature furnace was also used to heat the sample in the experiment. High-purity (99.999%) helium was used as the carrier gas with a flow rate of 1 mL/min. The outlet pressure of the cylinder was controlled at 0.2 MPa and the air flow rate at 80 mL/min. The split ratio was controlled at 150:1. Pona chromatographic columns were used in the experiment. The initial temperature of the chromatographic column box was 30 °C (for 15 min), and then the temperature was raised to 70 °C at a rate of 3 °C/min, and then to 300 °C (for 10 min) at a rate of 3 °C/min. As in Section 3.1, the composition of the test sample was determined by the retention time of the standard gas. Each sample was measured repeatedly to ensure that the difference between the two measurement results was not greater than the precision specified in the standard. Then, the arithmetic mean of the two measurement results was used as the analysis result. Therefore, the experimental results are reliable.

3.3. Carbon Isotopic Composition of the Natural Gas

The carbon isotope analysis of natural gas was completed on the Delta V Advantage isotope mass spectrometer connected with the Agilent 7890A gas chromatograph. The sample pretreatment and experimental process refer to the organic geochemical analysis standard of geological samples of the People's Republic of China (GB/T 18340.2-2010). First, the components of natural gas were separated using an Agilent 7890A gas chromatograph (the experiment used an HP-5MS column). Then, the hydrocarbon gas was sent into the isotope mass spectrometry oxidation furnace to be converted into CO_2. Finally, CO_2 was introduced into the Delta V Advantage isotope mass spectrometer to determine the carbon isotope composition. The initial temperature of the chromatographic column box was 40 °C (for 5 min), and then it rose to 200 °C (for 18 min) at 10 °C/min. The experiment used high-purity (99.999%) helium as the carrier gas at a flow rate of 2 mL/min. The split ratio of the methane carbon isotope analysis was 50:1 and the split ratio of the ethane-pentane isotope analysis was 10:1. The standard samples for experimental analysis were obtained from the national standard material sharing platform of China. The experimental results are based on the VPDB standard. The precision of the carbon isotope determination meets the requirements that the repeatability value (r) is lower than 0.4 and the reproducibility value (R) is lower than 0.5, which can be considered reliable.

4. Results

4.1. Components of the Natural Gas

The natural gas in the eastern belt around the Penyijingxi sag is absolutely dominated by alkane gases. The volume fraction of methane varies from 71.36% to 93.34%, with an average of 87.94% (Table 2). Natural gas has a dryness coefficient ranging from 0.76 to 0.95, averaging 0.91, and is dominated by wet gas. The wide range of its dryness coefficient indicates that the natural gas may be generated by source rocks at different stages of thermal evolution. The non-hydrocarbons in natural gas are mainly N_2 (volume fraction: 0.69–11.95%, with an average of 2.56%) and CO_2 (volume fraction: 0–1.49%, with an average of 0.45%). The gas composition varies among the zones: the gas in the Shixi bulge has a relatively low content of CH_4, with an average of less than 80%, and a relatively high content of N_2, averaging 6.75%. The gas in the Mobei bulge, Mobei Slope, and Mosuowan bulge has a relatively high content of CH_4, with an average greater than 87.58%, and a relatively low content of N_2, averaging less than 3.0% (Table 2).

Table 2. Natural gas compositions by zone in the eastern belt around the Penyijingxi sag, Junggar Basin.

Category	Carboniferous	Jurassic Sangonghe Formation			
	Shixi Bulge	Shixi Bulge	Mobei Bulge	Mobei Slope	Mosuowan Bulge
CH_4/%	71.36~88.82	75.10~91.12	84.56~93.34	73.88~91.20	87.49~89.51
	79.10	81.42	89.97	87.58	88.31
C_2H_6/%	3.70~7.92	4.38~9.11	3.10~6.69	4.02~10.90	4.25~4.70
	5.87	7.39	4.25	5.37	4.5
C_3H_8/%	1.07~4.36	1.47~4.72	0.93~3.01	1.07~5.96	1.47~1.98
	3.28	3.30	1.50	2.20	1.76

Table 2. *Cont.*

Category	Carboniferous	Jurassic Sangonghe Formation			
	Shixi Bulge	**Shixi Bulge**	**Mobei Bulge**	**Mobei Slope**	**Mosuowan Bulge**
$C_4H_{10}/\%$	1.08~4.55	0.90~4.04	0.50~1.95	0.60~4.38	0.84~1.43
	3.32	2.59	0.98	1.45	1.15
$C_5H_{12}/\%$	0.54~1.78	0.33~1.49	0.08~0.93	0.24~1.57	0.24~0.62
	1.32	1.00	0.36	0.54	0.45
$CO_2/\%$	0.00~0.37	0.63~1.07	0~1.49	0.31~0.70	0.41~0.62
	0.17	0.80	0.40	0.52	0.54
$C_1/C_{1\text{-}5}$	0.82~0.93	0.80~0.93	0.87~0.95	0.76~0.94	0.91~0.93
	0.85	0.8	0.93	0.90	0.92

Note : $\frac{71.36\sim88.82}{79.10(4)} = \frac{\text{Min}\sim\text{Max}}{\text{Ave}}$, see Appendix A for all data.

4.2. Carbon Isotopic Composition of the Natural Gas

The carbon isotopic compositions of the components of the natural gas in the eastern belt around the Penyijingxi sag were analyzed. The carbon isotope ratio of methane ($\delta^{13}C_1$) ranges from −45.57‰ to −31.19‰, with an average of −37.48‰, showing a single peak mainly within the range from −42.5‰ to −35.0‰ (Figure 2a). The carbon isotope ratio of ethane ($\delta^{13}C_2$) ranges from −31.69‰ to −24.66‰, with an average of −27.62‰. Similar to $\delta^{13}C_1$, it shows a single peak mainly within the interval from −35.0‰ to −27.5‰ (Figure 2b). The carbon isotope ratio of methane ($\delta^{13}C_3$) ranges from −28.76‰ to −23.56‰, with an average of −26.27‰, showing a single peak mainly within the range from −27.5‰ to −24.5‰ (Figure 2c). The carbon isotope ratio of methane ($\delta^{13}C_4$) ranges from −27.96‰ to −23.64‰, with an average of −26.41‰, showing a single peak mainly within the range from −26.5‰ to −24.5‰ (Figure 2d).

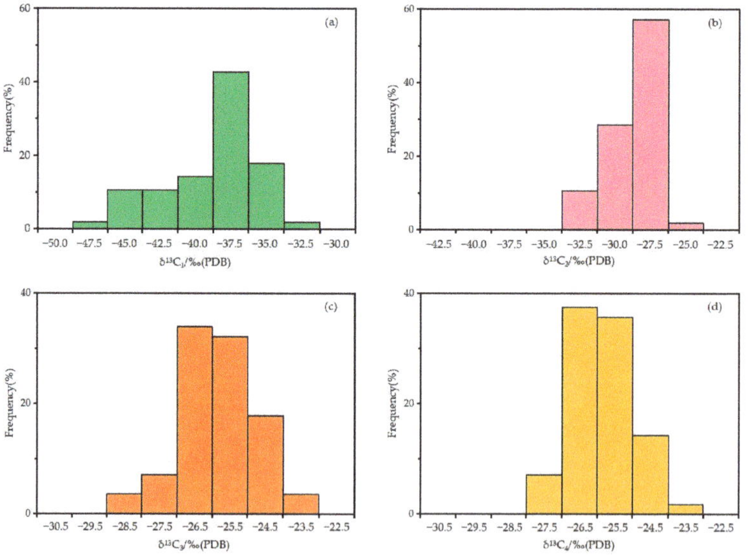

Figure 2. Histogram of $\delta^{13}C_1$ (**a**), $\delta^{13}C_2$ (**b**), $\delta^{13}C_3$ (**c**), and $\delta^{13}C_4$ (**d**) of the natural gas in the eastern belt around the Penyijingxi sag, Junggar Basin.

Due to their reversed isotope kinetic fractionation pattern, the $\delta^{13}C$ of the natural gas of organic origin increases gradually with the carbon number, forming a positive carbon isotopic series, while the $\delta^{13}C$ of the natural gas of inorganic origin forms a negative carbon isotopic series [9,10]. Gas samples with positive carbon isotopic series account for 41.38% of all the samples collected from the studied area. The remaining samples are all slightly and partially isotopically reversed, with the carbon isotopic series as $\delta^{13}C_1 < \delta^{13}C_2 < \delta^{13}C_3 < \delta^{13}C_4$ (Figure 3). The causes for carbon isotopic reversal include: (i) mixing of organic and inorganic alkane gases, (ii) mixing of coal- and oil-associated gases, (iii) microbial oxidation, and (iv) mixing of same-type alkane gases of different sources or same-source alkane gases of different periods [11,41]. In the studied area, natural gas reservoirs are generally located 3500 m below the surface or deeper. According to a geothermal gradient of 25°C/km [42], the reservoir temperature should be higher than 87.5°C, making it impossible for propane oxidizing bacteria to survive [43]. Therefore, we can exclude microbial oxidation from the causes. Other possible causes for carbon isotopic reversal are discussed below.

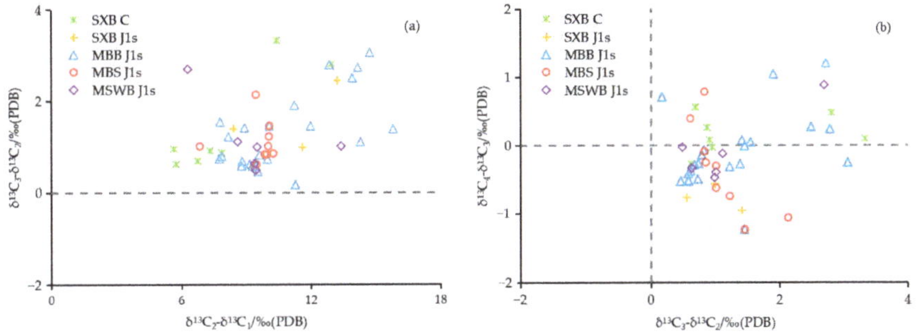

Figure 3. Cross-plots of $\delta^{13}C_2$-$\delta^{13}C_1$ vs. $\delta^{13}C_3$-$\delta^{13}C_2$ (**a**) and $\delta^{13}C_3$-$\delta^{13}C_2$ vs. $\delta^{13}C_4$-$\delta^{13}C_3$ (**b**) of the natural gas in the eastern belt around the Penyijingxi sag, Junggar Basin. SXB C means the Carboniferous system of Shixi bulge, SXB J1s means Jurassic Sangonghe formation of Shixi bulge, MBB J1s means Jurassic Sangonghe formation of Mobei bulge, MBS J1s means Jurassic Sangonghe formation of Mobei slope, MSWB J1s means Jurassic Sangonghe formation of Mosuowan bulge.

5. Discussion

5.1. Genesis Types of the Natural Gas

The carbon isotopic compositions of methane and ethane and the carbon isotopic series of their homologues are important indicators to identify whether the alkane gases are of inorganic or organic origin, and are commonly used to determine the genesis of natural gas [7]. The alkane gases in the natural gases from the Shixi, Mobei, and Mosuowan bulges and the Mobei slope have a positive carbon isotopic series ($\delta^{13}C_1 < \delta^{13}C_2 < \delta^{13}C_3$) that is reversed for butane (Figure 3). This partial reversal may be due to the mixing of natural gases of different genesis, migration, or secondary changes [11,12]. Primary alkane gases should have a positive carbon isotopic series ($\delta^{13}C_1 < \delta C^{13}{}_2 < \delta^{13}C_3 < \delta^{13}C_4$). In addition, while the isotopic composition of methane of inorganic origin is generally greater than −30‰, this value of the gases in the studied area is less than −30‰ (Figure 4). Therefore, it can be concluded that the natural gases in the eastern belt around the Penyijingxi sag are of organic origin, and that the reversed isotopic series of alkane gases should not be caused by the mixing of organic and inorganic alkane gases.

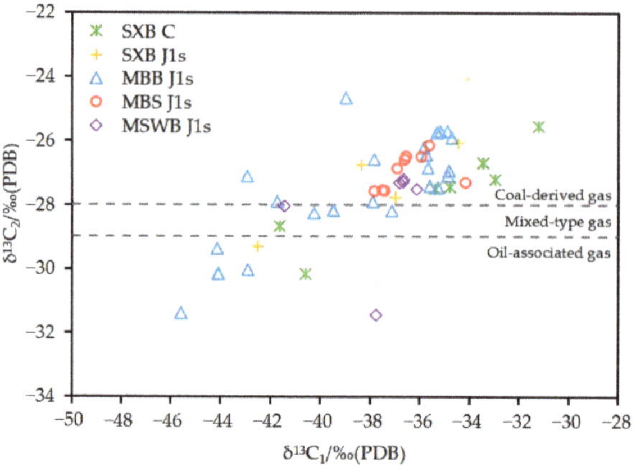

Figure 4. Genetic types identification chart for natural gases in the eastern belt around the Penyijingxi sag, Junggar Basin based on $\delta^{13}C_2$ & $\delta^{13}C_1$ (plate from ref. [11]).

The $\delta^{13}C_2$ value of alkane gases is a feature basically inherited from the parent material and is less influenced by the maturity of the source rocks. It is usually used as an important indicator for gas genesis identification [14,15]. In this paper, we identify gases with $\delta^{13}C_2 > -28‰$ as coal-derived gas, gases with $\delta^{13}C_2 < -29‰$ as oil-associated gas, and gases with $\delta^{13}C_2$ between $-28‰$ to $-29‰$ as mixed-type gas (Figure 4) [11]. The samples from the Shixi, Mobei, and Mosuowan bulges are distributed in all three intervals (Figure 4) but are mainly coal-derived gases. Only a few samples are oil-associated or mixed-type gas, which are the thermal degradation products of sapropelic-type and humic-type kerogens. The samples from the Mobei slope all fall in the coal-derived gas interval, which is generated by humic kerogen. Obviously, different from those in the Shixi, Mobei, and Mosuowan bulges, the Jurassic reservoir of the Mobei Slope only produces coal-derived gas. As revealed by the period analysis of the faults in the studied area, both the Hercynian and Yanshanian faults developed in the bulge zone, while only the Yanshanian faults developed in the slope zone [32]. This difference in vertical migration channels may result in the different gas types in bulge and slope zones.

Light hydrocarbons are important components of both natural gas and crude oil. Their variety becomes much wider as the number of carbon atoms increases, and their boiling points do not exceed 200°C in general. In natural gas genesis identification, indicators related to liquid and light C_5~C_8 hydrocarbons are commonly used for comparing the natural gases' type, maturity, and source [44]. The indicators for identifying organic matter type include the relative content of dimethyl cyclopentane (ΣDMCH) of various structures, n-heptane (nC_7), methylcyclohexane (MCH) in C_7 light hydrocarbons as well as the relative content of cycloalkanes, n-alkanes, and isomeric alkanes in $C_{5\sim7}$ hydrocarbons [45]. MCH, mainly from higher plants' lignin, cellulose, sugar, etc., has relatively stable thermodynamic properties and is a good parameter to indicate the type of terrestrial parent material. Its abundance is an important characteristic of light hydrocarbons in coal-derived gas. ΣDMCH, mainly from the lipid compounds of aquatic organisms, is affected by maturity. The high content of ΣDMCH indicates oil-associated gas. nC_7, mainly from algae and bacteria, is a good maturity indicator [46]. In the C_7 light hydrocarbons from the gases in the studied area, the relative content of MCH ranges from 29.38% to 53.35%, with an average of 41.98% (Figure 5). The relative content of ΣDMCH ranges between 4.63% and 45.79%, with an average of 14.30%. The high relative content of MCH and low relative content of ΣDMCH indicate that the natural gases in the studied area are mainly from type III (humic) kerogen and are dominated by coal-derived gas. This conclusion is consistent with the results of the alkane carbon isotopic analysis above.

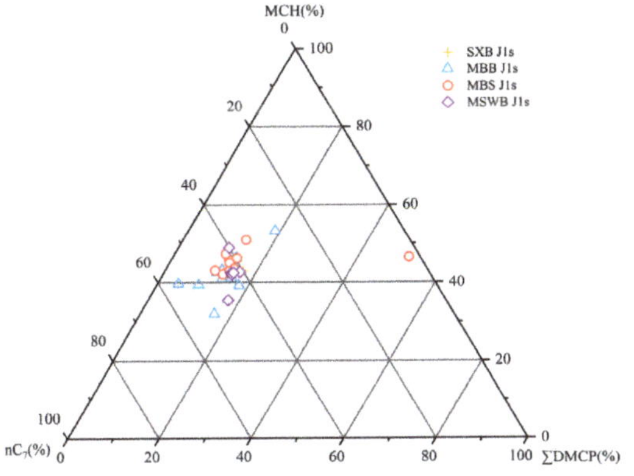

Figure 5. Characteristics of C_7 light hydrocarbons from natural gases in the eastern belt around the Penyijingxi sag, Junggar Basin.

5.2. Sources of Natural Gas

In the studied area, adjacent to the hydrocarbon-rich Penyijingxi sag in the Central Depression of the Junggar Basin, multiple sets of potential Carboniferous, Permian, and Jurassic source rocks have been developed (Table 3). According to the geochemical evaluation method of terrigenous source rocks (SY/T 5735-1995), the Carboniferous source rocks are of medium-poor quality, primarily composed of type III kerogen, and are mainly gas-producing. The Jiamuhe formation, Lower Urho formation, and Jurassic source rocks are of medium-good quality, composed of type III kerogen, and are mainly gas-producing. The Fengcheng formation source rocks are of good quality, composed of type II kerogen, and are mainly oil-producing. According to the 4th resource evaluation of the Junggar Basin, the total gas generation intensity of the source rocks in the Penyijingxi sag is 8000×10^6~$13{,}000 \times 10^6$ m^3/km^3 [28], which indicates that the above-mentioned source rocks have generated a large amount of natural gas and can provide sufficient gas to fill the eastern belt around the sag.

Table 3. Geochemical characteristics of the main source rocks in the Penyijingxi sag, Junggar Basin [47,48].

Stratum	TOC/%	$(S_1 + S_2)$/(mg/g)	Chloroform Bitume "A"/%	Hydrogen Index/(mg/g.TOC)	Kerogen Type
J2x	0.40~5.87	0.05~17.70	0.016~0.918	/	III
	1.42	2.03	0.267	/	
J1b	0.42~5.86	0.08~29.67	0.025~4.916	/	III
	1.68	3.03	0.555	/	
P2w	0.18~14.03	0.01~37.52	0.0007~0.8024	1.20~950.00	III
	1.69	2.06	0.0692	74.16	
P1f	0.03~4.43	0.1~59.84	0.0004~1.8933	3.33~1872.37	II
	0.93	4.66	0.2507	306.54	
P1j	0.1~14.04	0.01~17.60	0.0025~0.4539	1.64~507.89	III
	2.38	1.81	0.052	55.85	
C	0.03~19.8	0.01~37.52	0.001~0.3515	1.63~365.06	III
	1.63	0.84	0.031	52.45	

Note : $\frac{0.40\sim5.87}{1.42} = \frac{\text{Min}\sim\text{Max}}{\text{Ave}}$.

The Jurassic coal-bearing source rocks have high organic matter abundance and great hydrocarbon generation potential, but the vitrinite reflectance (Ro) is as low as 0.5% to 0.7% [40]. These source rocks enter their early gas generation stage only when their Ro is greater than 0.8% [49]. Therefore, the Jurassic source rocks are not the main hydrocarbon source for natural gases in the studied area. The Permian Fengcheng formation (P_1f) is a residual sea-lagoon deposit of a sea–land transition environment with type II kerogen (Table 3). It is in the mature-highly mature stage [48]. Given the results of the ethane carbon isotopic analysis above, it should be the source of oil-associated gas in the middle and lower assemblages of the eastern belt around the Penyijingxi sag. The Carboniferous, Jiamuhe, and Lower Urho formations contain abundant organic matter, have an average TOC value greater than 1.5% (Table 3), and are dominated by type III kerogen. Meanwhile, they are in the mature-highly mature stage [40]. According to the relationship between the hydrocarbon generation stage and the Ro value of source rocks [47], these three sets of gas-prone source rocks are at the peak of gas generation, and all of them could be the potential gas sources for the coal-derived gases in the studied area.

Mango has proposed the theory of light hydrocarbon generation based on the light hydrocarbon data of more than 2000 different types of crude oil and the steady-state catalytic kinetic model of heptane genesis [44,50]. According to his theory, all light hydrocarbons generated from the same source rocks have similar K_1 values (Equation (1)) and K_2 values (Equation (2)), which are related to their parent material, but not to the maturity.

$$K_1 = A_1/A_2 \tag{1}$$

where A_1 = 2-MH + 2,3-DMP and A_2 = 3-MH + 2,4-DMP. 2-MH means 2-methylhexane, 2,3-DMP means 2,3-dimethylpentane, 3-MH means 3-methylhexane, and 2,4-DMP means 2,4-dimethylpentane.

$$K_2 = P_3/(P_2 + N_2) \tag{2}$$

where P_2 = 2-MH + 3-MH, P_3 = 2,2-DMP + 2,4-DMP + 2,3-DMP + 3,3-DMP + 3-EP, and N_2 = *cis*-1,3-DMCP + *trans*-1,3-DMCP + 1,1-DMCP. 2,2-DMP means 2,2-dimethylpentane, 3,3-DMP means 3,3-dimethylpentane, 3-EP means 3-ethylpentane, cis-1,3-DMCP means cis-1,3-dimethylcyclopentane, and trans-1,3-DMC means trans-1,3-dimethylcyclopentane.

The K_1 values of the light hydrocarbons associated with the natural gas in the studied area are distributed along two different trend lines (Figure 6a) and are well correlated. The average K_1 values of the black trend line and the orange trend line (0.63 vs. 1.29) are significantly different, indicating two different sources of natural gas in the studied area. In the cross-plot of the K_1 vs. K_2 values, natural gas from different sources will be distributed in different regions [50]. The gas samples collected from the studied area are distributed in two areas: the oil-associated gas area on the left, represented by Well Mobei 2 (with $\delta^{13}C_2$ as −29.38‰) and the coal-derived gas area on the right, represented by Well Qianshao 1 (with $\delta^{13}C_2$ as −27.55‰) (Figure 6b). This pattern further indicates that the gas in the studied area should come from one set of sapropelic source rocks and one set of humic source rocks separately. To further determine the source of the coal-derived gas, the geochemical characteristics of light hydrocarbons from the natural gas in the studied area and from crude oil were compared.

By comparing biomarker compounds, Wu (2012) identified that the Cretaceous crude oil produced from the well Shixi 10 located in the Shixi bulge is from the Lower Urho formation [51]. In addition, he selected seven light hydrocarbons, such as *trans*-1,3-DMP/*trans*-1,2-DMP, from crude oil and determined their fingerprint features (Figure 7). By comparing the light hydrocarbons associated with natural gas in the studied area with the light hydrocarbons associated with crude oil from the Lower Urho formation source rocks, we find that the fingerprint features of the light hydrocarbons associated with natural gas in the Shixi bulge, Mobei bulge, Mobei slope, and Mosuowan bulge are highly similar to those of the above-mentioned crude oil-associated light hydrocarbons (Figure 7). Therefore, it can be inferred that the Lower Urho formation of the Penyijingxi sag should be the main source rock for the coal-derived gas in the studied area.

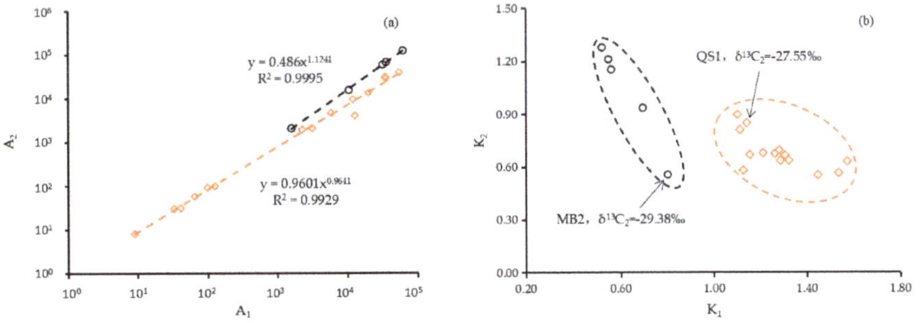

Figure 6. Cross-plots of A_1 vs. A_2 (**a**) and K_1 vs. K_2 (**b**) for the light hydrocarbons associated with the natural gas in the Eastern belt around the Penyijingxi sag, Junggar Basin (Plate from ref. [50]).

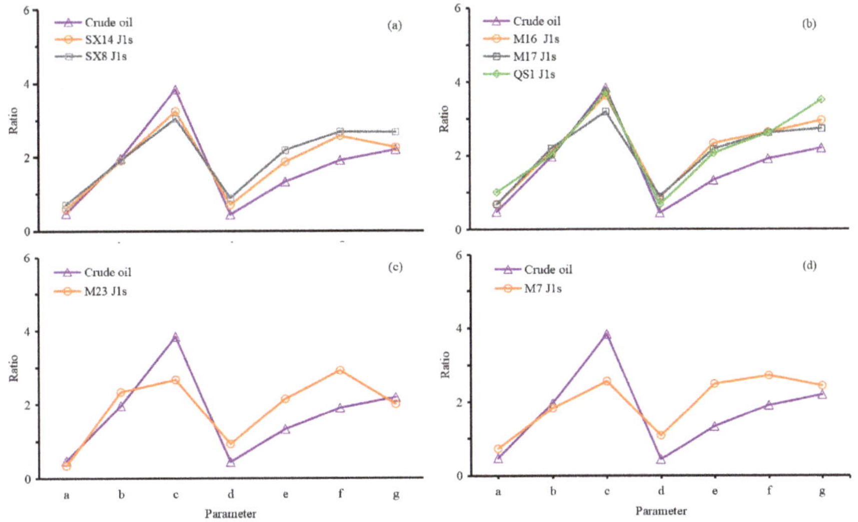

Figure 7. Fingerprint characteristics of light hydrocarbons associated with the natural gas in the eastern belt around the Penyijingxi sag, Junggar Basin and the crude oil (crude oil data from ref. [51]). a. *trans*-1,3-DMP/*trans*-1,2-DMP, b. CH/MCH, c. MCH/∑DMCH, d. n-heptane/(ECH + MCH), e. n-hexane/CH, f. 3-MP/3-MP, and g. 3-MH/2,3-DMP. (**a**) Shixi bulge, (**b**) Mobei Slope, (**c**) Mosuowan bulge, and (**d**) Mobei bulge. trans-1,3-DMP means 1-trans 3-dimethylpentane, trans-1,2-DMP means 1 trans 2-dimethylpentane, CH means cyclohexan, MCH means methylcyclohexane, ΣDMCH means Σdimethyl cyclohexane, ECH means ethylcyclohexane, 2-MP means 2-methylpentane, 3-MP means 3-methylpentane, 3-MH means 3-methylhexane, 2,3-DMP means 2,3-dimethylpentane.

The good linear relationship between the alkane gases' $\delta^{13}C_1$ and the Ro value of their source rocks, with carbon isotopes becoming heavier with thermal evolution, is useful for gas source comparison [14]. Many researchers have fitted the $\delta^{13}C$-Ro regression equations for coal-derived and oil-associated gases [52,53]. Considering the regional geochemical characteristics, we used the empirical regression Equations (3) and (4) proposed by Chen et al. (2021) to calculate the maturity degree of the source rocks of the natural gas. The results show (Table 4) that the Ro value of the oil-associated gas is between 0.75% to 1.55%, and the Ro value of the coal-derived gas is between 0.61% to 1.36%. As a result, it is assumed that oil-associated gas and coal-derived gas originate in the mature to highly mature stage from the source rocks of the Fengcheng formation and Lower Urho formation, respectively. This conclusion is consistent with the measured Ro values of the source rocks of the Fengcheng and Lower Urho formations [40,49]. The ratio of the methane carbon isotope in the alkane gases varies widely from −45.57‰ to −31.19‰ (Figure 4), indicating that natural gases are products of source rocks at different stages of thermal evolution. This conclusion is consistent with the results based on the dryness coefficient.

$$\text{Coal} - \text{derived gas}: \ \delta^{13}C_1 = 25\lg Ro - 37.5 \tag{3}$$

$$\text{Oil} - \text{associated gas}: \ \delta^{13}C_1 = 25\lg Ro - 42.5 \tag{4}$$

Table 4. Maturity degrees of natural gases in the eastern belt around the Penyijingxi sag, Junggar Basin.

Category		Carboniferous	Jurassic Sangonghe Formation			
		Shixi Bulge	Shixi Bulge	Mobei Bulge	Mobei Slope	Mosuowan Bulge
Oil-associated	$\delta^{13}C_1$-$\delta^{13}C_2$/‰	−10.41	−13.19	−14.72~−12.85	/	−13.19
	Ro/%	1.19	1.00	0.75~0.96	/	1.55
Coal-derived	$\delta^{13}C_1$-$\delta^{13}C_2$/‰	−7.86~−5.64	−11.59~−8.38	−15.81~−7.75	−10.25~−6.84	−10.2~−8.15
	Ro/%	1.22~1.79	0.93~1.33	0.61~1.30	0.97~1.36	1.01~1.21

5.3. Gas Migration and Accumulation

With longer migration distances, the alkane gas' $\delta^{13}C_1$ value will decrease and the C_1/C_2 ratio will increase. Therefore, they are highly sensitive geochemical parameters representing the migration characteristics of natural gas [25]. Furthermore, when the gas has undergone no or only weak secondary alteration processes, the differences in its components and carbon isotope compositions are mainly influenced by the maturity of the source rocks, with both $\delta^{13}C_1$-$\delta^{13}C_2$ and Ln(C_1/C_2) increasing with the maturity of the source rocks. On the contrary, when the gas has been subject to diffusion, migration, and dispersion, its $\delta^{13}C_1$-$\delta^{13}C_2$ will gradually increase and its Ln(C_1/C_2) will decrease [54]. During the Neogene-Quaternary period, the southern margin of the Junggar Basin tilted extensively, not only causing the disappearance of the Chepaizi-Mosuowan Paleo-uplift, but also making the central Junggar Basin a southward-tilted monocline [55]. As a result, hydrocarbons migrated and adjusted inevitably. Was there a large-scale lateral gas migration from south to north in the middle and lower assemblages in the eastern belt around the Penyijingxi sag during this period? We collected gas samples from adjacent wells of the same members within the Mobei Bulge for a comparative study.

At present, the proven gas reserves of both well areas M 7 (Mo 7) and MB 2 (Mobei 2) in the Mobei Bulge are concentrated in the Jurassic Sangonghe formation (Figure 1b). Although close to each other, the two well areas have significantly different gas components and carbon isotopic compositions. The average $\delta^{13}C_1$ and $\delta^{13}C_2$ of the gas in well area M 7 are −35.85‰ and −26.11‰, respectively, while the average of δ^{13} C_1 and $\delta^{13}C_2$ of the gas in well area MB 2 are −40.05‰ and −28.39‰, respectively. These data indicate that the source rocks for the gas in well area M 7 are more mature than those for the gas in well area MB 2. Eight typical wells were selected in the well areas M 7 and MB 2 to compare gas migration parameters (Figure 8). It was found that the $\delta^{13}C_1$ of the alkane gas and especially the single hydrocarbon component C_1/C_2 ratio tend to decrease from south to north in the Mobei bulge. If the lightening of alkane gas $\delta^{13}C_1$ is caused by the long-distance lateral migration of gas, then the C_1/C_2 ratio should be increasing rather than decreasing. Taking into account the maturity analysis of the gases in the well areas M 7 and MB 2 above, the difference in the gas components and carbon isotopic compositions between adjacent well areas in the eastern belt around the Penyijingxi sag should be caused by the varied maturity of the source rocks rather than the lateral migration of gas. In other words, there was no significant lateral gas migration due to tilting.

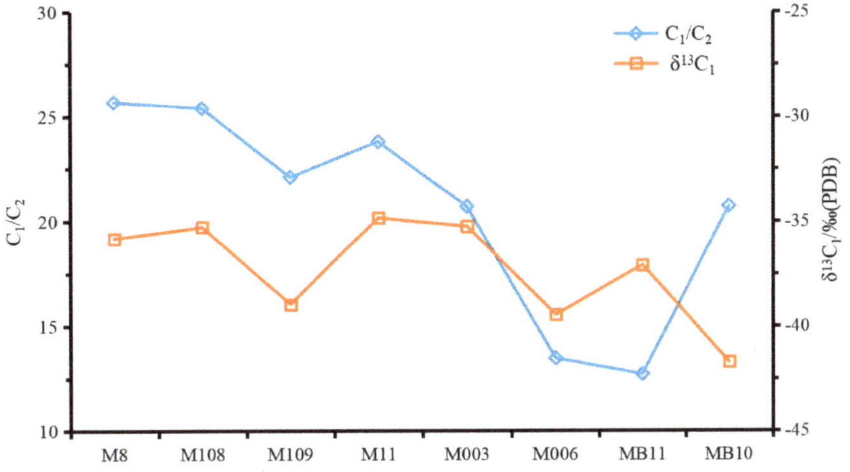

Figure 8. Geochemical parameters of the Mobei bulge in the eastern belt around the Penyijingxi sag, Junggar Basin.

Did large-scale gas migration occur between the superimposed gas-bearing formations in the eastern belt around the Penyijingxi sag? We explored this problem by conducting a case study of the typical well pen 4 (P 4) in the Mosuowan bulge (Figure 1b). From deep to shallow strata, coal-derived, mixed-type, and oil-associated gases appear successively in well pen 4, with the carbon isotopic series changing from reversed ones to positive ones (Figure 9). The maturity of the natural gas source rocks is obtained using regression Equations (3) and (4): the oil-associated gases with Ro values ranging from 1.42% to 1.55% and the coal-derived gases with Ro values ranging from 0.95% to 1.05%. The alkane gases from reservoirs at depths of 4676.00 m and 4514.00 m in well pen 4 are mixed-type gases (Figure 9a) with reversed carbon isotopic series (Figure 9b), and should be a mixture of late filled low-maturity (at a depth of 4514.00 m, with alkane gas $\delta^{13}C_1$ as −43.88‰) to mature (at a depth of 4676.00 m, with alkane gas $\delta^{13}C_1$ as −37.99‰) coal-derived gas and early filled highly matured oil-associated gas. The alkane gases from reservoirs at depths of 5032.45 m and 5100.57 m in well pen 4 are coal-derived gases (Figure 10a), but they also have a reversed carbon isotopic series. The results of the gas source and maturity analyses indicate it is caused by the mixing of the gases of the same genetic type formed at different stages. This conclusion is consistent with the view proposed by Zou et al. (2005) that hydrocarbons are continuously filling the Jurassic system of the central Junggar Basin [33]. The analysis above shows that gases from the upper and lower gas reservoirs of well pen 4 are mixed. Further, it is reasonable to infer that there is vertical gas migration in the studied area, and that the reversed carbon isotopic series of the mixed-type and coal-derived gases are caused by the mixing of coal-derived and oil-associated gases and the mixing of alkane gases of the same genetic type that formed at different stages, respectively.

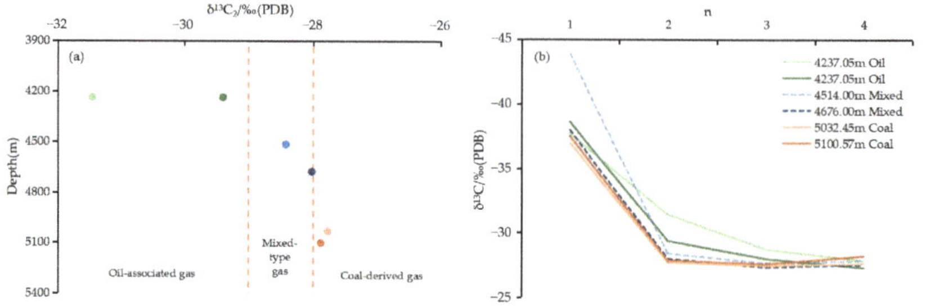

Figure 9. Relationship of $\delta^{13}C_2$ vs. depth (**a**) and carbon isotopic compositions of alkanes (**b**) of the natural gases from well pen 4 in the Mosuowan Bulge, Junggar Basin (data from ref. [56]).

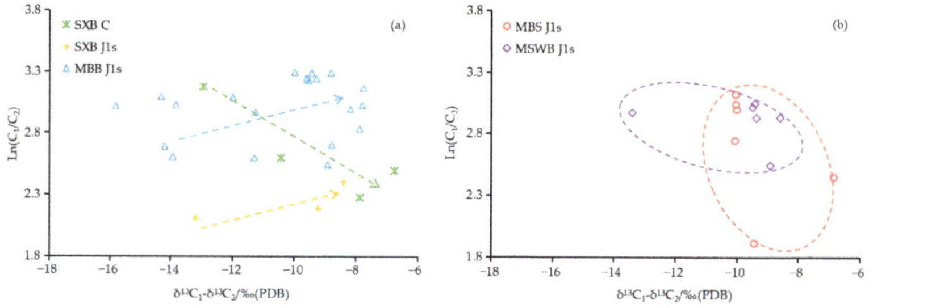

Figure 10. Relationship of $\delta^{13}C_1$-$\delta^{13}C_2$ vs. $Ln(C_1/C_2)$ in the eastern belt around the Penyijingxi sag, Junggar Basin (Plate from ref. [54]).

The $\delta^{13}C_1$-$\delta^{13}C_2$ and $Ln(C_1/C_2)$ of the gases from the Sangonghe formation of the Shixi Bulge and the Sangonghe formation of the Mobei Bulge in the studied area increase synchronously, reflecting the change of gas parameters with maturity (Figure 10a). The data about the gases from the Sangonghe formation of the Mobei Slope and Mosuowan Bulge show no regular pattern (Figure 10b), preventing us from determining the main controlling factors for the differences in their components and carbon isotopic compositions. The carboniferous gases in the Shixi Bulge have gradually increasing $\delta^{13}C_1$-$\delta^{13}C_2$ and gradually decreasing $Ln(C_1/C_2)$, showing obvious characteristics of residual gases after diffusion migration (Figure 10a). In addition, the gas–oil ratio of the reservoir decreases from 4794.59 m^3/t to 204.86 m^3/t along the direction indicated by the green arrow in Figure 10, further confirming the previous diffusion migration of Carboniferous gases, as the gas–oil ratio only decreases due to the loss of lighter components after oil and gas accumulation [57]. Diffusion migration is an important method of forming gas pools in sedimentary basins [25], therefore we speculated that there may be gas pools formed by diffusion migration of carboniferous natural gas in the shallow layers. Through a comprehensive analysis of natural gas composition, carbon isotope, and light hydrocarbon fingerprint parameters, this study explains the genesis and source of natural gas, as well as clarifies the migration characteristics of natural gas after reservoir formation. We can conclude that the gases in the middle and lower assemblages of the eastern belt around the Penyijingxi sag were mainly produced in the processes where hydrocarbons were generated by the source rocks of the Lower Urho and Fengcheng formations and then migrated along faults under greater pressure to fill the near (Carboniferous) or distant (Jurassic Sangonghe formation) reservoirs. The Yanshanian faults only cut through the Triassic–Jurassic systems, resulting in only coal-derived gas production in the slope zone. The Hercynian faults cut through the Carboniferous–Triassic systems and "relay" gases with the Yanshanian faults, enabling the bulge areas to produce coal-derived, oil-associated, and mixed-type gases. The tilting during Himalayan movements did not cause significant lateral migration of the gases in the Sangonghe formation, while diffusion migration of the

Carboniferous gases occurred after reservoir formation. We believe that: (i) the shallow layer (e.g., the Cretaceous) can be considered one of the key strata series for searching for secondary hydrocarbon reservoirs in the next stage of hydrocarbon exploration and (ii) the geochemical parameters and analysis process used in this paper have certain reference values for studying the origin of natural gas in similar petroliferous basins (e.g., the Tarim Basin and the Georgina Basin).

6. Conclusions

The natural gases in the middle and lower assemblages of the eastern belt around the Penyijingxi sag, Junggar Basin consist of a high percentage of methane (71.36–93.34%) with the dryness coefficient ranging from 0.76 to 0.95, averaging 0.91. There are also varying amounts of non-hydrocarbons, such as CO_2 (<1.49%) and N_2 (0.69–11.95%). The carbon isotopic composition of methane ($\delta^{13}C_1$) ranges widely from −45.57‰ to −31.19‰, indicating that the natural gases may be products of source rocks at different stages of thermal maturity.

The contained alkanes show an overall carbon isotopic composition trend as $\delta^{13}C_1 < \delta^{13}C_2 < \delta^{13}C_3 < \delta^{13}C_4$ and have $\delta^{13}C_1$ values < −30‰, indicating that the natural gases are of organic origin. The methane and ethane isotopic compositions and the characteristics of light hydrocarbons show that the natural gases in the studied area are dominated by coal-derived gas and contain a small amount of oil-associated and mixed-type gas. According to the gas source comparison results, it is basically confirmed that the coal-derived gas is from the mature to highly mature source rocks of the Lower Urho formation, and the oil-associated gas is from the mature to highly mature source rocks of the Fengcheng formation.

Gas once migrated vertically in the gas-bearing formations, leading to the mixing of oil-associated and coal-derived gases, as well as the mixing of alkane gases of the same genetic type formed at different stages and possibly causing a reversed carbon isotopic series. While the components and carbon isotopic composition of the natural gases in the Jurassic Sangonghe formation vary with the maturity of the source rocks, these features of the Carboniferous gases are mainly affected by the gas diffusion migration after reservoir formation. Natural gas migration characteristics indicate the shallow layer (e.g., the Cretaceous) in the eastern belt around the Penyijingxi sag may be a favorable area for future oil and gas exploration, which is suitable for searching for secondary hydrocarbon reservoirs.

Author Contributions: Methodology, J.Q. and X.D.; Validation, M.Z. (Minghui Zhou) and T.G.; Investigation, K.L.; Resources, H.L.; Writing–original draft, K.L.; Writing–review & editing, K.L.; Supervision, M.Z. (Ming Zhao) and X.D.; Project administration, M.Z. (Ming Zhao); Funding acquisition, H.L. All authors have read and agreed to the published version of the manuscript.

Funding: This work was supported by major projects of PetroChina Science and Technology (2021DJ0206).

Data Availability Statement: Restrictions apply to the availability of these data. With the permission of Xinjiang Oilfield, it can be obtained from the authors.

Conflicts of Interest: The authors declare no conflict of interest.

Appendix A

Table A1. Natural gas compositions in the eastern belt around the Penyijingxi Sag, Junggar Basin.

Location	Well	Formation	Depth/m	Chemical Composition/%							Carbon Isotopic Composition/‰ (PDB)			
				N_2	CO_2	CH_4	C_2H_6	C_3H_8	C_4H_{10}	C_5H_{12}	$\delta^{13}C_1$	$\delta^{13}C_2$	$\delta^{13}C_3$	$\delta^{13}C_4$
MBB	M121	J1s	4222.00	1.68	0.02	93.34	3.10	0.93	0.57	0.21	nd	nd	nd	nd
MBB	M121	J1s	4254.50	1.60	0.15	91.58	4.02	1.37	0.81	0.28	nd	nd	nd	nd
MBB	M109	J1s	4158.00	0.69	0.67	93.30	3.42	0.98	0.59	0.21	nd	nd	nd	nd
MBB	M113	J1s	4205.00	1.07	0.14	92.70	3.51	1.19	0.84	0.35	nd	nd	nd	nd
MBB	M115	J1s	4204.00	1.43	0.20	88.16	5.39	2.23	1.70	0.63	nd	nd	nd	nd
MBB	M116	J1s	4195.00	2.68	0.23	91.11	3.40	1.19	0.80	0.32	nd	nd	nd	nd
MBB	M117	J1s	4237.50	1.89	0.30	90.63	3.98	1.48	1.04	0.42	nd	nd	nd	nd
MBB	M119	J1s	4258.25	2.04	0.18	90.88	3.98	1.41	0.87	0.31	nd	nd	nd	nd
MBB	M119	J1s	4236.00	1.69	0.19	92.17	3.55	1.23	0.75	0.25	nd	nd	nd	nd
MBB	M003	J1s	3915.00	1.08	0.59	92.40	3.61	1.20	0.74	0.24	−41.07	−29.26	−27.76	−26.95
MBB	M003	J1s	3975.00	2.86	0.29	88.77	4.29	1.65	nd	nd	−35.25	−27.46	−25.92	−25.87
MBB	MB2	J1s	3921.00	2.68	0.41	92.98	2.52	0.77	nd	nd	−42.89	−30.04	−27.26	−27.02
MBB	MB2	J1s	3921.00	2.72	0.51	91.22	3.29	1.09	nd	nd	−44.12	−29.38	−26.33	−26.58
MBB	MB5	J1s	3726.20	2.71	0.00	88.16	5.19	1.73	nd	nd	−34.80	−26.93	−26.14	−26.28
MBB	MB10	J1s	3666.00	3.27	0.41	88.24	4.26	1.69	nd	nd	−41.72	−27.90	−26.59	−26.62
MBS	M16	J1s	4047.25	1.60	0.40	89.90	4.30	1.47	1.05	0.53	−37.56	−27.52	−26.73	−26.76
MBS	QS1	J1s	3944.50	1.35	0.70	89.90	4.34	1.53	1.10	0.52	−37.40	−27.55	−26.72	−25.94
MBS	QS1	J1s	3944.75	1.65	0.46	90.66	4.33	1.48	0.92	0.27	−35.60	−26.14	−25.53	−25.14
MBS	QS2	J1s	3981.00	1.09	0.45	91.18	4.31	1.47	0.94	0.33	−37.49	−27.54	−26.71	−26.80
MBS	QS4	J1s	4014.25	1.56	0.31	90.79	4.39	1.48	0.91	0.33	−37.82	−27.57	−26.72	−26.97
MSWB	M101	J1s	4204.00	2.95	0.62	89.51	4.25	1.47	0.84	0.24	−36.64	−27.21	−26.72	−26.75
SXB	S006	J1s	3577.00	0.92	0.68	91.12	4.38	1.47	0.90	0.33	nd	nd	nd	nd
SXB	S006	C	4373.00	4.35	0.21	88.82	3.70	1.07	1.08	0.54	−41.62	−28.68	−25.88	−25.40
SXB	S007	C	4408.50	11.95	0.09	71.36	5.32	4.36	4.55	1.78	−40.58	−30.17	−26.85	−26.75
SXB	SX1	C	4438.00	5.21	0	78.97	6.52	3.96	3.95	1.39	−33.43	−26.69	−26.00	−25.44
SXB	SX1	C	4473.00	5.49	0.37	77.25	7.92	3.73	3.70	1.55	−35.36	−27.50	−26.63	−26.37
SXB	S015	J1s	nd	4.68	1.07	75.10	9.11	4.04	4.04	1.49	−42.5	−29.31	−26.87	−26.47
SXB	SX8	J1s	nd	1.25	0.63	84.04	7.61	2.98	1.97	0.77	−34.44	−26.06	−24.66	−25.61
SXB	SX14	J1s	nd	3.63	0.82	75.41	8.45	4.72	3.45	1.40	−36.98	−27.75	−27.20	−27.96
MBB	MB11	J1s	3710.75	1.31	0.82	84.98	6.69	3.01	1.93	0.75	−37.11	−28.19	−26.78	−26.7
MBB	MB2	J1s	3907.00	2.81	0.07	91.56	3.40	1.13	0.64	0.23	−35.65	−26.84	−26.16	−26.44
MBB	M003	J1s	3971.50	3.39	0.60	89.31	4.58	1.31	0.64	0.18	−37.82	−26.58	−24.68	−23.64

Table A1. *Cont.*

Location	Well	Formation	Depth/m	Chemical Composition/%							Carbon Isotopic Composition/‰ (PDB)			
				N_2	CO_2	CH_4	C_2H_6	C_3H_8	C_4H_{10}	C_5H_{12}	$\delta^{13}C_1$	$\delta^{13}C_2$	$\delta^{13}C_3$	$\delta^{13}C_4$
MBB	MB9	J1s	3761.25	1.82	0.35	90.57	4.42	1.37	0.97	0.50	−42.92	−27.11	−25.73	−26.00
MBB	MB9	J1s	3778.00	1.71	1.29	84.56	5.75	2.67	1.90	0.93	−45.57	−31.39	−28.67	−27.47
MBB	M005	J1s	3890.25	2.36	0.42	86.76	6.41	2.38	1.59	0.08	−44.08	−30.16	−27.67	−27.4
MBB	M006	J1s	3759.25	2.01	0.45	85.84	6.38	2.71	1.95	0.66	−39.46	−28.18	−28.01	−27.3
MBB	MB5	J1s	3726.20	3.96	0.22	89.21	4.46	1.12	0.81	0.23	−35.58	−27.41	−26.19	−26.5
MBB	M108	J1s	4179.00	2.74	0.30	89.66	3.53	1.10	0.75	0.37	−35.28	−25.74	−25.28	−25.8
MBB	M109	J1s	4185.00	1.81	0.22	91.97	4.16	1.10	0.50	0.12	−38.96	−24.66	−23.56	nd
MBB	M11	J1s	4139.00	3.60	1.49	88.58	3.30	1.01	0.68	0.32	−35.15	−25.71	−25.13	−25.64
MBB	M11	J1s	4177.00	1.65	0.27	91.82	3.86	1.33	0.77	0.22	−34.84	−27.09	−26.35	−26.61
MBB	M7	J1s	4227.50	2.40	0.47	90.92	3.54	1.19	0.80	0.36	−35.72	−26.43	−25.85	−26.37
MBB	M7	J1s	4260.00	3.65	0.50	89.99	3.33	1.15	0.71	0.31	−37.88	−27.90	−27.17	−27.66
MBB	M8	J1s	4233.00	1.47	0.51	91.43	3.56	1.07	0.70	0.36	−35.82	−26.21	−25.39	−25.46
MBB	M8	J1s	4265.50	1.09	0.42	87.70	5.86	2.23	1.53	0.62	−34.69	−25.91	−25.33	−25.74
MBS	M16	J1s	4041.00	2.69	0.62	87.44	5.62	1.77	0.90	0.32	−36.54	−26.48	−25.03	−26.26
MBS	M171	J1s	4472.85	1.45	0.40	90.45	4.52	1.58	0.98	0.37	−36.59	−26.58	−25.57	−26.19
MBS	M17	J1s	4161.50	1.86	0.57	91.20	4.02	1.07	0.60	0.24	−36.9	−26.86	−25.64	−26.38
MBS	M12	J1s	4235.00	3.25	0.66	80.44	6.96	4.14	2.72	0.95	−34.14	−27.30	−26.29	−26.59
MBS	M17	J1s	4192.00	1.55	0.66	73.88	10.90	5.96	4.38	1.57	−35.93	−26.49	−24.36	−25.42
MSWB	M101	J1s	4209.00	3.01	0.41	88.76	4.35	1.56	1.07	0.48	−36.81	−27.31	−26.32	−26.79
MSWB	M102	J1s	4251.00	2.71	0.52	88.13	4.70	1.92	1.21	0.48	−36.11	−27.51	−26.40	−26.52
MSWB	M103	J1s	4250.50	2.59	0.58	87.65	4.69	1.98	1.43	0.62	−36.66	−27.28	−26.647	−26.98
MSWB	P5	J1s	4250.00	3.25	0.57	87.49	4.49	1.86	1.20	0.45	−41.43	−28.04	−27.03	−27.42

Note: nd means no data, C means Carboniferous, J1s means Jurassic Sangonghe Formation, MBB means Mobei bulge, MSWB maens Mosuowan bulge, SXB means Shixi bulge, MBS means Mobei slope.

References

1. Glasby, G.P. Abiogenic origin of hydrocarbons: An historical overview. *Resour. Geol.* **2006**, *56*, 83–96. [CrossRef]
2. Dai, J.; Gong, D.; Ni, Y.; Huang, S.; Wu, W. Stable carbon isotopes of coal-derived gases sourced from the Mesozoic coal measures in China. *Org. Geochem.* **2014**, *74*, 123–142. [CrossRef]
3. Song, Y.; Xu, Y.C. Origin and identification of natural gases. *Pet. Explor. Dev.* **2005**, *32*, 24–29.
4. Milkov, A.V. Methanogenic biodegradation of petroleum in the West Siberian Basin (Russia): Significance for formation of giant Cenomanian gas pools. *AAPG Bull.* **2010**, *94*, 1485–1541. [CrossRef]
5. Hutcheon, I.; Krouse, H.R.; Abercrombie, H.J. Controls on the origin and distribution of elemental sulfur, H_2S, and CO_2 in Paleozoic hydrocarbon reservoirs in Western Canada. *ACS Symp. Ser.* **1995**, *612*, 426–438.
6. Arrouvel, C.; Prinzhofer, A. Genesis of natural hydrogen: New insights from thermodynamic simulations. *Int. J. Hydrogen Energy* **2021**, *46*, 18780–18794. [CrossRef]
7. Milkov, A.V.; Giuseppe, E. Revised genetic diagrams for natural gases based on a global dataset of >20,000 samples. *Org. Geochem.* **2018**, *125*, 109–120. [CrossRef]
8. Dai, J.X.; Huang, S.P. Discussion on the carbon isotopic reversal of alkane gases from the Huanglong Formation in the Sichuan Basin, China. *Acta Pet. Sin.* **2010**, *31*, 710–717.
9. Des Marais, D.J.; Donchin, J.H.; Nehring, N.L.; Truesdell, A.H. Molecular carbon isotopic evidence for the origin of geothermal hydrocarbons. *Nature* **1981**, *292*, 826–828. [CrossRef]
10. Tilley, B.; Muehlenbachs, K. Isotope reversals and universal stages and trends of gas maturation in sealed, self-contained petroleum systems. *Chem. Geol.* **2013**, *339*, 194–204. [CrossRef]
11. Dai, J.; Xia, X.; Qin, S.; Zhao, J. Origins of partally reversed alkane $\delta^{13}C$ values for biogenic gases in China. *Org. Geochem.* **2004**, *35*, 405–411. [CrossRef]
12. Hosgormez, H.; Etiope, G.; Yalin, M.N. New evidence for a mixed inorganic and organic origin of the Olympic Chimaera fire (Turkey): A large onshore seepage of abiogenic gas. *Geofluids* **2008**, *8*, 263–273. [CrossRef]
13. Ogbesejana, A.B.; Bello, O.M.; Ali, T. Origin and depositional environments of source rocks and crude oils from Niger Delta Basin: Carbon isotopic evidence. *China Geol.* **2020**, *3*, 602–610. [CrossRef]
14. Schoell, M. The hydrogen and carbon isotopic composition of methane from natural gases of various origins. *Geochim. Cosmochim. Acta* **1980**, *44*, 649–661. [CrossRef]
15. Kotarba, M. Isotopic geochemistry and habitat of the natural gases from the Upper Carboniferous Žacleř coal-bearing formation in the Nowa Ruda coal district (Lower Silesia, Poland). *Org. Geochem.* **1990**, *16*, 549–560. [CrossRef]
16. Battani, A.; Sarda, P.; Prinzhofer, A. Basin scale natural gas source, migration and trapping traced by noblegases and major elements: The Pakistan Indus basin. *Earth Planet. Sci. Lett.* **2000**, *181*, 229–249. [CrossRef]
17. Prinzhofer, A.; Mello, M.R.; Freitas, L.; Takaki, T. New geochemical characterization of natural gas and its use in oil and gas evaluation. *AAPG Mem.* **2000**, *73*, 107–119.
18. Allan, U.S. Model for hydrocarbon migration and entrapment within faulted structures. *AAPG Bull.* **1989**, *73*, 803–811.
19. Hillman, J.I.; Cook, A.E.; Daigle, H.; Nole, M.; Malinverno, A.; Meazell, K.; Flemings, P.B. Gas hydrate reservoirs and gas migration mechanisms in the Terrebonne Basin, Gulf of Mexico. *Mar. Pet. Geol.* **2017**, *86*, 1357–1373. [CrossRef]
20. Wanner, P.; Hunkeler, D. Isotope fractionation due to aqueous phase diffusion–What do diffusion models and experiments tell?–A review. *Chemosphere* **2019**, *219*, 1032–1043. [CrossRef]
21. Losh, S.; Walter, L.; Meulbroek, P.; Martini, A.; Whelan, J. Reservoir Fluids and Their Migration into the South Eugene Island Block 330 Reservoirs, Offshore Louisiana. *AAPG Bull.* **2002**, *86*, 1463–1488.
22. Prinzhofer, A.; Battani, A. Gas isotopes tracing: An important tool for hydrocarbons exploration. *Oil Gas Sci. Technol.* **2003**, *58*, 299–311. [CrossRef]
23. Schloemer, S.; Krooss, B.M. Molecular transport of methane, ethane and nitrogen and the influence of diffusion on the chemical and isotopic composition of natural gas accumulations. *Geofluids* **2010**, *4*, 81–108. [CrossRef]
24. Prinzhofer, A.; Neto, E.; Battani, A. Coupled use of carbon isotopes and noble gas isotopes in the Potiguar basin (Brazil): Fluids migration and mantle influence. *Mar. Pet. Geol.* **2010**, *27*, 1273–1284. [CrossRef]
25. Prinzhofer, A.; Mello, M.R.; Takaki, T. Geochemical Characterization of Natural Gas: A Physical Multivariable Approach and Its Applications in Maturity and Migration Estimates. *AAPG Bull.* **2000**, *84*, 1152–1172.
26. Bourg, I.C.; Sposito, G. Isotopic fractionation of noble gases by diffusion in liquid water: Molecular dynamics simulations and hydrologic applications. *Geochim. Cosmochim. Acta* **2008**, *72*, 2237–2247. [CrossRef]
27. Sun, P.; Wang, Y.; Leng, K.; Li, H.; Ma, W.; Cao, J. Geochemistry and origin of natural gas in the eastern Junggar Basin, NW China. *Mar. Pet. Geol.* **2016**, *75*, 240–251. [CrossRef]
28. Yang, H.B.; He, W.J.; Jia, X.Y. *The Fourth Evaluation of Oil and Gas Resources in Junggar Basin*; Internal Report of Exploration and Development Research Institute of Xinjiang Oilfield Company: Karamay, China, 2016.
29. IEA. *Offshore Energy Outlook*; IEA: Paris, France, 2018.
30. Tao, K.; Cao, J.; Wang, Y.; Ma, W.; Xiang, B.; Ren, J.; Zhou, N. Geochemistry and origin of natural gas in the petroliferous Mahu sag, northwestern Junggar Basin, NW China: Carboniferous marine and Permian lacustrine gas systems. *Org. Geochem.* **2016**, *100*, 62–79. [CrossRef]

31. Huayao, Z.; Fang, H.; Bin, Z.B.C. History of hydrocarbon-filling and remigrating in hinterland of the Junggar Basin. *Chin. J. Geol.* **2005**, *40*, 499–509.
32. Ma, W.; Wei, Y.; Li, X.; Tao, S.; Li, Y.; Zhao, Z. Accumulation process and control factors of Jurassic-Cretaceous distant-source and secondary-filled reservoirs in the hinterland of Junggur Basin. *Acta Sci. Nat. Univ. Pekin.* **2018**, *54*, 1195–1204.
33. Zhu, G.; Su, J.; Yang, H.; Wang, Y.; Fei, A.; Liu, K.; Zhu, Y.; Hu, J.; Zhang, B. Formation mechanisms of secondary hydrocarbon pools in the Triassic reservoirs in the northern Tarim Basin. *Mar. Pet. Geol.* **2013**, *46*, 51–66. [CrossRef]
34. Lei, J.; Worden, R.H.; Cai, C.F. Thermochemical sulfate reduction and fluid evolution of the Lower Triassic Feixianguan Formation sour gas reservoirs, northeast Sichuan Basin, China. *AAPG Bull.* **2014**, *98*, 947–973.
35. Volk, H.; George, S.C.; Kempton, R.H.; Liu, K.; Ahmed, M.; Ambrose, G.J. Petroleum migration in the Georgina Basin: Evidence from the geochemistry of oil inclusions and bitumens. In Proceedings of the Central Australian Basins Symposium (CABS). Northern Territory Geological Survey, Alice Springs, Northern Territory, Australia, 16–18 August 2007; pp. 282–303.
36. Ali, A.; Younas, M.; Ullah, M.; Hussain, M.; Toqeer, M.; Bhatti, A.S.; Khan, A. Characterization of secondary reservoir potential via seismic inversion and attribute analysis: A case study. *J. Pet. Sci. Eng.* **2019**, *178*, 272–293. [CrossRef]
37. Jahn, B.M.; Wu, F.; Chen, B. Massive granitoid generation in Central Asia: Nd isotope evidence and implication for continental growth in the Phanerozoic. *Episodes* **2000**, *23*, 82–92. [CrossRef] [PubMed]
38. Windley, B.F.; Alexeiev, D.; Xiao, W.; Kröner, A.; Badarch, G. Tectonic models for accretion of the Central Asian Orogenic Belt. *J. Geol. Soc.* **2007**, *164*, 31–47. [CrossRef]
39. Liu, H.L.; Yin, H.; Imin, A.; Ding, X. Origin and preservation of deep natural gas in the lower assemblage of the western Junggar Basin. *J. China Univ. Min. Technol.* **2022**, *51*, 148–159.
40. Chen, J.P.; Wang, X.L.; Deng, C.P.; Liang, D.G.; Zhang, Y.Q.; Zhao, Z.; Wang, Y.T. Geochemical features of source rocks and crude oil in the Junggar Basin, northwestem China. *Acta Geol. Sin.* **2016**, *90*, 37–67.
41. Saadati, H.; Al-Iessa, H.J.; Alizadeh, B.; Tarhandeh, E.; Jazayeri, M.H.; Bahrami, H.; Rashidi, M. Geochemical characteristics and isotopic reversal of natural gases in eastern Kopeh-Dagh, NE Iran. *Mar. Pet. Geol.* **2016**, *78*, 76–87. [CrossRef]
42. Qiu, N.S.; Yang, H.B.; Wang, X.L. Tectonic-thermal evolution in the Junggar Basin. *Chin. J. Geol.* **2002**, *37*, 423–429.
43. Milkov, A.V. Worldwide distribution and significance of secondary microbial methane formed during petroleum biodegradation in conventional reservoirs. *Org. Geochem.* **2011**, *42*, 184–207. [CrossRef]
44. Mango, F.D. The light hydrocarbons in petroleum: A critical review. *Org. Geochem.* **1997**, *26*, 417–440. [CrossRef]
45. Hu, T.L.; Ge, B.X.; Zhang, Y.G.; Liu, B. Development and application of fingerprint parameters of source rock adsorbed hydrocarbon and natural gas light hydrocarbon. *Pet. Geol. Exp.* **1990**, *12*, 375–394,450.
46. Leythaeuser, D.; Schaefer, R.; Cornford, C.; Weiner, B. Generation and migration of light hydrocarbons (C_2–C_7) in sedimentary basins. *Org. Geochem.* **1979**, *1*, 191–204. [CrossRef]
47. Qian, H.T.; Su, D.X.; Ablimit, I.; Wang, X.Y.; Li, Z.H.; Wang, G.D. Petroleum geological characteristics and exploration potential in slope area of Well Pen-1 Western Depression in Junggar Basin. *Nat. Gas Geosci.* **2021**, *32*, 551–561.
48. Chen, J.P.; Sun, Y.G.; Zhong, N.N.; Huang, Z.K.; Deng, C.P. Lacustrine facies hydrocarbon expulsion efficiency of hydrocarbon source rock and hydrocarbon expulsion model. In Proceedings of the Fourteenth China Mineral Rock Geochemistry Academic Annual Conference, Nanjing, China, 21 April 2013.
49. Harry, D.J. Three common source rock evaluation errors made by geologists during prospect or play appraisals. *AAPG Bull.* **2009**, *93*, 341–356.
50. Mango, F.D. The origin of light hydrocarbons in petroleum: A kinetic test of the steady-state catalytic hypothesis. *Geochim. Cosmochim. Acta* **1990**, *54*, 1315–1323. [CrossRef]
51. Wu, M. *Geochemical Study on Hydrocarbon Accumulation in Deep Strata of the Central Junggar Basin, Northwest China*; Nanjing University: Nanjing, China, 2012.
52. Schoell, M. Genetic characterization of natural gases. *AAPG Bull.* **1983**, *67*, 2225–2238.
53. Chen, J.; Wang, X.; Chen, J.; Ni, Y.; Xiang, B.; Liao, F.; He, W.; Yao, L.; Li, E. New equation to decipher the relationship between carbon isotopic composition of methane and maturity of gas source rocks. *Sci. China Earth Sci.* **2021**, *64*, 470–493. [CrossRef]
54. Prinzhofer, A.; Pernaton, R. Isotopically light methane in natural gas: Bacterial imprint or diffusive fractionation? *Chem. Geol.* **1997**, *142*, 193–200. [CrossRef]
55. Hou, L.; Wang, J.; Kuang, L.; Zhang, G.; Liu, L.; Kuang, J. Provenance Sediments and Its Exploration Significance-A Case from Member 1 of Qingshuihe Formation of Lower Cretaceous in Junggar Basin. *Earth Sci. Front.* **2009**, *16*, 337–348. [CrossRef]
56. Kou, C.H. *Origins and Accumulation Model of Hydrocarbon in Deep Strata of the Central Junggar Basin, Northwestern China*; Nanjing University: Nanjing, China, 2017.
57. Cheshmeh Sefidi, A.; Ajorkaran, F. A novel MLP-ANN approach to predict solution gas-oil ratio. *Pet. Sci. Technol.* **2019**, *37*, 2302–2308. [CrossRef]

processes

MDPI

Article

Coupling Relationship between Diagenesis and Hydrocarbon Charging in Middle Permian–Lower Triassic in the Eastern Slope of Mahu Sag in Junggar Basin, Northwest China

Wenlong Jiang [1,*], Ping Song [1], Hailei Liu [1], Baoli Bian [1], Xueyong Wang [1], Wenjian Guo [1] and Nan Wang [2]

1 Research Institute of Exploration and Development, Xinjiang Oilfield Company, PethoChina, Karamay 834000, China
2 Shixi Oilfield Operation District, Xinjiang Oilfield Company, PetroChina, Karamay 834000, China
* Correspondence: jwl@petrochina.com.cn

Abstract: In this study, a variety of test and analysis methods, such as cast thin sections, fluorescent thin sections, scanning electron microscopy, fluid inclusions, etc., were comprehensively used and combined with logging data, sedimentary systems, burial history and other research results, to systematically study the diagenesis characteristics of Middle Permian Lower Wuerhe formation–Lower Triassic Baikouquan formation reservoirs and their control on hydrocarbon accumulation. The coupling relationship between the hydrocarbon accumulation process and reservoir secondary pores was established. The result shows that besides the development of primary intergranular pores, the reservoir develops secondary pores, such as particle dissolution pores, cement dissolution pores and fractures. The development of secondary dissolution pores, such as particle dissolution pores, carbonate and zeolite cement dissolution pores, is mainly controlled by the range and scale of organic acids produced by the thermal evolution of source rocks. It is considered that being located in the updip direction of source rocks, being involved in the development of the unconformity surface and the faults connecting source rocks, and being involved in the development of alkaline cements (such as laumontite) are the three dominant conditions for the development of secondary dissolution pores in the study area.

Keywords: diagenesis; secondary pores; organic acid dissolution; hydrocarbon accumulation; eastern slope of Mahu sag; Junggar Basin

Citation: Jiang, W.; Song, P.; Liu, H.; Bian, B.; Wang, X.; Guo, W.; Wang, N. Coupling Relationship between Diagenesis and Hydrocarbon Charging in Middle Permian–Lower Triassic in the Eastern Slope of Mahu Sag in Junggar Basin, Northwest China. *Processes* **2023**, *11*, 345. https://doi.org/10.3390/pr11020345

Academic Editors: Ping Gao, Yidong Cai, Yingfang Zhou and Quan Gan

Received: 1 November 2022
Revised: 8 January 2023
Accepted: 18 January 2023
Published: 20 January 2023

1. Introduction

The study of diagenesis is an important basis for finding different types of reservoirs. In recent years, good progress has been made in the study of diagenesis, mainly in the following three aspects: the influence of the interaction between fluid and rock on diagenesis [1,2], the relationship between hydrocarbon charging and reservoir densification [3,4], and the influence of high temperature and pressure on diagenesis evolution [5,6]. At the same time, diagenesis research also developed from qualitative to quantitative. Some scholars' statistics show that at least 1/3 of the global clastic reservoir space is formed by mineral dissolution [7]. More and more scholars realize that a large number of secondary pores are produced by organic acid dissolution. For example, Shi et al. and Zhao et al. found that compared with other types of feldspar, the secondary pores formed by the dissolution of potassium feldspar are the largest [8,9]. Yang et al. found that the main product of potassium feldspar dissolved in organic acids is kaolinite [10].

The exploration work of the study area is dominated by fault stratum traps in the early stage, and the exploration wells are all located near the fan delta plain or the transitional zone between the fan delta plain and fan delta front. The lithology is dominated by coarse-grained thick blocky sandy conglomerates with poor sorting and poor physical properties. Therefore, although many exploration wells in the study area obtained oil and gas, the yield is generally low and no substantive breakthrough has been achieved. In recent years, through the transformation of exploration ideas, the exploration of Triassic Baikouquan formation in the western slope of the Mahu sag, which is adjacent to the study area, has achieved fruitful results and many wells obtained high-yield industrial oil and gas flow. Therefore, more and more attention has been paid to the eastern slope of Mahu sag with the same favorable conditions for oil and gas migration and accumulation.

In this article, the diagenesis characteristics of the Middle Permian Lower Wuerhe formation–Lower Triassic Baikouquan formation reservoirs and their control on hydrocarbon accumulation will be studied to establish the coupling relationship between diagenesis and hydrocarbon charging in order to provide a theoretical basis for exploration work in the study area.

2. Geological Setting

The Junggar Basin is located in the northern part of Xinjiang province. It is sandwiched between the Zaire Mountains, the Qinggelidi Mountains, the Kelameili Mountains and the Yilinhebiergen Mountains. It is slightly triangular and has an area of about 13.6×10^4 km^2. It is a superimposed basin that has undergone multiple tectonic movements from Paleozoic to Quaternary. The study area is located in the northwestern margin of the Junggar Basin (Figure 1a), which is the most abundant area of oil–gas accumulation. The Carboniferous to Quaternary are well developed, and the maximum sedimentary thickness is nearly 15 km [11]. Permian and Triassic, Triassic and Jurassic, Jurassic and Cretaceous are regional unconformable contact relations (Figure 1b). The overlap and denudation of the Permian strata are obvious. Vertically, oil and gas reservoirs are distributed in 14 formations of 5 systems, i.e., Carboniferous, Permian, Triassic, Jurassic and Cretaceous [12]. The main objective strata of this study are Middle Permian Lower Wuerhe formation and Lower Triassic Baikouquan formation (Figure 1c).

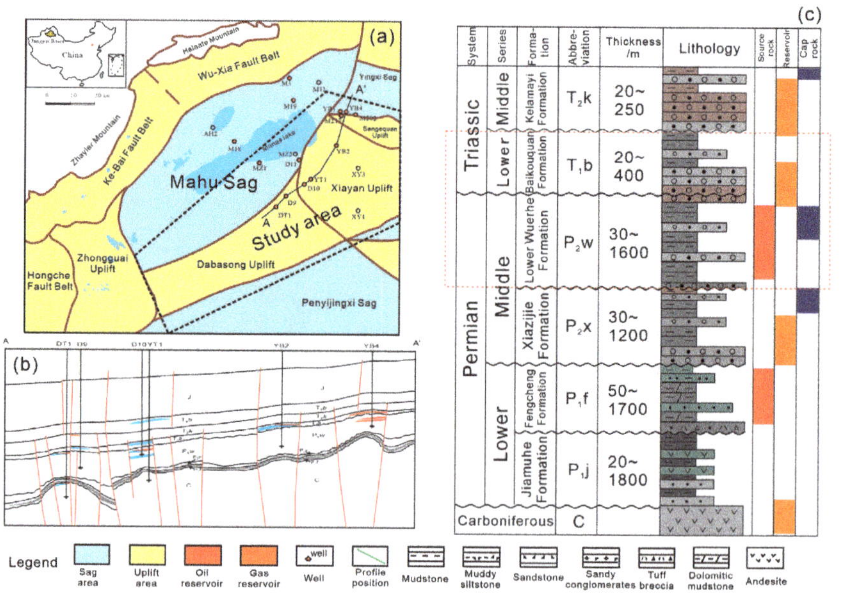

Figure 1. Distribution of structural units in the northwestern Junggar Basin. (**a**) study area tectonic unit and exploration well distribution; (**b**) typical reservoir profile of study area; (**c**) stratigraphy histogram of study area.

The study area has undergone multi-stage tectonic movements, such as Hercynian, Indosinian, Yanshanian and Himalayan tectonic movements. The sedimentary strata were relatively stable in the Indosinian movement (Triassic–early Jurassic), where the formation thickness changes little and gradually overlaps in the direction of uplift. In the early Yanshanian movement (middle Jurassic), it was relatively stable, forming delta facies and lake facies. In the middle and late Yanshanian movements (late Jurassic–early Cretaceous), the activity was intensified. It was manifested in the strata above the middle Jurassic system, especially the Toutunhe formation, which was obviously denuded and thinned to the high part and was in angular unconformity contact with the overlying Cretaceous. The late Yanshanian and Himalayan tectonic movements had little influence on the Cretaceous, but during the Himalayan (Neogene-Quaternary), the regional southward tilting caused the strata above the Jurassic in the study area to form a monocline structural feature uplifted to the northwest [13].

3. Methodology

In this study, an Olympus CX23 optical microscope was used to observe the microscopic mineral and pore structures of samples. The resolutions of the secondary electron image of the Tescan scanning electron microscope were 1.0 nm (15 kV) and 1.5 nm (1 kV), and the resolution of the analysis mode is 3.0 nm (15 kV, 5 nA, WD = 8 mm). This microscope was used for the ultra-high resolution observation of the pore type, pore structure, mineral type, and sample structure. The CM300 overburden porosimeter was used to determine the physical properties of the samples. The effective porosity test range is 0.01–40%, the permeability test range is 0.0005 mD–15 D. The experimental analyses were conducted at the Institute of Experiment and Analysis at the Xinjiang Oilfield Company (Karamay City, Xinjiang province, China).

The fluid inclusions were investigated via mono-polarizer and fluorescent observations using an Olympus Dual Channel Fluorescent-Transmission Light Microscope equipped with a telephoto lens (8 mm, 100×). The digenetic occurrence of inclusions and the fluorescent colors of the hydrocarbon inclusions were observed. Doubly polished thin sections were prepared for fluid inclusion microthermometric analysis using a Linkam THMS-600 heating–freezing stage. The homogenization temperature (Th) of fluid inclusions were obtained by cycling to the liquid phase. The measurements were determined using a heating rate of 10 °C/min. The measured temperature precisions for the homogenization is ±1 °C. The laboratory testing temperature was 25 °C, and the humidity was 65%. The experimental analyses were conducted at the Analtical Laboratory of Beijing Research Institute of Uranium Geology.

PetroMod-1D basin modeling software was used to reconstruct the burial history, thermal history and hydrocarbon generation history.

4. Results

4.1. Compositions, Porosity and Permeability

The sandy conglomerates reservoir of the Middle Permian Lower Wuerhe formation–Lower Triassic Baikouquan formation in the study area is characterized by low porosity and permeability, and its vertical differentiation characteristics are also obvious. The reservoir properties of each formation are as follows.

The size of detrital grains in the reservoir of the Baikouquan formation is mainly distributed between 0.25–2 mm. There is also a large amount of distribution between 2–10 mm, up to 10 mm. It has the characteristics of low textural maturity and low composition maturity. The particle sorting is poor, and the roundness is mainly subangular–subround. The particles are mostly line contact and particle support. Reservoir porosity is normal distribution (Figure 2a), mainly concentrated in 6~12%, an average of 9.38%. Permeability is also of a normal distribution but relatively concentrated, mainly distributed in (0.1~1) $\times 10^{-3}$ μm^2, except for the influence of fracture factors; the average is 2.35 $\times 10^{-3}$ μm^2 (Figure 2b), which is a typical low porosity and low permeability reservoir. The analysis shows that the main reason for the great change of reservoir physical properties in the Baikouquan formation is the difference of facies and grain size. The high porosity sandy conglomerates reservoirs are mainly distributed in the underwater distributary channel subfacies of fan delta front, and the micro-fractures developed near the fault zone also have a great effect on the improvement of reservoir physical properties.

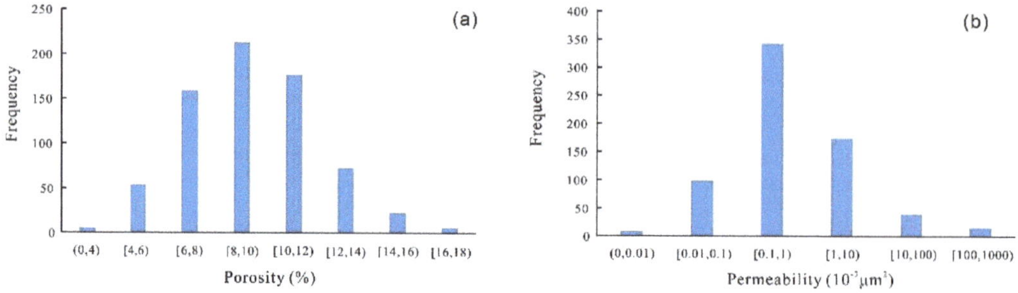

Figure 2. Distribution frequency histogram of porosity (**a**) and permeability (**b**) of Baikouquan formation.

The sizes of detrital grains in the reservoir of the Lower Wuerhe formation are mainly larger than 2 mm, up to 20 mm. It also has the characteristics of low textural maturity and low composition maturity. The particle sorting is poor, and the roundness is mainly subrounded. The contact between particles is mainly line contact, followed by point–line contact and point contact. Compared with the Baikouquan formation, the porosity and permeability of sandy conglomerates in the Lower Wuerhe formation decrease: the porosity distribution is relatively concentrated between 6% and 10%, with an average of 8.53% (Figure 3a). Permeability distribution range is relatively wide, mainly distributed in the range of $(0.01{\sim}10) \times 10^{-3}$ μm^2 (Figure 3b). The reservoir properties of sandy conglomerates in the Lower Wuerhe formation are greatly affected by the cementation of laumontite and the dissolution of volcanic clastic. For example, the reservoirs with porosity greater than 16% in Well YB1 are mainly laumontite-dissolved reservoirs.

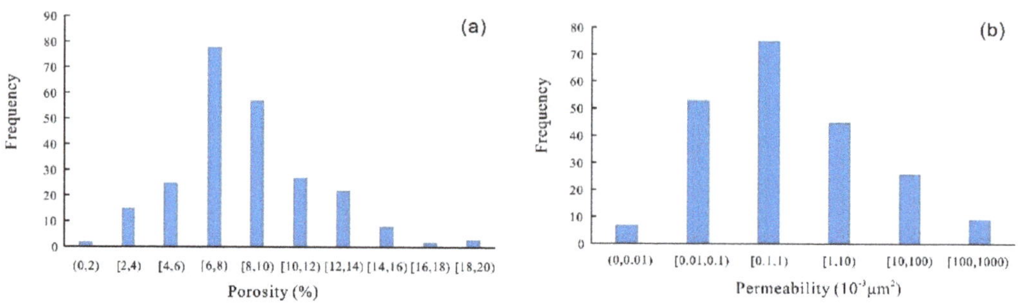

Figure 3. Distribution frequency histogram of porosity (**a**) and permeability (**b**) of Lower Wuerhe formation.

4.2. Compaction

The compaction effect in the study area is obvious, which is the main pore reduction effect of the reservoir. Under the microscope, the contact relationship between the clastic particles is mainly point–line contact, and the concave–convex contact is locally developed. With the deepening of the sandy conglomerates burial, the compaction effect is enhanced so that the clastic particles are closely arranged in a certain direction (Figure 4a). The rigid particles (such as quartz and feldspar particles) are broken under pressure, and the feldspar is mostly broken along the cleavage crack. With the further strengthening of compaction, the contact mode of particles transits from point contact to line contact until the concave–convex contact (Figure 4b). Finally, the phenomenon of pressure solution appears with sutured contact, as shown in Figure 4c,d.

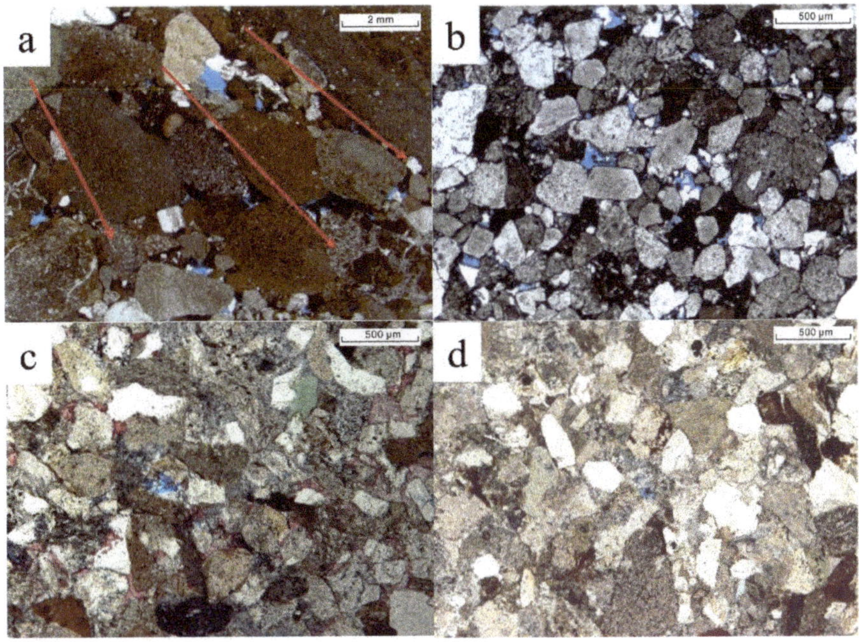

Figure 4. Cast thin sections of compaction in study area. (**a**) well YT1, 4821.9 m, Lower Wuerhe formation, sandy conglomerates, grains in close contact, line contact–concavo–convex contact; (**b**) well YT1, 4810 m, Lower Wuerhe formation, medium sandstone, grains in close contact, line contact–concavo–convex contact; (**c**) well MZ4, 3580 m, grains in close contact, sutured contact; (**d**) well M18, 3438.3 m, grains in close contact, sutured contact.

4.3. Cementation

Previous studies have confirmed that the initial sedimentary water of the Baikouquan formation in the study area is acidic, and the Lower Wuerhe formation is alkaline [14–17]. Therefore, the Baikouquan formation mainly develops neutral-acidic cements, such as kaolinite, quartz secondary overgrowth rims and illite. Silicate and carbonate cements can also be formed in some areas. The sandy conglomerates of the Lower Wuerhe formation mainly forms cements, such as silicate and carbonate in alkaline sedimentary environment, especially zeolite cement.

4.3.1. Carbonate Cementation

Carbonate cementation is common in the reservoir of the study area, and cements are mainly calcite, followed by iron calcite. Calcite is most widely distributed and common in the Baikouquan formation, while iron calcite basically disappears in the Lower Wuerhe formation. The volume fraction of calcite in the Baikouquan formation varies greatly, from 2.39% to 15.31%, which in the Lower Wuerhe formation is relatively stable, with a minimum of 2.64% and a maximum of 9.78%. However, there is strong iron calcite cementation (nearly 7%) in the Lower Wuerhe formation reservoir of Well YT1, and the iron calcite is blue-purple under microscopic observation due to the large amount of iron content. From the microscope, it can be seen that the calcite cement in the reservoir is granular and mosaic. When carbonate cementation is strong, reservoir properties are generally poor. However, it can provide conditions for dissolution (Figure 5a).

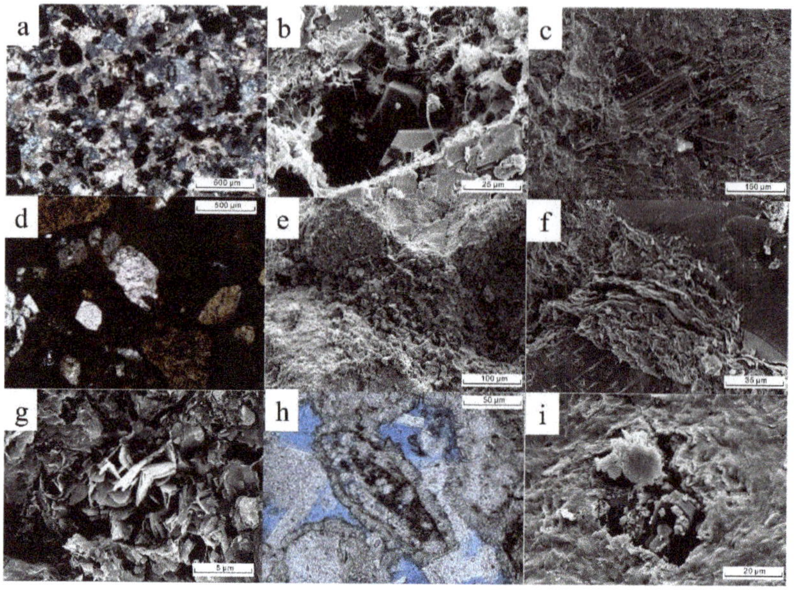

Figure 5. Thin sections and SEM images of cementation in study area. (**a**) thin section of well YT1, 5044 m, Lower Wuerhe formation, strong calcite cementation; (**b**) SEM image of well MZ2, 4199.36 m, Baikouquan formation, intergranular filling of authigenic quartz and filamentous illite; (**c**) SEM image of well M217, 4000.39 m, Lower Wuerhe formation, intergranular filling of zeolite; (**d**) thin section of well M217, Lower Wuerhe formation, heulandite cementation; (**e**) SEM image of well D11, 4280.63 m, Baikouquan formation, intergranular filling of worm-like kaolinite; (**f**) SEM image of well D11, 4294.47 m, Baikouquan formation, intergranular filling of lineated flaky illite; (**g**) SEM image of well M217, 4004.56 m, Lower Wuerhe formation, intragranular dissolved pores filling of leaf-shaped chlorite; (**h**) cast thin section of well YT1, 4872.5 m, Lower Wuerhe formation, authigenic sodium feldspar in intergranular pores and chlorite film developed; (i) SEM image of well MZ2, 4283.04 m, Baikouquan formation, intragranular dissolved pores filling of sodium feldspar crystal.

4.3.2. Siliceous Cementation

The volume fraction of siliceous cement in the study area is relatively low, with an average of 1.2%, which is mainly composed of quartz secondary overgrowth rims and filled in pores as automorphic granular. Under the scanning electron microscope (SEM), small automorphic quartz crystals are commonly found in intergranular pores or intragranular dissolved pores at the edge of clastic particles (Figure 5b), which play a role in reducing porosity. However, the formation of a certain number of siliceous cements can enhance the anti-compaction strength of the sandstone and prevent the destruction of residual primary intergranular pores by compaction, which has certain positive significance for reservoirs.

4.3.3. Zeolite Cementation

The zeolite cement in the study area is composed of laumontite and heulandite, mainly laumontite (Figure 5c,d). Laumontite generally is columnar; it often grows along the intergranular pores and intragranular dissolved pores. It mainly formed in the middle and late diagenesis stage. The distribution of zeolite cements is controlled by volcaniclastic materials in the source area, and they often coexist with calcite in intergranular pores to plug pores. However, the zeolite cementation is also conducive to the formation of secondary pores by later dissolution, thereby improving reservoir physical properties.

4.3.4. Authigenic Clay Mineral Cementation

According to the results of X-ray diffraction (XRD), thin section and SEM, it is found that the clay minerals of the Baikouquan formation in the study area are mainly composed of illite–smectite mixed layer, chlorite, illite and kaolinite, while the Lower Wuerhe formation is mainly composed of illite–smectite mixed layer, chlorite and illite. Kaolinite has completely disappeared. In the Baikouquan formation and Lower Wuerhe formation, the content of flaky illite is high, followed by the filamentous illite transformed from kaolinite (Figure 5b), while the honeycomb illite transformed from montmorillonite is rare. In SEM, kaolinite is a worm-like filling in the reservoir pores of the Baikouquan formation or distributed on the surface of the particles (Figure 5e); illite is a curved flake filling in the pores or distributed on the surface of the particles (Figure 5f) and chlorite is an irregular flake, filamentous filling in the pores (Figure 5g). The authigenic clay minerals mainly come from the connate water, the alteration of unstable components of clastic rocks and the transformation of clay from overlying mudstone.

4.3.5. Feldspar Cementation

The type of feldspar cement in the study area is mainly sodium feldspar, which is mostly formed by small automorphic crystals and filled in pores in fine lath or granular form (Figure 5h,i).

4.4. Dissolution

The burial depth of the Baikouquan formation and Lower Wuerhe formation in the study area is large, the overlying load compaction is strong and the cementation is common, resulting in a large volume loss of primary pores, especially in the superimposed structural compressive fault zone. As a result, the primary porosity of sandy conglomerates in the Lower Wuerhe formation is generally less than 10%, and that in the Baikouquan formation is less than 15% [18,19]. Therefore, dissolution and micro-fractures are important genetic mechanisms for the effectiveness of sandy conglomerates' pore structure. Through the observation of cast thin sections, it is found that the dissolved materials in the study area are mainly debris, feldspar, calcite, laumontite, etc. The dissolution of debris particles usually has three characteristics: only part of the minerals in the particles have local dissolution, forming spotted, honeycomb and striped intragranular dissolution pores (Figure 6a,b); the particles are strongly dissolved, leaving only part of the residue or the entirety of it dissolved to form a mold pore (Figure 6c–e); and the edge of the particles dissolved irregularly or expanded to form intergranular enlarged dissolved pores (Figure 6f,g). In addition, intergranular interstitial materials and calcite, laumontite, sodium feldspar and other cements also exist dissolution. The laumontite dissolution pores is one of the important factors to improve reservoir physical properties in the Lower Wuerhe formation (Figure 6h,i).

4.5. Petrography and Homogeneous Temperature of Fluid Inclusion

The fluid inclusions in the Baikouquan formation and the Lower Wuerhe formation in the study area mainly occur in diagenetic microcracks of quartz grain, calcite veins and analcime veins. Two-phase hydrocarbon inclusions are common and the content is about 5–10%. Among the observed inclusions, there are many two-phase hydrocarbon inclusions in which methane bubbles are mostly elliptical and some of them are irregular in shape. The content of hydrocarbon inclusions can reach 15~20% (Figure 7).

The appearance of hydrocarbon inclusions with different fluorescence can provide strong evidence for the migration and charging of oil and gas in different periods. Microscopic fluorescence observation found that the fluorescence color of hydrocarbon inclusions in the study area is diverse, mainly light blue (Figure 8a,b), blue (Figure 8c), light yellow (Figure 8d,e) and yellow (Figure 8f), indicating the change in maturity of oil and gas. Judging from the fluorescence color, there are at least two stages of hydrocarbon charging in the reservoir in the study area. The first stage should be the yellow fluorescent inclusions representing the early charging, and the other stage should be the blue fluorescent inclusions representing the late charging.

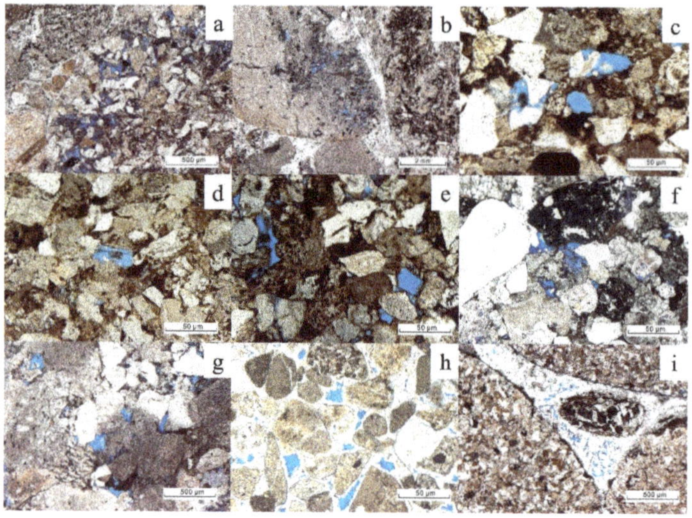

Figure 6. Cast thin sections of dissolution in study area. (**a**) Well YB4, 3913.77 m, Lower Wuerhe formation, intragranular dissolved pore; (**b**) Well YB4, 3915.31 m, Lower Wuerhe formation, intragranular dissolved pore; (**c**) Well YT1, 4538.73 m, Baikouquan formation, mold pore; (**d**) Well YT1, 4538.05 m, Baikouquan formation, mold pore; (**e**) Well YT1, 4533.83 m, mold pore; (**f**) Well YB4, 3692.21 m, Baikouquan formation, remaining intergranular pores; (**g**) Well YB4, 3693.04 m, remaining intragranular pores; (**h**) Well YB1, 4070.9 m, Lower Wuerhe formation, laumontite-dissolved pore; (**i**) well YB4, 3869.37 m, Lower Wuerhe formation, laumontite-dissolved pore.

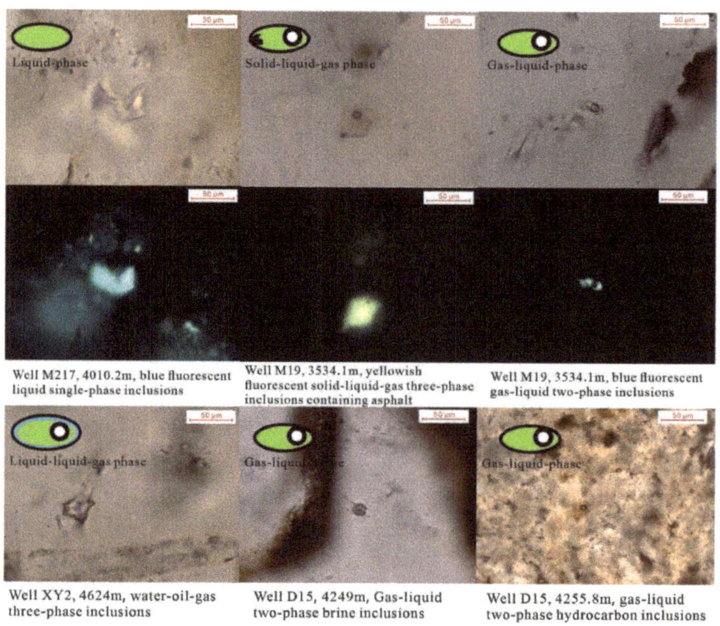

Figure 7. Microscopic photos of multi-stage hydrocarbon inclusions in reservoir.

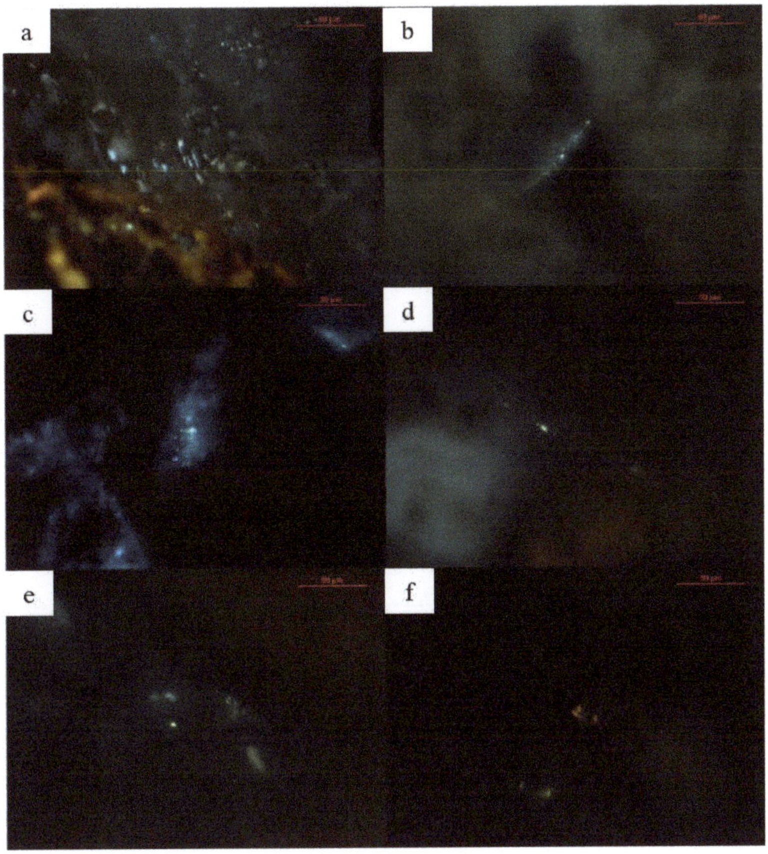

Figure 8. Microscopic photos of different fluorescent hydrocarbon inclusions in reservoir. (**a**) Well YB4, 3871 m, distributed along the micro-crack of analcime cement, showing blue, yellow, yellow-green or blue-green fluorescent hydrocarbon inclusions; (**b**) well M19, 3526.7 m, distributed in belts along microcracks after diagenesis of quartz grains, showing yellow-light blue fluorescent hydrocarbon inclusions; (**c**) well DT1, 5695.4 m, distributed along feldspar dissolved pores, showing blue fluorescent hydrocarbon inclusions; (**d**) well D13, 4207.5 m, distributed in belts along the micro-fractures after the diagenesis of quartz grains, showing light yellow fluorescent hydrocarbon inclusions; (**e**) well D15, 4249 m, distributed in belts along micro-fractures after diagenesis of quartz grains, showing yellow, blue or yellow-green fluorescent hydrocarbon inclusions; (**f**) well M19, 3534.1 m, distributed in belts along micro-fractures after diagenesis of quartz grains, showing yellow fluorescent hydrocarbon inclusions.

5. Discussion

5.1. Timing of Hydrocarbon Accumulation

By studying the homogenization temperature of fluid inclusions in the reservoirs of the Baikouquan formation and the Lower Wuerhe formation in the study area, it is found that the homogenization temperature of brine inclusions associated with yellow fluorescent hydrocarbon inclusions is concentrated between 70~80 °C, and the homogenization temperature of brine inclusions associated with blue hydrocarbon inclusions is between 140~150 °C (Figure 9). Combined with the study of burial temperature history of well DT1 in the slope zone (Figure 10), it is considered that the first stage of oil charging is mature crude oil generated in the early Jurassic, which is characterized by yellow fluorescent hydrocarbons. High-intensity and large-scale migration and charging occurred in the study area, and large-scale accumulation occurred in Mahu sag and its periphery. The second stage of hydrocarbon charging is in the early–middle Cretaceous, which is blue-white fluorescent high maturity crude oil and is also widely charged and accumulated. These results are consistent with previous studies [20,21].

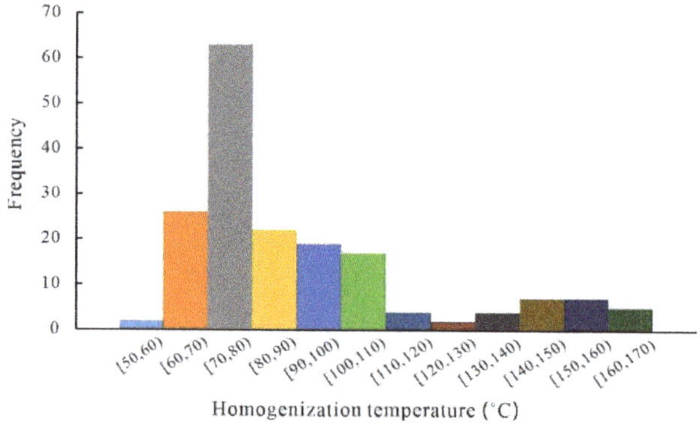

Figure 9. Homogeneous temperature histogram of brine inclusion.

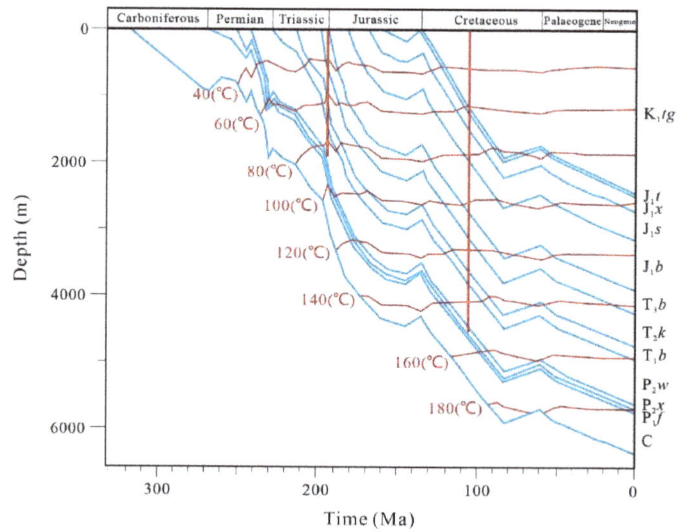

Figure 10. Burial history of well DT1.

5.2. *Organic Acid Dissolution and Its Control on Accumulation*

Hydrocarbon source rocks can produce a large amount of organic acids before generating a large amount of oil. The organic acids formed have a great influence on the diagenesis of reservoirs, which has been recognized by many scholars [22–25]. The kerogen in the source rock will remove oxygen-containing functional groups (such as carboxyl and phenol) and form a large number of organic acids (such as oxalic acid and acetic acid) under the thermal action of 80~120 °C. These organic acids are easy to form complexes with Al^{3+}, which increases the activity of Al^{3+} and promotes the dissolution of aluminosilicate and calcite. When the temperature rises to 120~160 °C, the carboxylic acid anion will undergo thermal decarboxylation and crack into hydrocarbons and CO_2, which will increase the concentration of CO_2 in the strata water and reduce the concentration of organic acids [22,25]. At the same time, the concentration of acid in strata water will increase due to the dissolution of CO_2 in water to form carbonic acid. As a result, alkaline minerals will continue to be dissolved.

As shown in Figure 11, yellow fluorescent mature hydrocarbon inclusions and blue fluorescent highly mature hydrocarbon inclusions can be observed simultaneously. It is speculated that it may be due to the formation of new dissolution pores by organic acids dissolving alkaline minerals during the migration of mature oil in the first stage, which provides space for the capture of highly mature hydrocarbon in the second stage, resulting in the occurrence of yellow fluorescent inclusions and blue fluorescent inclusions in the same part of the reservoir. It provides strong evidence for organic acid dissolution minerals to improve reservoir physical properties.

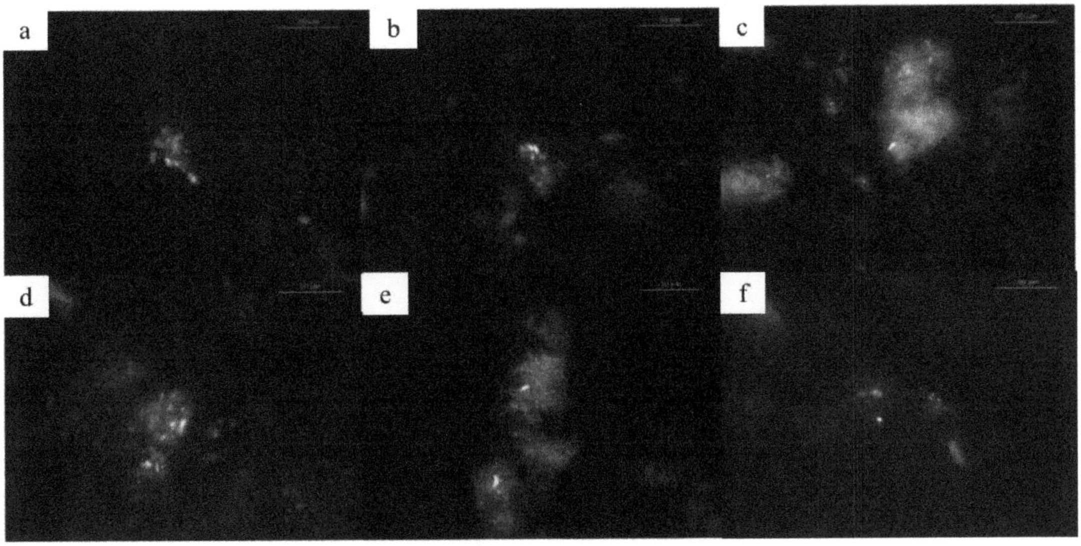

Figure 11. Fluorescence photos of hydrocarbon inclusions coexisting with different maturity. (**a**) Well M19, 3534.1 m, distributed along the microcracks of quartz grain, showing yellow or blue-green fluorescence; (**b**) well M217, 4010.02 m, distributed along micro-fractures of analcime cement, showing blue-green or yellow-green fluorescence; (**c**) well M217, 4010.02 m, distributed in belts along micro-fractures of quartz grain, showing blue-green fluorescence; (**d**) well YB4, 3871 m, distributed in belts along the micro-fractures of quartz particles, showing yellow-green or bluish-green fluorescence; (**e**) well YB4, 3914 m, distributed in belts along micro-fractures of quartz particles, showing yellow or yellow-green fluorescence; (**f**) well D15, 4249 m, distributed in belts along micro-fractures of quartz particles, showing yellow, blue or yellow-green fluorescence.

A large number of dissolution pores are formed in the reservoirs of the Baikouquan formation and the Lower Wuerhe formation in the study area, which greatly improves the reservoir physical properties. There are three main favorable conditions for this phenomenon: (1) The study area is located in the upward direction of the hydrocarbon generation center, which is conducive to the migration of organic acids and makes it easier to react with alkaline minerals. (2) Development of faults and unconformable surfaces: faults connecting source rocks are developed in the study area, providing a vertical channel for the migration of organic acids. In addition, the unconformity surface between the Baikouquan formation and the Lower Wuerhe formation and stratigraphic pinch-out line are developed in the study area [26], which provides an important channel for the lateral migration of organic acids. (3) The laumontite cementation particularly developed in the study area. Previous studies on the distribution of zeolite in this area suggest that the study area is close to the ancient lake center, and the high salinity and alkalinity in the early diagenesis are the two major factors for the development of laumontite, resulting in the formation of laumontite development zone in the study area [27]. The more laumontite cement, the greater the probability of acidic fluid contacting laumontite through fractures, unconformity surfaces and residual intergranular pores, as well as being easier to form dissolution pores.

The above-mentioned studies indicate that the dissolution during diagenesis in the study area is closely related to the hydrocarbon charging process. It can be seen from Figure 12 that after the organic acid reaches the unconformity surface with hydrocarbon along the faults connecting to source rock, it will migrate to the high part of the structure, so the sandstone reservoir in the high part will first develop secondary pores. The concentration of organic acids in the high part is relatively higher and dissolution is more intense. As shown in Figure 12, there are faults connecting source rocks between wells MZ1 and MZ2. Organic acids migrate vertically with hydrocarbon through faults to the unconformity surface, and then migrate laterally along the unconformity surface (or sandstone) to the high part of the structure. Therefore, a large number of secondary pores dissolved by organic acids are developed in the reservoir of well MZ2, which is located in the high structural position. There is no fault connecting source rocks in the downdip direction of well MZ1, so there is almost no organic acid passing through. As a result, the reservoir is basically not dissolved and the physical properties is poor. In addition, the sedimentary subfacies of the Lower Wuerhe formation in well XY3 is in the fan delta plain, and the reservoir is very tight. So even if there is organic acid fluid passing through, dissolution is not easy because there is no chance for organic acids to contact with alkaline cements.

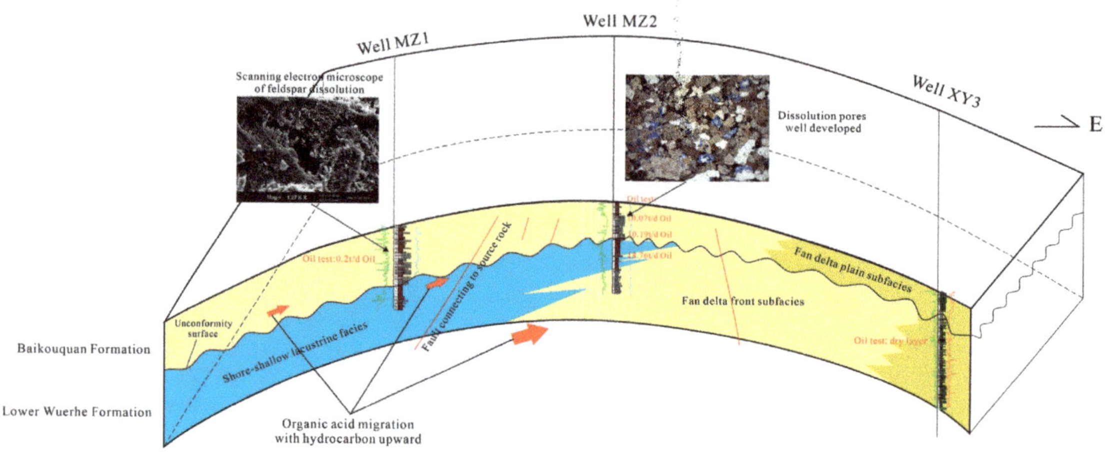

Figure 12. Relationship between diagenesis and hydrocarbon charging.

In summary, it suggests that the sandstone of fan delta front subfacies rich in alkaline cements is the basis for the formation of secondary pores, the dissolution of organic acids is the main factor controlling the development of secondary pores, and the dominant migration channel is a necessary condition for the formation of a large number of secondary pores.

6. Conclusions

(1) The sandy conglomerates reservoirs of the Lower Wuerhe formation–Baikouquan formation in the study area are characterized by low porosity and permeability, and both of them have low textural maturity and compositional maturity and poor size classification. However, the reservoir physical properties of the Lower Wuerhe formation are more affected by laumontite cementation and volcaniclastic dissolution.

(2) There are two stages of large-scale hydrocarbon charging in the study area: the first stage is the mature crude oil generated in the early Jurassic, which is characterized by yellow fluorescent inclusions; the second stage of hydrocarbon charging is in the early–middle Cretaceous, characterized by blue-white fluorescence inclusions.
(3) By restoring the process of hydrocarbon accumulation in the study area, the coupling relationship between diagenesis and hydrocarbon charging is established. The sandstone of fan delta front subfacies in the study area is rich in alkaline mineral. Therefore, when organic acids enter the reservoir along the dominant migration path with hydrocarbon, it can have strong dissolution, providing more reservoir space for subsequent hydrocarbon accumulation.

Author Contributions: Conceptualization, W.J.; methodology, H.L. and W.J.; validation, P.S., H.L. and B.B.; formal analysis, P.S.; investigation, H.L.; resources, P.S. and N.W.; data curation, P.S. and W.G.; writing—original draft preparation, W.J.; writing—review and editing, W.J.; visualization, X.W. All authors have read and agreed to the published version of the manuscript.

Funding: This research was funded by Major projects of PetroChina science and technology [2021DJ0206].

Data Availability Statement: Data is unavailable due to confidentiality.

Conflicts of Interest: The authors declare that they have no known competing financial interests or personal relationships that could have appeared to influence the work reported in this paper.

References

1. Yang, L.L.; Xu, T.F.; Liu, K.Y.; Peng, B.; Yu, Z.C.; Xu, X.M. Fluid-rock interactions during continuous diagenesis of sandstone reservoirs and their effects on reservoir porosity. *Sedimentology* **2017**, *64*, 1303–1321. [CrossRef]
2. Hu, H.W.; Li, H.Y.; Yu, H.B.; Xiao, S.G.; Xu, W. Quantitative analysis of source-to-sink system controls on sand-body distribution of the Paleogene in Chengbei low uplift and surrounding areas, Bohai Bay Basin. *J. Palaeogeogr.* **2020**, *22*, 266–277.
3. Nader, F.H.; Champenois, F.; Barbier, M.; Adelinet, M.; Rosenberg, E.; Houel, P.; Delmas, J.; Swennen, R. Diagenetic effects of compaction on reservoir properties: The case of early callovian"DalleNacrée"formation (Paris basin, France). *J. Geodyn.* **2016**, *101*, 5–29. [CrossRef]
4. Li, M.; Hou, Y.D.; Luo, J.L.; Chen, J.P.; Luo, X.R.; Jia, Y.N. Burial, diagenesis, hydrocarbon charging evolution process and quantitative analysis of porosity evolution: A case study from He 8 tight sand gas reservoir of the Upper Paleozoic in Eastern Ordos Basin. *Oil Gas Geol.* **2016**, *37*, 882–892.
5. Li, W.; He, S.; Zhang, B.Q.; He, Z.L.; Chen, M.F.; Zhang, D.W.; Li, T.Y.; Gao, J. Characteristics of paleo-temperature and paleo-pressure of fluid inclusions in shale composite veins of Longmaxi Formation at the western margin of Jiaoshiba anticline. *Acta Pet. Sin.* **2018**, *39*, 402–415.
6. Meng, B.; Jin, M.B.; Wu, B.X. Jurassic Yan'an Formation reservoir rocks in the Maling oil field, Ordos Basin. *Sediment. Geol. Tethyan Geol.* **2019**, *39*, 73–83.
7. Morad, S.; Ketzer, J.M.; DeRos, L.F. Spatial and temporal distribution of diagenetic alterations in siliciclastic rocks: Implications for mass transfer in sedimentary basins. *Sedimentology* **2000**, *47*, 95–120. [CrossRef]
8. Shi, J.A.; Jin, H.J.; Xue, L.H. Analysis on mechanism of feldspar dissolution and its influencing factors in feldspar-rich sandstone reservoir. *Acta Sedimentol. Sin.* **1994**, *12*, 67–75.
9. Zhao, G.Q.; Li, K.M.; Zhao, H.L.; Liu, R.E.; Sun, F.J.; Di, Y.J.; Li, Y.T. Feldspar corrosion and secondary pore formation in the Upper Paleozoic gas reservoir, Ordos Basin. *Pet. Explor. Dev.* **2005**, *32*, 53–55+75.
10. Yang, J.J.; Huang, Y.M.; Zhang, W.Z.; Liu, G.X. Experimental approach of dissolution of feldspar sand stone by acetic acid. *Pet. Explor. Dev.* **1995**, *22*, 82–86.
11. Li, P.L. Petroleum geological characteristics and exploration orientation of large oilfields in Junggar Basin. *Acta Pet. Sin.* **2005**, *26*, 7–9.
12. Shi, J.N.; Zou, H.Y. Study on the conduit system types and hydrocarbon accumulation rules in Junggar Basin. *J. Oil Gas Technol.* **2009**, *31*, 7–13.
13. Zou, H.Y.; Hao, F.; Zhang, B.Q.; Chen, J.P. Fluid-conduit frame work and its control on petroleum accumulation in the Junggar Basin. *J. China Univ. Geosci.* **2005**, *30*, 609–616.
14. Zhu, N.; Cao, Y.C.; Xi, K.L.; Wu, S.T.; Zhu, R.K.; Yan, M.M.; Ning, S.K. Diagenesis and physical properties evolution of sandy conglomerate reservoirs: A case study of Triassic Baikouquan formation in northern slope zone of Mahu despression. *J. China Univ. Min. Technol.* **2019**, *48*, 1102–1118.
15. Feng, C.; Lei, D.W.; Qu, J.H.; Huo, J.Z. Controls of paleo-overpressure, faults and sedimentary facies on the distribution of the high pressure and high production oil pools in the lower Triassic Baikouquan Formation of the Mahu Sag, Junggar Basin, China. *J. Pet. Sci. Eng.* **2019**, *176*, 232–248. [CrossRef]

16. Tang, W.B.; Zhang, Y.Y.; Pe-Piper, G.; Piper, D.J.W.; Guo, Z.J.; Li, W. Permian to early Triassic tectono-sedimentary evolution of the Mahu sag, Junggar Basin, western China: Sedimentological implications of the transition from rifting to tectonic inversion. *Mar. Pet. Geol.* **2021**, *123*, 1–15. [CrossRef]
17. Song, F.; Su, N.N.; Kong, X.W.; Liu, C.J.; Song, C.F. Sedimentary characteristics of humid alluvial fan and its control on hydrocarbon accumulation: A case study on the northern margin of the Junggar Basin, west China. *Pet. Sci. Eng.* **2020**, *187*, 106729. [CrossRef]
18. Chen, B.; Wang, Z.T.; Kang, L.; Zhang, S.C.; Shi, J.A. Diagenesis and pore evolution of Triassic Baikouquan formation in Mabei Region, Junggar Basin. *J. Kirin Univ. Nat. Sci.* **2016**, *46*, 23–35.
19. Zou, N.N.; Zhang, D.Q.; Jiang, Y.; Zhu, Y.C.; Zhang, S.C.; Lu, X.C.; Shi, J.A. Diagenesis and porosity evolution of permian Lower Wuerhe formation reservoir in the Madong Area, Junggar Basin. *Geol. Sci. Technol. Inf.* **2015**, *34*, 42–48.
20. Li, J.Z.; Wang, X.J.; Yang, F.; Song, Y.; Ablmt, I.; Bian, B.L.; Liu, H.L.; Wang, X.Y.; Gong, D.Y. Hydrocarbon accumulation pattern and exploration prospect of the structural traps in lower play of the western Central Depression in the Junggar Basin. *Oil Gas Geol.* **2022**, *43*, 1059–1072.
21. Tang, Y.; Song, Y.; He, W.J.; Zhao, L.; Yang, H.B.; Zhao, C.Y.; Zheng, M.L.; Sun, S.; Fei, L.Y. Characteristics of composite hydrocarbon accumulation in a superimposed basin, Junggar Basin. *Oil Gas Geol.* **2022**, *43*, 132–148.
22. Heydari, E.; Wade, W.J. Massive recrystallization of low-Mg Calcite at high temperatures in hydrocarbon source rocks: Implications for organic acids as factors in diagenesis. *AAPG Bull.* **2002**, *86*, 1285–1303.
23. Surdam, R.C.; Crossey, L.J.; Hagen, E.S.; Heasler, H.P. Organic-inorganic interactions and sandstone diagenesis. *AAPG Bull.* **1989**, *73*, 1–23.
24. Crossey, L.J.; Frost, B.R.; Surdam, R.C. Secondary porosity in laumontite-bearing sandstones: Part 2. Aspects of Porosity Modification. *AAPG Bull.* **1984**, *59*, 225–237.
25. Huang, S.J.; Zhang, M.; Zhu, S.Q.; Wu, W.H.; Huang, C.G. Control of origin of pores over relationship of porosity to permeability in sandstone reservoir: A case study from Yanchang sandstone of Triassic of eastern Gansu, Ordos Basin. *J. Chengdu Univ. Technol. Sci. Technol. Ed.* **2004**, *31*, 648–653.
26. Wu, K.Y.; Zha, M.; Liu, G.D. The unconformity surface in the Permian of Junggar basin and the characters of oil-gas migration and accumulation. *Pet. Explor. Dev.* **2002**, *29*, 53–57.
27. Guo, M.Z.; Shou, J.F.; Xu, Y.; Guo, H.J.; Zou, Z.W.; Han, S.H. Distribution and controlling factors of Permian zeolite cements in Zhongguai-Northwest margin of Junggar Basin. *Acta Pet. Sin.* **2016**, *37*, 695–705.

processes

MDPI

Article

Petroleum Geology and Exploration of Deep-Seated Volcanic Condensate Gas Reservoir around the Penyijingxi Sag in the Junggar Basin

Baoli Bian [1], Ablimit Iming [1], Tianze Gao [2], Hailei Liu [1], Wenlong Jiang [1], Xueyong Wang [1] and Xiujian Ding [2,*]

[1] Research Institute of Exploration and Development, Petro China Xinjiang Oilfield Company, Karamay 834000, China
[2] School of Geosciences, China University of Petroleum, Qingdao 266580, China
* Correspondence: dingxj@upc.edu.cn

Abstract: Many types of volcanic rock oil and gas reservoirs have been found in China, showing great petroleum exploration potential. Volcanic reservoir also is one of the key fields of exploration in the Junggar Basin and mainly concentrated in the middle and shallow layers, while the deep volcanic rock and natural gas fields have not been broken through. Based on comprehensive analysis of core observation, single well analysis, reservoir description, source rocks evaluation, combined with seismic data and time-frequency electromagnetic technology, multiple volcanic rock exploration targets were identified, and industrial oil and gas flow was obtained in the well SX 16 of the Penyijingxi Sag, western Junggar Basin. It is believed that the deep Permian source rocks have relatively higher natural gas generation potential and volcanic breccia usually have large reservoir space. And the mudstone of the Upper Wuerhe Formation played as the role of caprock. The success of exploration well SX16 has achieved a major breakthrough in natural gas exploration in the Penyijingxi Sag, which has essential guiding significance for the exploration of deep volcanic rocks and large-scale gas exploration in the Junggar Basin.

Keywords: condensate gas; volcanic rocks; Penyijingxi Sag; Junggar basin; Carboniferous

Citation: Bian, B.; Iming, A.; Gao, T.; Liu, H.; Jiang, W.; Wang, X.; Ding, X. Petroleum Geology and Exploration of Deep-Seated Volcanic Condensate Gas Reservoir around the Penyijingxi Sag in the Junggar Basin. *Processes* **2022**, *10*, 2430. https://doi.org/10.3390/pr10112430

Academic Editors: Ping Gao, Yidong Cai, Yingfang Zhou and Quan Gan

Received: 16 October 2022
Accepted: 15 November 2022
Published: 17 November 2022

1. Introduction

Both deep-seated petroleum reservoirs and unconventional petroleum are hotspot for exploration and development [1–3]. And deep-seated petroleum exploration is an important strategic replacement field for oil and gas exploration in China [4,5]. With the deepening of theoretical understanding of high-quality reservoir formation mechanism, deep oil and gas hydrocarbon generation and reservoir formation, deep oil and gas exploration has gradually attracted extensive attention. Volcanic rock is an essential field of oil and gas exploration in China [6,7]. Many types of volcanic rock oil and gas reservoirs have been found in the Songliao, Bohai Bay, Hailar, Erlian, and Sichuan basins of China [8–10], showing great petroleum exploration potential.

In China, as early as the 1970s to 1980s, it was recognized that volcanic rocks can be used as reservoir rocks and volcanic reservoirs such as Huangshatuo, Oulituozi, and Rehetai were successively discovered. The Junggar Basin is one of the large petroliferous basins in northwestern China [11,12]. More than sixty years of petroleum exploration show that the Junggar Basin is rich in oil and poor in natural gas [13]. Volcanic reservoir is one of the key fields of oil and gas exploration in Junggar Basin [14,15]. In the early stage of exploration, many oil reservoirs were found in the northwest margin of the basin [16]. Volcanic oil and gas reservoirs are mainly concentrated in the middle and shallow layers, the deep seated volcanic rock and natural gas fields have not been broken through. In order to find large-scale oil and natural gas reservoirs, based core observation, single well analysis, reservoir description and hydrocarbon source rock evaluation, combined with seismic data and time-frequency electromagnetic technology, several volcanic rock targets were identified for the deep volcanic rocks in the Penyijingxi Sag of the Junggar Basin. The exploration well was successfully deployed, and high-yield industrial oil and gas flow was obtained, revealing the great potential for oil and gas exploration in deep volcanic rocks, which has essential guiding significance for the exploration of deep volcanic rocks and large-scale gas exploration in the Junggar Basin.

In summary, many types of volcanic rock oil and gas reservoirs have been found in China, showing great petroleum exploration potential. Volcanic reservoir also is one of the key fields of exploration in the Junggar Basin and mainly concentrated in the middle and shallow layers, while the deep volcanic rock and natural gas fields have not been broken through. The study has essential guiding significance for the exploration of deep volcanic rocks and large-scale gas exploration in the Junggar Basin.

2. Exploration History of Volcanic Rocks in the Junggar Basin

Junggar basin is located in the west of China, with Junggar Boundary Mountain in the northwest, Altai Mountain in the northeast and North Tianshan Mountain in the south [17,18]. It is a slightly triangular closed basin with an area of 130,000 km^2 (Figure 1a,b). It is one of the four major oil-gas basins with oil and gas resources of more than 10 billion tons in China [11,12]. It has been rich in oil and poor in gas for a long time [13,19].

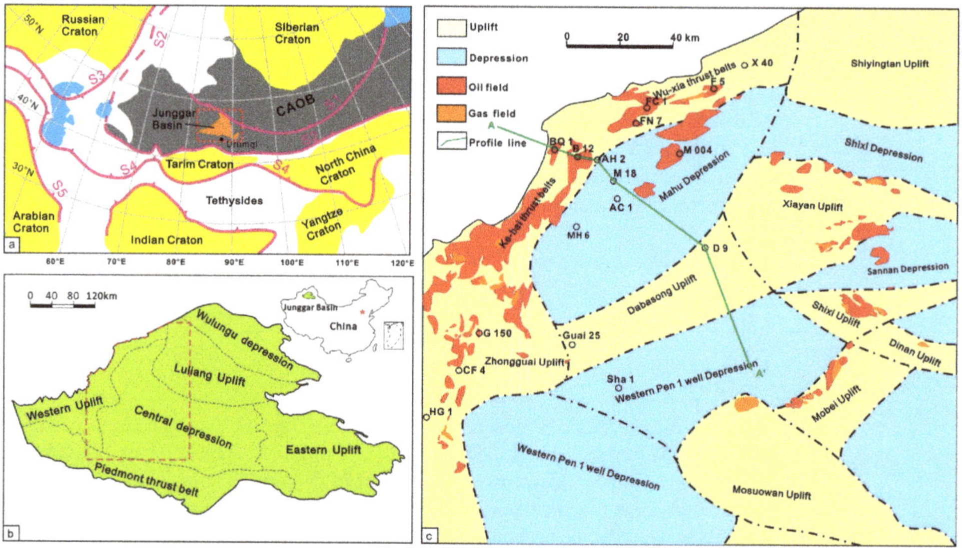

Figure 1. Distribution of structural units, oil fields, gas fields, and sample wells distributions in the northwestern Junggar Basin (After reference [20]). (**a**) geotectonic setting of the Junggar Basin; (**b**) tectonic unit of the Junggar Basin; (**c**) study area tectonic unit and petroleum, exploration well distribution. Reprinted with permission from Ref. [20].

Before 1990, the volcanic exploration in Junggar Basin mostly belonged to the early exploration stage, and the exploration discoveries were mainly oil reservoirs, which were all located in the northwest margin region of the Junggar Basin [20]. From 1990 to 2000, with the continuous improvement of geological knowledge and exploration technology, targeted exploration was carried out in Shixi area and Chepaizi area of Junggar basin [21]. The exploration discoveries were mainly oil and some natural gas. Since 2000, volcanic exploration in Junggar basin has stepped into an accelerated stage, and breakthroughs have been made successively in the Liuzhong Area of the northwest margin [22]. In 2007 and 2008, large-scale breakthroughs in natural gas have been made in the northwest margin, and significant progress and breakthroughs have been made. The early volcanic exploration in Junggar Basin mainly focused on the middle and shallow layers such as Shixi uplift. Due to the recognition accuracy and engineering technology, the deep layer has not been explored yet.

With the increase of exploration degree, deep volcanic rocks have gradually become the focus of exploration and experienced a tortuous exploration process [23]. In 2006, the Moshen-1 well was deployed and the design well depth was 7380 m. After more than one year of drilling, nearly 400 m tuff was drilled in the Carboniferous system. The gas logging showed good results. Gas invasion occurred during the drilling process of the Carboniferous system, but oil and gas flow could not be obtained during the oil test. Although the exploration of Moshen-1 well did not obtain industrial oil and gas flow, the good gas logging shows increased confidence of deep volcanic rock exploration. In 2015, Datan 1 well was implemented, with a drilling depth of 6226 m. The Carboniferous-Jurassic system has obtained extremely active oil and gas display, with a longitudinal span of more than 2500 m, and no industrial oil flow has been obtained from igneous rocks. In view of the active oil and gas display of well Datan 1, well Yantan 1 was deployed in 2016. Frequent interbedding of tuff andesite and basalt was encountered in Carboniferous, unfortunately there was no oil and gas display in igneous rocks.

In 2016, the exploration focus was adjusted from Dabasong Uplift to the surround of the Penyijingxi Sag, and a comprehensive study on the deep volcanic rocks in the Penyijingxi Sag was carried out. The study shows that the Permian hydrocarbon source rocks have a high degree of deep organic matter thermal evolution and large-scale gas generation potential. And well SX 16 exploration well was deployed for deep volcanic rocks in 2019. Formation testing was conducted in the 4800 m deep andesitic breccia well section of Carboniferous system, and high-yield oil and gas flow was obtained. Since then, a major breakthrough has been made in the exploration of deep volcanic rocks and deep natural gas in the basin.

3. Data and Method

Based on comprehensive analysis of core observation, microscopic slice observation and physical property, combined with logging and seismic data, volcanic reservoirs were distinguished and predicted. Source rocks distribution was defined by seismic data and petroleum filling intensity was defined by the data of grain with oil inclusions. With seismic data and time-frequency electromagnetic technology, multiple volcanic rock exploration targets were identified.

4. Result of Reservoir and Source Rock Characteristics

4.1. Types and Characteristics of Volcanic Reservoir Space

The volcanic reservoir in the Penyijingxi Sag is mainly developed in andesitic and basaltic volcanic rocks, mainly including andesitic cryptoexplosive breccia, andesitic volcanic breccia, andesitic tuff, stomatal andesite, massive andesite, stomatal basalt, and massive basalt. The reservoir space and association law of andesitic volcanic rocks in Shixi area are systematically summarized by using core and cast thin section analysis methods (Figure 2).

Andesitic cryptoexplosive breccia is mainly developed near the volcanic channel. It is formed by blasting the early volcanic rocks in situ during the later magma rising process. The magama filled between the self-broken breccias has poor crystallinity and is easy to dissolve, forming a large number of intergranular solution pores. It is characterized by large dissolution pores, and its shape is limited by the boundary of the self-broken breccias (Figure 2a). Andesitic volcanic breccia and andesitic tuff are formed by volcanic eruption. The volcanic debris is composed of crystal debris and rock debris, and the matrix is volcanic dust. Because volcanic dust is composed of volcanic glass with poor stability, it is prone to devitrification to form intergranular micropores. It is easy to dissolve under the action of fluid, forming a large number of dissolution pores in the matrix (Figure 2b,c).

Stomatal andesite and stomatal basalt are developed in the upper part of lava flow unit, some primary pores are filled by amygdaloid, and secondary dissolution occurs in the later stage, forming a large number of dissolution pores in amygdaloid (Figure 2d,e). In addition, the curved condensation shrinkage cracks between the amygdaloid body and the pore wall are also common in the stomatal andesite and the stomatal basalt, which are formed by the condensation shrinkage of the amygdaloid body (Figure 2f). Massive andesite and massive basalt are developed in the middle and lower part of the lava flow unit, with similar reservoir space characteristics. They are characterized by massive structure, less porosity and lack of primary pores. Under the action of weathering fluid, phenocrysts and matrix can undergo dissolution to a certain extent, forming intracrystalline solution pores (Figure 2g) and matrix dissolved pores with sieve-like distribution (Figure 2h). In addition, structural fractures of massive basalt and andesite are relatively developed (Figure 2i). Structural fractures are formed by regional tectonic stress and deformation. The fractures are generally flat or in micro saw tooth shape (tracking tensional joints), and the filling content is generally not high (Figure 2i). In summary, there are many types of pores in the volcanic reservoir, showing good reservoirs developed in deep-seated volcanic.

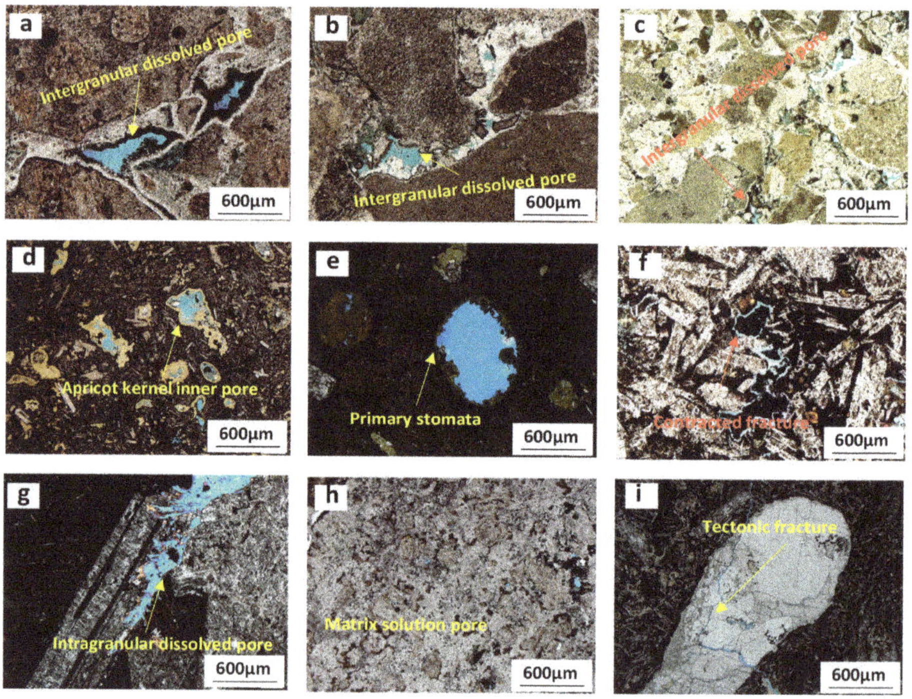

Figure 2. Types and characteristics of volcanic reservoir space around the Penyijingxi Sag. (**a**) Interparticle dissolution pore of andesitic cryptoexplosive breccia. (**b**) Interparticle dissolution pore of andesitic volcanic breccia. (**c**) Matrix solution pore of andesitic tuff. (**d**) Pore inner stomatal basalt amygdaloid. (**e**) Pore inner andesite amygdala. (**f**) Basalt amygdaloid contraction joint. (**g**) Intracrystalline solution pore of andesite. (**h**) Andesite matrix solution pore. (**i**) Structural fracture.

4.2. Physical Property Characteristics of Volcanic Reservoir

The reservoir physical property statistics of volcanic reservoirs with different lithology in the study area are shown in Figure 3. The matrix porosity of andesitic volcanic breccia formed by eruption is between 10 and 20%, and the maximum permeability can reach 100×10^{-3} μm^2, which is the best reservoir. Stomatal andesite, stomatal basalt, and andesitic cryptoexplosive breccia have similar physical properties. Their matrix porosity is between 10 and 20% and permeability is 0.1×10^{-3}~1×10^{-3} μm^2, which is also a relatively favorable volcanic reservoir. The matrix porosity of massive andesite is between 3 and 10% and the seepage rate is 0.01×10^{-3} μm^2~1×10^{-3} μm^2, the reservoir physical property is poor. The porosity of massive basalt matrix is between 1 and 5% and the permeability is 0.01×10^{-3} μm^2~0.1×10^{-3} μm^2, which basically cannot form an effective reservoir. Compared with clastic rocks and dolomitic rocks, the Carboniferous deep volcanic rocks are less affected by compaction, and the widely developed unconformity at the top of the Carboniferous system will significantly improve the physical properties of the reservoir. It is comprehensively considered that the deep volcanic rocks develop large-scale reservoirs.

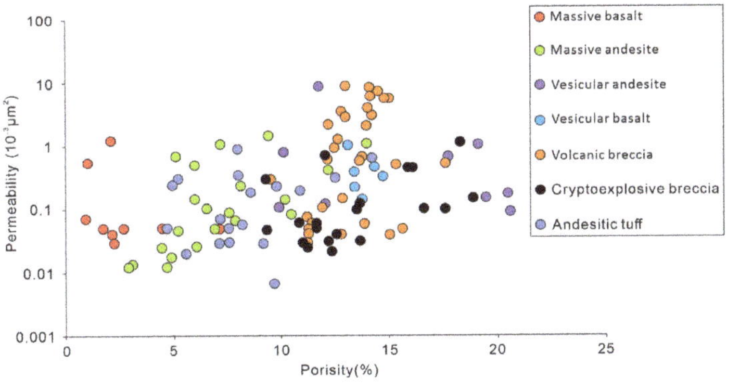

Figure 3. Crossplot of porosity and permeability of volcanic rocks in Shixi area of the Junggar Basin.

4.3. Distribution and Evolution of Petroleum Source Rocks

At present, the Permian source rocks were widely drilled in the Mahu Sag and no well has been drilled in the Penyijingxi Sag. The Permian source rocks in the Mahu Sag have high organic matter abundance and are mainly medium-good hydrocarbon source rocks, with organic matter type II, and are developed from low to high maturity [24]. Through the comparison between Mahu Sag and Penyijingxi Sag, it is considered that the depositional environment of Permian is similar as a whole. It is concluded that two sets of source rocks of the Fengcheng Formation and lower Wuerhe Formation are developed in the Penyijingxi Sag, which is similar to the Mahu Sag.

According to the seismic data, the formation thickness of the Fengcheng Formation and Lower Wuerhe Formation is identified. The mudstone of the Fengcheng Formation and Lower Wuerhe Formation is widely distributed, with a thickness of more than 200 m.

In the Mahu Sag, the burial depth of source rocks of the Fengcheng Formation in the Mahu Sag is mostly less than 6500 m, and decade's years petroleum exploration had confirmed that the source rocks are in the stage of oil generation, and has not entered in the stage of condensate gas generation stage.

From the Mahu Sag to the Penyijingxi Sag, the burial depth of source rocks gradually increases, and the buried depth of Fengcheng Formation source rocks in the Penyijingxi Sag is more than 9000 m. Combined with the burial depth of source rocks and the thermal evolution simulation of organic matter, it is considered that the evolution degree of source rocks in the Lower Wuerhe Formation is dominated by high maturity, mainly in the condensate generation oil and gas and wet gas stages (Figure 4a), and the evolution degree of Fengcheng Formation source rocks is mainly high mature and over mature, mainly in the stage of generating wet gas and dry gas (Figure 4b).

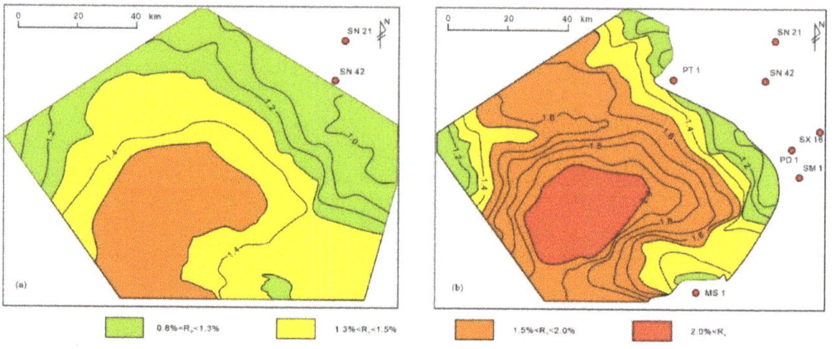

Figure 4. Plane distribution of organic matter maturity (R_o) of Permian main source rocks. (**a**) R_o of the Lower Wuerhe Formation; (**b**) R_o of the Fengcheng Formation.

5. Discussion on Petroleum Accumulation and Natural Gas Exploration

5.1. Distribution Model of Volcanic Rocks Reservoir

The Penyijingxi Sag is surrounded by Dabasong Uplift, Xiayan Uplift, Shixi Uplift, Mobei Uplift, and Mosuowan Uplift. According to the lithological characteristics of Upper Carboniferous revealed by drilling, Xiayan Uplift and Dabasong Uplift in the west of Penyijingxi Sag are mainly basic basalt, and tuff, intermediate acid andesite and dacite with a certain thickness are developed. Affected by the ancient landform, some low-lying areas also developed sedimentary volcaniclastic rocks and sedimentary rocks.

The identification of volcanic lithofacies is the basis for establishing volcanic institutions and making high-quality reservoir predictions [25]. Based on core observation, thin section identification, logging data and seismic, the lithology and lithofacies of volcanic reservoir in the Shixi Uplift was classified and identified [25]. The volcanic rocks in Shixi Uplift, Mobei Uplift and Mosuowan Uplift in the east of Penyijingxi Sag are mainly medium acid andesite, dacite and rhyolite, and volcanic breccia and tuff are developed. Under the influence of volcanic mechanism, volcanic lava and pyroclastic rocks are interbedded, as shown in Figure 5. The volcanoes in Shixi Uplift and Mobei Uplift are continental eruptions and do not develop sedimentary pyroclastic rocks and sedimentary rocks. However, thick sedimentary rocks are developed in the upper part of Mosuowan Uplift, and there are sedimentary pyroclastic rocks and sedimentary rock intercalations in the lower volcanic rocks. On the plane, the lithology of the Upper Carboniferous volcanic rocks around the Penyijingxi Sag has obvious cyclicity.

5.2. Oil and Gas Filling Intensity

The industrial oil and gas flow is obtained from the formation testing of well SX 16. And the gas test has the characteristics of low gas/oil ratio, high oil production and high gas production.

The formation testing results of well SX 16 showed that condensates are the primary petroleum type. Through the observation of reservoir thin sections, it is found that the hydrocarbon phases are diverse, with residual black oil (Figure 6a), black brown heavy oil inclusions (Figure 6b), light yellow fluorescent condensate oil and gas (Figure 6c) and black brown gas hydrocarbon inclusions (Figure 6d). Grain with oil inclusions (GOI) is an essential parameter to evaluate the oil and gas filling intensity. It is generally considered that the GOI of oil layer is mostly greater than 5%, the GOI of the oil-water intermediate layer is 1~5%, the GOI of SX16 well is 8%, and the grain with gas inclusion is 4%, revealing a relative stronger oil and gas charging intensity.

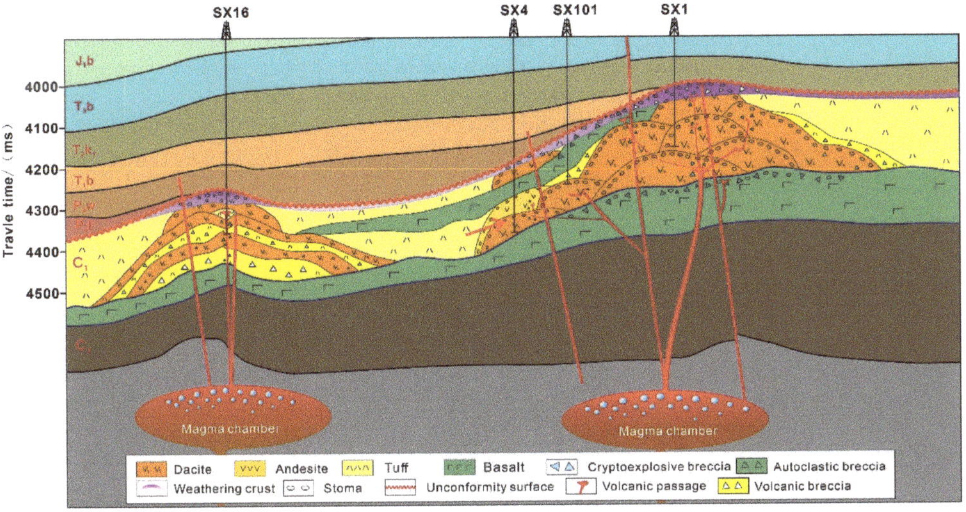

Figure 5. Regional distribution of volcanic rocks in the Penyijingxi Sag and its surroundings.

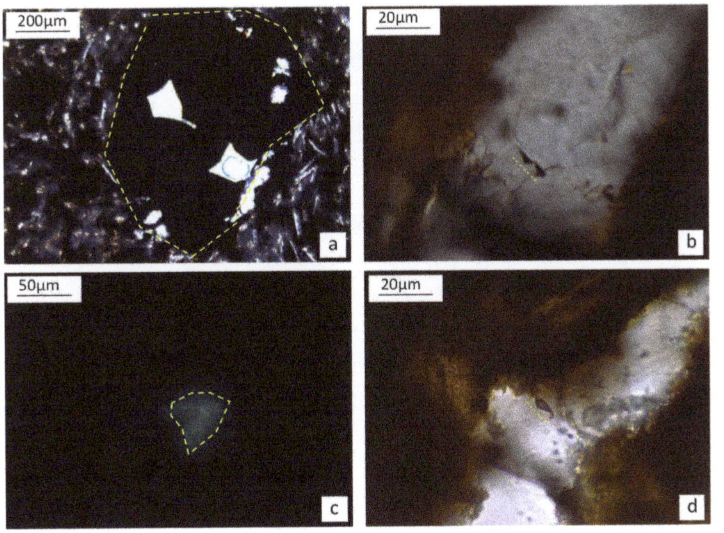

Figure 6. Microscopic characteristics and inclusions characteristics of volcanic reservoir in well SX16. (**a**), well SX 16, 4806.9 m, black oil, plane polarized light; (**b**), well SX 16, 4806.9 m, heavy oil inclusion, plane polarized light; (**c**), well SX 16, 4806.9 m, light yellow, condensate, fluorescent light; (**d**), well SX 16, 4806.9 m, gas hydrocarbon inclusion, plane polarized light.

5.3. Natural Gas Preservation and Caprock Condition

Most deep-seated oil and gas reservoirs in Junggar basin have experienced multi-stage tectonic movements, and most of hydrocarbon are light oil and natural gas, thus caprock and petroleum reservoir preservation are often one of the key controlling factors. For instance, the oil and gas show of well Datan-1 ranges from the Carboniferous to the Jurassic, with a depth range of more than 2500 m. However, there is and no industrial oil flow has been obtained in the Carboniferous volcanic rocks of well DT 1.

The regionally distributed mudstone of the Upper Wuerhe Formation, Fengcheng Formation are developed on the Carboniferous volcanic rocks, played as the regional cap rocks of volcanic oil and gas reservoirs. The Carboniferous igneous rock of the well SX16 is overlaid with mudstone of the Upper Wuerhe Formation (P_3w) with a cumulative thickness of more than 100 m, which has good preservation conditions and is conducive to the preservation of oil and gas reservoirs (Figure 7).

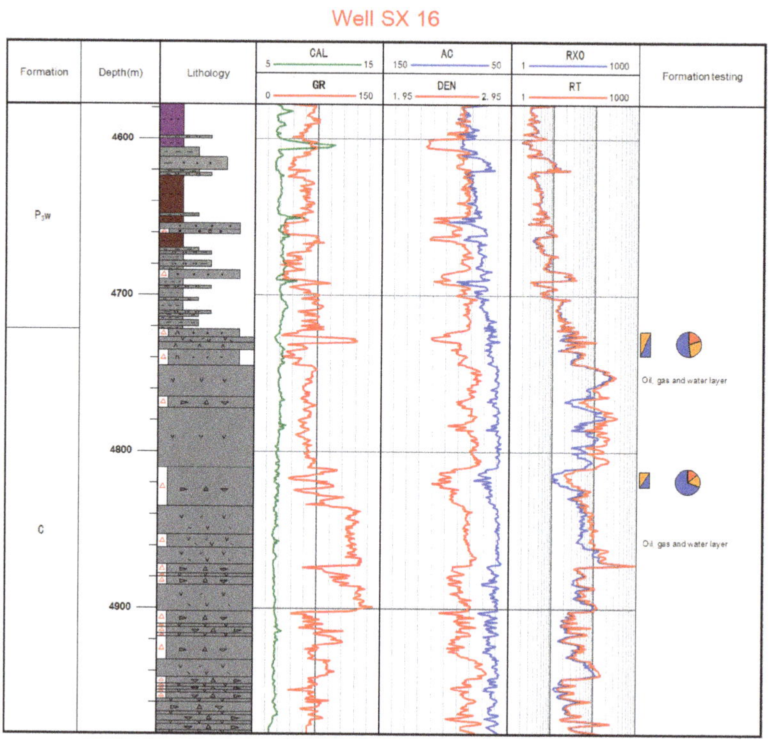

Figure 7. Formation testing results and comprehensive histogram of upper caprocks of well SX 16 around the Penyijingxi Sag.

5.4. Key Techniques for Deep Volcanic Gas Reservoir Exploration

Compared with the medium and shallow layers, the deep oil and gas exploration technology is much more difficult. The improvement of exploration technology has advantageously guaranteed the breakthrough of oil and gas exploration.

Due to the limitations of previous geophysical exploration acquisition techniques and the shielding effect of middle and shallow Jurassic coal rock strata, the existing 3D seismic data have relatively low deep resolution and signal-to-noise ratio (SNR), which cannot meet the requirements for deep volcanic oil and gas reservoir exploration. The progress of 3D seismic acquisition and processing technology for deep volcanic rocks is the key to exploration.

The new seismic imaging technology named "wide frequency, wide azimuth and high-density seismic exploration" have a significant effect on improving the imaging quality [26]. Recently, the "wide frequency, wide azimuth and high-density seismic exploration" new technology had been used to discover exploration target (Figure 8).

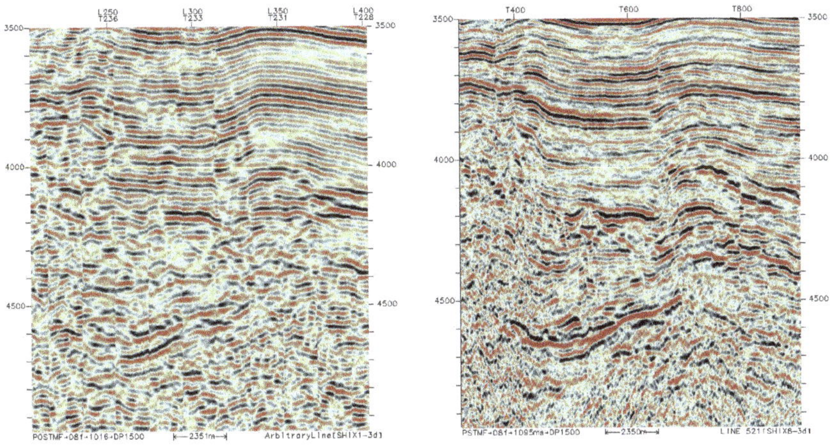

Figure 8. Comparison diagram of seismic profile through well SX16 with conventional (left) and new (right) technology.

Time frequency electromagnetic method (TFEM) is an artificial source electromagnetic detection method based on the principle of electromagnetic induction to observe and study the response field in time domain and frequency domain [27]. This method can effectively eliminate the influence of complex surface conditions and large horizontal variation of underground geological structure, and obtain resistivity and polarizability parameters. It can cooperate with seismic to identify reservoir target lithology, implement geological structure, predict oil, and gas potential, etc. In recent years, this technology has been widely applied in many regions at China and abroad, and has become an indispensable and important means in oil and gas exploration.

Through statistical analysis of the physical property differences between the lithology of different strata and the specific lithology of Carboniferous in the Junggar basin, the resistivity anomaly information template of different types of Carboniferous volcanic rocks is established. Then the abnormal information of time-frequency electromagnetic resistivity and polarizability in SX16 well area is inversed and interpreted (Figure 9). The Carboniferous volcanic rocks were characterized in detail, which effectively improved the drilling success rate of deep and complex volcanic reservoirs and reduced the drilling risk. Through the TFEM exploration application, the distribution of volcanic rocks is determined, which provides a favorable support for the exploration of deep volcanic gas reservoirs.

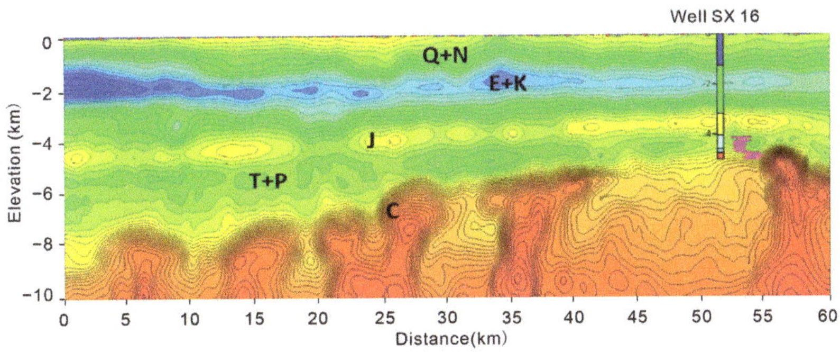

Figure 9. TFEM (time frequency electromagnetic method) profile through well SX16.

5.5. Deep Volcanic Gas Reservoir Exploration Inspiration

The source rocks in the Penyijingxi Sag are characterized by large thickness and wide area, but the source rocks have not been drilled in the study area, and the gas generation capacity of the source rocks is questionable. The great breakthrough which has been made in the exploration of natural gas in deep volcanic rocks confirms that the Permian source rock in the Penyijingxi Sag has the large-scale natural gas generation potential.

The Carboniferous distributed in the Penyijingxi Sag develops large-scale volcanic rocks, and the buried depth gradually increases from north to south. The statistical analysis of physical properties shows that with the increase of depth, the volcanic rocks are supported by the rock framework and volcanic breccia, and the physical properties change relatively little. The exploration breakthrough of well SX16 reveals that the volcanic reservoir has good physical properties, large rock mass thickness, and high gas production, and may become an important exploration area for large-scale natural gas exploration.

6. Conclusions

Based on comprehensive analysis of core observation, single well analysis, reservoir description, and source rocks evaluation, it could be concluded that the source rocks have relatively higher natural gas generation potential and supply a relative stronger oil and gas charging intensity. The deep volcanic rocks may develop large-scale weathered crust reservoirs and the Upper Wuerhe Formation played as the role of caprock and controlling natural gas reservoirs preservation.

Based on comprehensive analysis of petroleum geology elements and the new seismic imaging technology named "wide frequency, wide azimuth and high-density seismic exploration", a major breakthrough has been made in the exploration of deep volcanic rocks and deep natural gas in the basin. This is the first breakthrough of deep seated gas reservoirs exploration in volcanic reservoir. The success has achieved a major breakthrough in natural gas exploration in the Penyijingxi Sag, which has essential guiding significance for the exploration of deep volcanic rocks and large-scale gas exploration in the Junggar Basin.

Author Contributions: Conceptualization, B.B. and A.I.; methodology, T.G.; validation, H.L., W.J. and X.W.; formal analysis, X.W.; investigation, B.B.; resources, B.B.; data curation, B.B.; writing—original draft preparation, B.B.; writing—review and editing, X.D.; visualization, H.L. All authors have read and agreed to the published version of the manuscript.

Funding: This research was funded by Major projects of PetroChina science and technology [2021DJ0206].

Acknowledgments: We thank Xinjiang Oil Field Company, PetroChina, for providing access to the cored rock samples.

Conflicts of Interest: The authors declare that they have no known competing financial interests or personal relationships that could have appeared to influence the work reported in this paper.

References

1. Zou, C.; Zhang, G.; Tao, S.; Hu, S.; Li, X.; Li, J.; Dong, D.; Zhu, R.; Yuan, C.; Hou, L.; et al. Geological features, major discoveries and unconventional petroleum geology in the global petroleum exploration. *Pet. Explor. Dev.* **2010**, *37*, 129–145.
2. Tolmachev, O.; Urunov, A.; Muminova, S.; Dvoichenkova, G.; Davydov, I. Review of unconventional hydrocarbon resources: Production technologies and opportunities for development. *Min. Miner. Depos.* **2020**, *14*, 113–121. [CrossRef]
3. Matkivskyi, S. Increasing hydrocarbon recovery of Hadiach field by means of CO2 injection as a part of the decarbonization process of the energy sector in Ukraine. *Min. Miner. Depos.* **2022**, *16*, 114–120. [CrossRef]
4. Wang, Y.; Jia, D.; Pan, J.; Wei, D.; Tang, Y.; Wang, G.; Wei, C.; Ma, D. Multiple-phase tectonic superposition and reworking in the Junggar Basin of northwestern China-Implications for deep-seated petroleum exploration. *AAPG Bull.* **2018**, *102*, 1489–1521. [CrossRef]
5. Zhou, X.; Lü, X.; Zhu, G.; Cao, Y.; Yan, L.; Zhang, Z. Origin and formation of deep and superdeep strata gas from Gucheng-Shunnan block of the Tarim Basin, NW China. *J. Pet. Sci. Eng.* **2019**, *177*, 361–373. [CrossRef]
6. Chen, Z.; Yan, H.; Li, J.; Zhang, G.; Zhang, Z.; Liu, B. Relationship between tertiary volcanic rocks and hydrocarbons in the Liaohe basin, People's Republic of China. *AAPG Bull.* **1999**, *83*, 1004–1014.

7. Hu, T.; Chen, Z.; Dong, X.; Yao, W.; Liang, Z.; Wu, K.; Guan, J.; Gao, M.; Pang, Z.; Li, S.; et al. Oil Origin, Charging History and Crucial Controls in the Carboniferous of western Junggar Basin, China: Formation Mechanisms for Igneous Rock Reservoirs. *J. Pet. Sci. Eng.* **2021**, *203*, 108600. [CrossRef]
8. Zhao, W.-Z.; Zou, C.-N.; Feng, Z.-Q.; Hu, S.-Y.; Zhang, Y.; Li, M.; Wang, Y.-H.; Yang, T.; Yang, H. Geological features and evaluation techniques of deep-seated volcanic gas reservoirs, Songliao Basin. *Pet. Explor. Dev.* **2008**, *35*, 129–142. [CrossRef]
9. Wenzhi, Z.; Caineng, Z.; Jianzhong, L.; Zhiqiang, F.; Guangya, Z.; Suyun, H.; Lichun, K.; Yan, Z. Comparative study on volcanic hydrocarbon accumulations in western and eastern china and its significance. *Pet. Explor. Dev.* **2009**, *36*, 1–11. [CrossRef]
10. Zou, C.-N.; Zhao, W.-Z.; Jia, C.-Z.; Zhu, R.-K.; Zhang, G.-Y.; Zhao, X.; Yuan, X.-J. Formation and distribution of volcanic hydrocarbon reservoirs in sedimentary basins of china. *Pet. Explor. Dev.* **2008**, *35*, 257–271. [CrossRef]
11. Cao, J.; Wang, X.; Sun, P.A.; Zhang, Y.; Tang, Y.; Xiang, B.; Lan, W.; Wu, M. Geochemistry and origins of natural gases in the central Junggar Basin, northwest China. *Org. Geochem.* **2012**, *53*, 166–176. [CrossRef]
12. Pan, J.; Wang, G.; Qu, Y. Origin and charging histories of diagenetic traps in the Junggar Basin. *AAPG Bull.* **2021**, *105*, 275–307. [CrossRef]
13. Cao, J.; Zhang, Y.; Hu, W.; Yao, S.; Wang, X.; Zhang, Y.; Tang, Y. The Permian hybrid petroleum system in the northwest margin of the Junggar Basin, northwest China. *Mar. Pet. Geol.* **2005**, *22*, 331–349. [CrossRef]
14. Yin, J.; Chen, W.; Xiao, W.; Yuan, C.; Sun, M.; Tang, G.; Yu, S.; Long, X.; Cai, K.; Geng, H.; et al. Petrogenesis of Early-Permian sanukitoids from West Junggar, Northwest China: Implications for Late Paleozoic crustal growth in Central Asia. *Tectonophysics* **2015**, *662*, 385–397. [CrossRef]
15. Yang, X.F.; He, D.F.; Wang, Q.C.; Tang, Y.; Tao, H.F.; Li, D. Provenance and tectonic setting of the Carboniferous sedimentary rocks of the East Junggar Basin, China: Evidence from geochemistry and U–Pb zircon geochronology. *Gondwana Res.* **2012**, *22*, 567–584. [CrossRef]
16. Yao, W.; Chen, Z.; Dong, X.; Hu, T.; Liang, Z.; Jia, C.; Pan, T.; Yu, H.; Dang, Y. Storage space, pore–throat structure of igneous rocks and the significance to petroleum accumulation: An example from Junggar Basin, western China. *Mar. Pet. Geol.* **2021**, *133*, 105270. [CrossRef]
17. Ding, X.; Gao, C.; Zha, M.; Chen, H.; Su, Y. Depositional environment and factors controlling β-carotane accumulation: A case study from the Jimsar Sag, Junggar Basin, northwestern China. *Palaeogeogr. Palaeoclimatol. Palaeoecol.* **2017**, *485*, 833–842. [CrossRef]
18. Ding, X.; Qu, J.; Imin, A.; Zha, M.; Su, Y.; Jiang, Z.; Jiang, H. Organic matter origin and accumulation in tuffaceous shale of the lower Permian Lucaogou Formation, Jimsar Sag. *J. Pet. Sci. Eng.* **2019**, *179*, 696–706. [CrossRef]
19. Tao, K.; Cao, J.; Wang, Y.; Ma, W.; Xiang, B.; Ren, J.; Zhou, N. Geochemistry and origin of natural gas in the petroliferous Mahu Sag, northwestern Junggar Basin, NW China: Carboniferous marine and Permian lacustrine gas systems. *Org. Geochem.* **2016**, *100*, 62–79. [CrossRef]
20. Imin, A.; Zha, M.; Ding, X.; Bian, B.; Liu, Y.; Zheng, M.; Han, C. Identification of a Permian foreland basin in the western Junggar Basin (NW China) and its impact on hydrocarbon accumulation. *J. Pet. Sci. Eng.* **2020**, *187*, 106810. [CrossRef]
21. Yao, W.; Chen, Z.; Hu, T.; Liang, Z.; Jia, C.; Wu, K.; Pan, T.; Yu, H.; Dang, Y. Storage space, pore structure, and primary control of igneous rock reservoirs in Chepaizi Bulge, Junggar Basin, western China: Significance for oil accumulation. *J. Pet. Sci. Eng.* **2020**, *195*, 107836. [CrossRef]
22. Yao, Z.; Yang, F.; Jianatayi, D.; Wang, W.; Gao, Y.; Sun, L.; Wang, W.; Mu, B.-Y.; Pan, P.-B. Application of multi-attribute matching technology based on geological models for sedimentary facies: A case study of the 3rd member in the Lower Jurassic Badaowan Formation, Hongshanzui area, Junggar Basin, China. *Pet. Sci.* **2022**, *19*, 12. [CrossRef]
23. Gong, D.; Song, Y.; Wei, Y.; Liu, C.; Wu, Y.; Zhang, L.; Cui, H. Geochemical characteristics of Carboniferous coaly source rocks and natural gases in the Southeastern Junggar Basin, NW China: Implications for new hydrocarbon explorations. *Int. J. Coal Geol.* **2018**, *202*, 171–189. [CrossRef]
24. Yiming, A.; Ding, X.; Qian, L.; Liu, H.; Hou, M.; Jiang, Z. Gas generation Potential of Permian Oil-Prone Source Rocks and Natural Gas Exploration Potential in the Junggar Basin, NW China. *Appl. Sci.* **2022**, *12*, 11327. [CrossRef]
25. Qiao, S.; Shan, X.; Li, A.; Yi, J.; Yiming, A.; Bian, B. Characteristics and identification of Carboniferous volcanic lithofacies in Shixi area, Junggar Basin. *Glob. Geol.* **2022**, *25*, 146–158.
26. Zhang, L.; Li, A.; Yu, C. Application of broadband, wide-azimuth, and high-density 3D seismic exploration. *Oil Geophys. Prospect.* **2017**, *52*, 1236–1245.
27. Dong, W.; Zhao, X.; Liu, F.; Zhao, G. The time-frequency electromagnetic method and its application in western China. *Appl. Geophys.* **2008**, *5*, 127–135.

processes

Article

The Whole-Aperture Pore Structure Characteristics and Their Controlling Factors of the Dawuba Formation Shale in Western Guizhou

Kun Yuan [1,2], Wenhui Huang [1], Xianglin Chen [2], Qian Cao [3], Xinxin Fang [4,*], Tuo Lin [2,*], Chunshuang Jin [2], Shizhen Li [2], Chao Wang [2] and Ting Wang [2]

1 School of Energy Resources, China University of Geosciences, Beijing 100083, China; kunyuanogscgs@sina.com (K.Y.); huangwh@cugb.edu.cn (W.H.)
2 Oil and Gas Resources Survey, China Geological Survey, Beijing 100083, China; chenxianglin2022@163.com (X.C.); jincs2002@163.com (C.J.); lishz2006@sina.com (S.L.); wangchaopku@126.com (C.W.); wangting@mail.cgs.gov.cn (T.W.)
3 Sichuan Keyuan Testing Center of Engineering Technology, Chengdu 610091, China; cissiy0923@163.com
4 Institute of Geomechanics, Chinese Academy of Geological Sciences, Beijing 100081, China
* Correspondence: freestarxin@163.com (X.F.); everdeer@163.com (T.L.)

Abstract: Since shale gas mainly occurs in shale pores, research on pore structure characteristics is the key to understanding the shale gas accumulation mechanism. The pore structure of the Lower Carboniferous Dawuba Formation shale in the Qianxi area, represented by the well QSD-1, which obtains a daily shale gas flow of 10,000 m^3 and represents an important breakthrough in the investigation of marine shale gas in the Upper Paleozoic region of southern China, is characterized by high-pressure mercury compression experiments and low-temperature gas adsorption (N_2 and CO_2) experiments with whole pore size. The main controlling factors affecting the pore development of the shale are discussed. (i) The micropores and mesopores are more developed in the shale, and the macropores are the second most developed in the shales of the Dawuba Formation in the Qianxi area, among which the mesopores and macropores contribute most to the pore volume and the micropores and mesopores contribute most to the pore-specific surface area. (ii) The microfractures and interlayer pores of clay minerals are developed in the shales of the Dawuba Formation, which are the main storage spaces for hydrocarbon gases. (iii) The main factors affecting the adsorption capacity of the shales of the Dawuba Formation in the Qianxi area are the organic carbon content and clay mineral content of the shales, both of which have an obvious positive correlation with the variation of pore structure.

Keywords: taphrogenic trough; Qianshuidi 1 well; carboniferous strata; pore structure; total pore volume

Citation: Yuan, K.; Huang, W.; Chen, X.; Cao, Q.; Fang, X.; Lin, T.; Jin, C.; Li, S.; Wang, C.; Wang, T. The Whole-Aperture Pore Structure Characteristics and Their Controlling Factors of the Dawuba Formation Shale in Western Guizhou. *Processes* **2022**, *10*, 622. https://doi.org/10.3390/pr10040622

Academic Editor: Carlos Sierra Fernández

Received: 9 February 2022
Accepted: 18 March 2022
Published: 22 March 2022

1. Introduction

Since the "shale gas revolution", China has carried out a lot of exploration work on shale gas resources [1–9]. However, there are few breakthroughs in Carboniferous strata in South China with great oil and gas potential [10–14], and there are few studies on the pore development and its main controlling factors of Carboniferous shale [13,14]. In 2021, well QSD-1 was deployed in the Liupanshui area of Qianxi, which was tested in the Carboniferous Dawuba Formation and obtained 10,000 m^3/D shale gas flow, achieving a major breakthrough in the investigation of shale gas in the Lower Carboniferous area of South China. Drilling revealed that the shale thickness of the Dawuba Formation was nearly a kilometer, and 53 sets of gas reservoirs were comprehensively interpreted, with the highest total hydrocarbon value of 63.4%. As an example of typical shale gas drilling in the area, this paper takes the core samples from the Dawuba Formation in Well QSD-1 as the research object, characterizes the whole pore size of the shale pore structure, and discusses its influencing factors, in order to provide reference for shale gas investigation in the Qianxi area.

Shale gas occurs mainly as free gas in pores and natural fractures, adsorbed gas in organic matter and clay minerals, dissolved gas in residual oil, and in water in shale reservoirs [15,16]. The size, volume, shape, connectivity and permeability of the pores significantly influence shale gas enrichment and exploration [17]. The pore structure and fracture systems have been characterized by researchers based on various models [18–20]. For example, the low temperature CO_2 and N_2 adsorption are widely used to characterize the pore distributions of micropores (<2 nm) and mesopore size (2–50 nm), respectively. The high-pressure mercury injection method is used to characterize the pore distributions of macropores (>50 nm).

The shale characteristics of the Lower Paleozoic Silurian strata represented by the Sichuan Basin have been extensively studied [21]. In contrast, due to the large difference in mineral composition and strong heterogeneity of shale in the Upper Paleozoic marine rocks in southern China, the pore structure characteristics and their main controlling factors are quite different from those in the Lower Paleozoic strata [22].

In order to understand the pore structure characteristics of shale in the Dawuba Formation in Qianxi area and to understand the enrichment mechanism of shale gas in the Lower Carboniferous, qualitative and quantitative studies of the shale in the Dawuba Formation were carried out by FE-SEM, low temperature gas adsorption (N_2 and CO_2) and high-pressure mercury injection. The pore structure of the shale in the Dawuba Formation was characterized from different scales. The pore structure parameters such as specific surface area, pore size distribution and pore volume of the shale were obtained, and the main controlling factors were analyzed.

2. Experimental Scheme and Sample Selection

2.1. Sample Selection

The shale samples selected in this paper were taken from the cores of different layers of the Dawuba Formation in Well QSD-1, which was located in the Qiannan depression, southwestern China (Figure 1). The study area is located in the west part of the Qiannan depression, and confined by the Ziyun and Guiyang fault. Under the tectonic processes of multiple stresses, the Carboniferous, Permian and Triassic strata have undergone strong structural deformation and denudation of different extents. In the Indosinian-Yanshan and Himalayan stages, strong folds and thrust nappes resulted in massive shortening deformation.

According to the lithology and electrical characteristics, and combined with the outcrop situation of the surrounding strata, the shale samples can be divided into four sections from bottom to top [23,24]. (1). The first member of the Dawuba Formation is the thickest and most uniform lithology member of this formation, which is mainly composed of thick black mudstone. (2). The shale in the second member of the Dawuba Formation is thin interbedded sandwiched in marl, and the shale lamellation is well developed. (3). The third member of the Dawuba Formation is grayish-black calcareous shale, mudstone, and dark gray argillaceous limestone mixed with grayish-black carbonaceous mudstone, and the interbedded structure of calcareous mudstone and micritic limestone is developed. (4). The fourth member of the Dawuba Formation is characterized by dark gray and grayish-black marl with carbonaceous shale and calcareous mudstone (Figure 2).

In order to ensure the distinguishability of the samples, nine samples from different members of the Dawuba Formation were selected for the high-pressure mercury injection experiment and the low-temperature gas adsorption experiment. At the same time, the organic carbon content, maturity, mineral composition and rock density of the samples were tested (Table 1) in order to fully reflect the pore structure characteristics of Dawuba Formation shale.

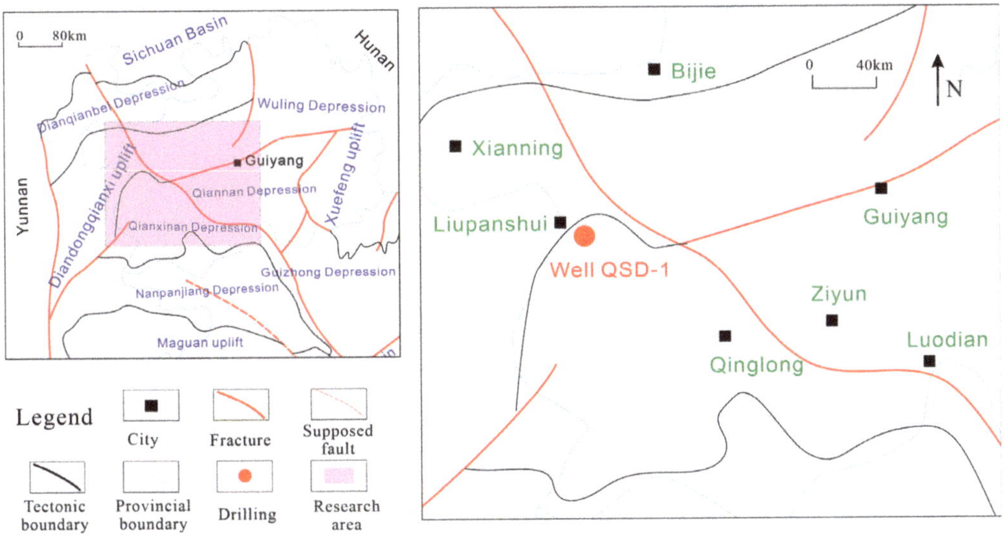

Figure 1. Study area and well location of Qianshuidi 1.

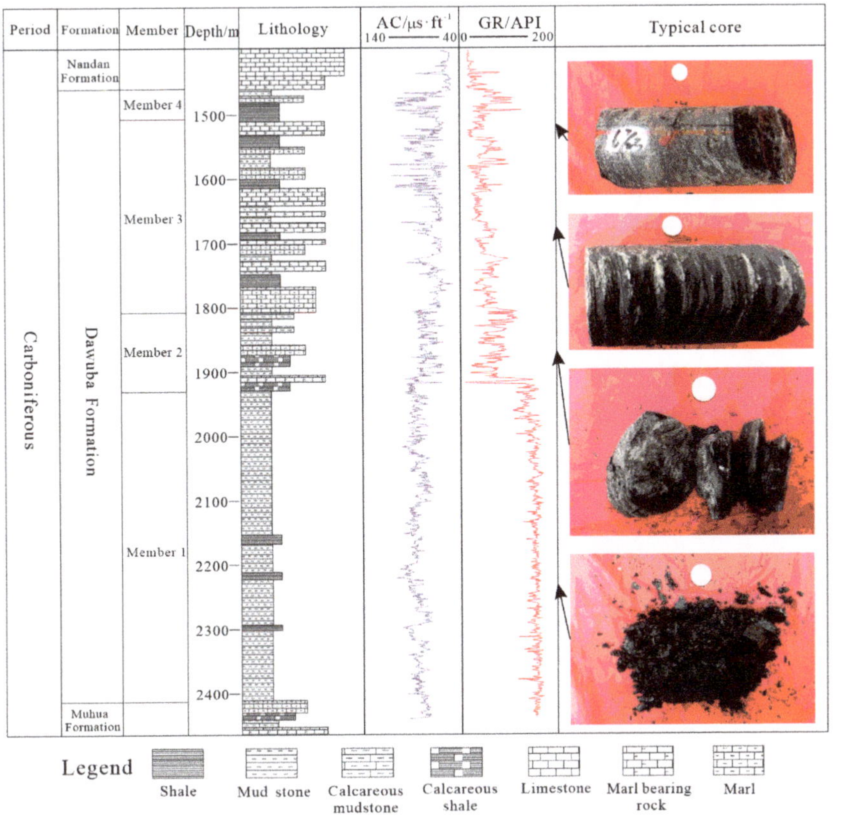

Figure 2. Stratigraphic characteristics and division scheme of the Dawuba Formation in Well QSD-1.

Table 1. Shale sample parameters of the Dawuba Formation in the Qianxi area.

Group	Member	Depth/m	TOC/%	Ro/%	Main Mineral Component Content/%					Rock Density/g·cm^3
					Quartz	Calcite	Dolomite	Pyrite	Clay Mineral	
Dawuba Formation	Member 4	1648.10	0.76	2.17	12	76	3	1	5	2.65
	Member 3	1649.95	0.7	2.28	5	83	2	1	7	2.66
		1656.50	0.53	2.18	9	74	7	1	9	2.66
		1660.20	0.58	1.95	10	56	19	2	13	2.67
		1682.25	0.99	2.32	16	37	30	2	15	2.40
		1687.40	0.96	2.19	19	27	31	2	21	1.78
		1695.59	0.88	2.18	12	34	33	1	20	2.21
	Member 2	1931.25	1.08	2.29	8	12	37	3	40	2.69
	Member 1	1940.90	1.22	2.57	27	0	4	0	69	2.49

2.2. Mineral Composition Analysis

A commonly used and statistically valid method to evaluate minerals in shale reservoirs is X-ray diffraction. The diffract grams were recorded in detail from 4° to 70° at a scanning speed of 1° 2θ/min by a Rigaku automated powder diffractometer (D/MAX-RA) equipped with a Cu X-ray source (40 Kv, 35 mA). The samples from the Dawuba shale were analyzed for whole-bulk and clay fraction mineralogy by quantitative X-ray diffraction following two independent processes. First, the bulk mineral composition of the powder sample was determined at this stage to only include the total clay content. The whole-rock samples were analyzed over an angular range of 4° to 70° 2θ at a scanning speed of 1° 2θ/min. Second, the individual clay mineral content of the clay fractions separated from the rock powder sample was determined. The clay minerals were analyzed over an angular range of 3° to 65° 2θ at a scanning speed of 1.5° 2θ/min. The quantification of the minerals was based on calculations of the integrated area using JADE 5 software.

2.3. SEM Observation

A ZEISS Sigma 300 (Oberkochen, Germany) SEM and Bruker Quantax 200G (Berlin, Germany) spectrometer were used to visualize the shale minerals and micropores of the Dawuba Formation. The organic matter, porosity, fracture distribution, size and pore connectivity were quantitatively described [25]. The porosity of the dry samples was determined from the grain density obtained from helium pycnometry (skeletal density) and the bulk volume of the plugs was calculated from mercury immersion (bulk density). The total porosity was calculated from the difference between the bulk and skeletal densities. A Gatan model 685.C (Pleasanton, CA, USA) was used for argon ion polishing in the SEM observations. Ar ion polishing is a common processing procedure carried out before observing under SEM. The samples were milled through a two-step process: (1) source operated at 2.5 kV for 2 h, followed by (2) source operated at 1.0 kV for 1 h. Each ion milling step used a 40% focus and 5° tilt angle, and the samples were continuously rotated 360° to ensure the entire surface was milled and common induced features such as curtaining were minimized.

2.4. High-Pressure Mercury Injection Experiment

A high-pressure mercury injection experiment is mainly used to analyze the macropore development in shale. The AutoPore 9500 mercury injection instrument from Micrometrics Instrument Co., Ltd. (Shanghai, China) is used for testing and analysis. The maximum external pressure is 400 MPa (60,000 psia). The experimental steps are based on the national standard. Before the test, the cylindrical samples (diameter 2.5 cm, length 3 cm) were first dried (dried at 150° or higher temperature for an hour), and mercury was injected into the sample tube under vacuum conditions. The test analysis was carried out at low-pressure and high-pressure stages, respectively. During the low-pressure test, the pressure was increased incrementally. When the maximum external pressure was reached, the pressure was reduced to atmospheric pressure, and the sample tube was moved into the high-pressure station for analysis. At the same time, mercury was filled into the dilatometer under vacuum conditions. The continuous pressure gradually increased the mercury to fill the pores, so as to obtain the relationship curve between the press-in amount and the pressure [26,27]. The distribution of the pore structure characteristic parameters of the different samples could then be obtained by analysis. To minimize the potential effects of the formation of artificially induced microfractures in the cylindrical sample processing, samples without any microfractures on the surface were used. However, it is hard to be certain that no artificially induced microfractures occurred in these samples.

2.5. Low-Temperature N_2 and CO_2 Adsorption Experiments

N_2 and CO_2 were selected as the adsorbed gas, and the isothermal adsorption curve was obtained by measuring the adsorption capacity of the sample under different pressures. The distribution of the pore structure characteristic parameters such as the pore size, specific surface area and pore volume of the sample was calculated by BET and BJH models [27]. Among them, the liquid nitrogen adsorption–desorption test was used to characterize the mesoporous distribution characteristics of the sample. The instrument was an ASAP 2460 automatic specific surface and pore size analyzer. The relative pressure range of the isothermal adsorption–desorption experiment of the instrument is 0.00–0.995. The pore size range that can be tested is 0.35 nm–400 nm, and the minimum specific surface area that can be tested is 0.0005 m^2/g.

The samples were ground to powder with the particle size of 60–80 mesh. The experimental temperature is 77.3 k when the experimental adsorbate is liquid nitrogen, and 273.15 k when the experimental adsorbate is carbon dioxide. Before the experimental test, the appropriate amount of sample was weighed and put into a specific long tube (high-temperature and low-temperature resistance), and the vacuum degassing treatment was carried out at 150 °C for 4–6 h in order to remove the gas adsorbed on the surface of the material. After the degassing was complete, the sample was cooled to room temperature and fixed on the instrument to backfill helium and nitrogen (carbon dioxide) in turn. Then, the corresponding gas adsorption amount was determined by gradually changing the pressure at the test temperature, and the adsorption and desorption tests were carried out to obtain the adsorption–desorption isotherm curves of different samples.

3. Results

3.1. Mineral Composition Characteristics

The overall performance of the Dawuba Formation in Well QSD-1 was dominated by carbonate minerals and clay minerals, followed by quartz minerals, and by fewer iron and magnesium minerals. The carbonate rock content was 4–93%, average 53%; the clay mineral content was 5–82%, average 33%; the quartz mineral content was 4–27%, average 11%; some samples contained iron–magnesium minerals and plagioclase: the pyrite content was 1–3%, average 2%, and the plagioclase content was 1–3%, average 1%.

Vertically, the fourth and third members of the Dawuba Formation are dominated by carbonate minerals (58–93%, average 70%), clay minerals (6–15%, average 10%) and felsic minerals (6–11%, average 7%). Carbonate minerals (32–61%, average 49%) and clay minerals (37–48%, average 40%) are the main minerals in the second member of the Dawuba Formation, followed by felsic minerals (average 8%) and iron–magnesium minerals (average 3%). The first member of the Dawuba Formation is dominated by clay minerals (54–82%, average 69%), followed by felsic minerals (16–33%, average 27%), and carbonate minerals (4–13%, average 6%) (Figure 3).

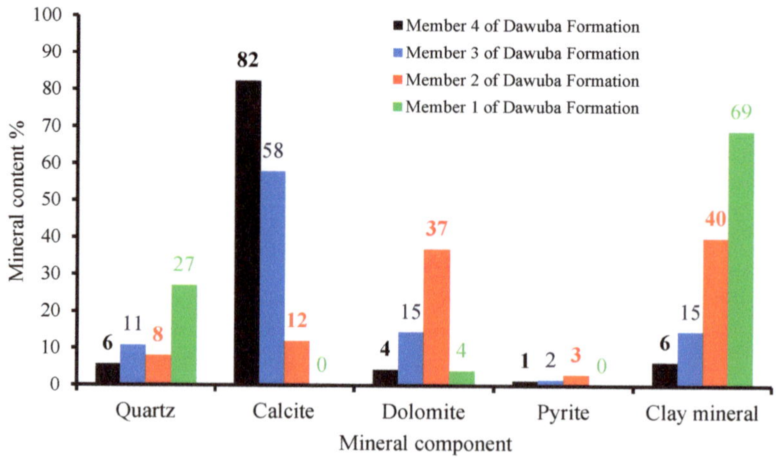

Figure 3. Mineral component content of the Dawuba Formation in Well QSD-1.

3.2. Pore Types and Morphological Characteristics

The SEM observation results show that several pore types are mainly observed in the Dawuba Formation of Well QSD-1: primary intergranular pore, organic matter pore, dissolved pore and intergranular pore of carbonate minerals, diagenetic fracture and interlayer pore fracture of clay minerals (Figure 4). It was found that the internal/intergranular pores of inorganic minerals such as clay minerals, quartz and feldspar, and the internal and marginal pores of organic matter differ greatly in size and morphology. The SEM observation shows that the distribution of different organic matter is quite different, and most of them are distributed in the shale matrix independently in the form of lumps and ribbons. Some organic matter grows together with the surrounding matrix minerals, filling in inorganic mineral particles (crystals) such as siliceous minerals, clay minerals or pyrite, and their boundaries are mostly mixed with the boundary between the mineral particles/crystals. There are a certain number of pores in the internal and edge of the organic matter, with large morphological differences. Ellipsoid, needle-like or irregular shapes are distributed, and the pore scale is relatively small. The content of the clay minerals in the samples is relatively high, and the pores in the crystal (grain) are most developed in the illite and illite–montmorillonite mixed layer. The intergranular pores of the clay minerals mostly exist in the natural faults between the laminated structures, particles and the internal plates of the particles. The bending lamellar pore aggregates are developed in the mineral crystals such as the illite and illite–montmorillonite mixed layer. Due to compaction, some clay minerals show a flake shape, and there are a large number of flat or linear pores between these flake clay minerals [28]. Compared with organic matter pores, the microfractures and clay mineral interlayer pore fractures in the Dawuba Formation shale are more developed, providing a good reservoir space for hydrocarbon gas occurrence [29].

3.3. Experimental Curve of Mercury Injection

It can be seen that the shape of the mercury injection–withdrawal curve of the samples in each member of the Dawuba Formation is quite different. The "hysteresis loop" formed by the inconsistency of the mercury injection curve and the mercury withdrawal curve is also the performance of the distribution of parameters such as the shape and connectivity of the reservoir space in the samples [30], reflecting the different characteristics of the strata in each member of the Dawuba Formation. The mercury withdrawal efficiency reflects the capillary effect recovery of the non-wetting phase, which indicates that the throat volume accounts for the percentage of the total volume of pores and throats in the core [31]. The larger the mercury withdrawal efficiency, the more uniform the size distribution of the pores and throats in the core. The maximum mercury saturation reflects the connectivity of the pore-fracture system and the degree of pore development.

The mercury injection test results show that there are macropores in the first, second and third members of the Dawuba Formation, the macropores in the second and third members are more developed than those in the first member and the pore connectivity is better in the first member. The pores in the first member are relatively developed and the pore size distribution is more uniform (Figure 5a).

The first member of the Dawuba Formation: when the pressure increases to about 10 MPa, the mercury curve shows a slow increasing trend, the pore sorting is relatively good and the pore size distribution is relatively uniform. The hysteresis loop of the mercury injection and withdrawal is wide, the volume difference of mercury injection and withdrawal is small and the efficiency of the mercury injection and withdrawal is relatively low, indicating that the corresponding throat is thin and the connectivity is poor in the mercury injection test range.

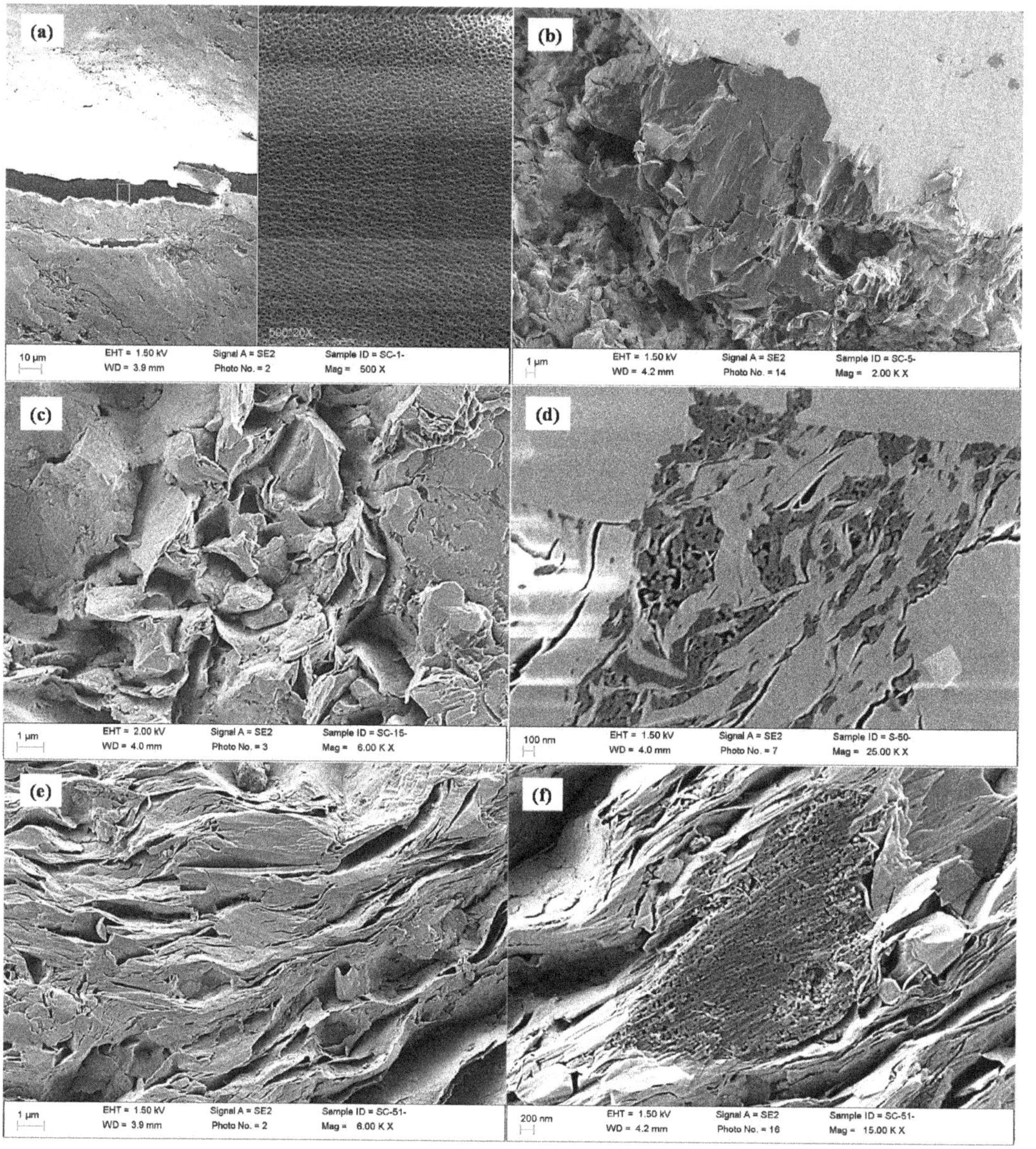

Figure 4. Microscopic characteristics of shale pores in the Dawuba Formation of Well QSD-1. (**a**) Black shale, member 4 of the Dawuba Formation, 1490 m, banded organic matter developed, micropores developed inside; (**b**) black shale, member 3 of the Dawuba Formation, 1531 m, organic matter is distributed in mineral grains, pores are developed in its interior and edge and some pores are connected to each other; (**c**) black shale, member 3 of the Dawuba Formation, 1660 m, clay mineral intergranular pore development; (**d**) black shale, member 2 of the Dawuba Formation, 1930 m, mixed distribution of organic matter and clay minerals; (**e**) black shale, member 1 of the Dawuba Formation, 1940 m, clay mineral interlayer fractures developed in a narrow strip; (**f**) black shale, member 1 of the Dawuba Formation, 1940 m, clay mineral interlayer fractures and micropores developed.

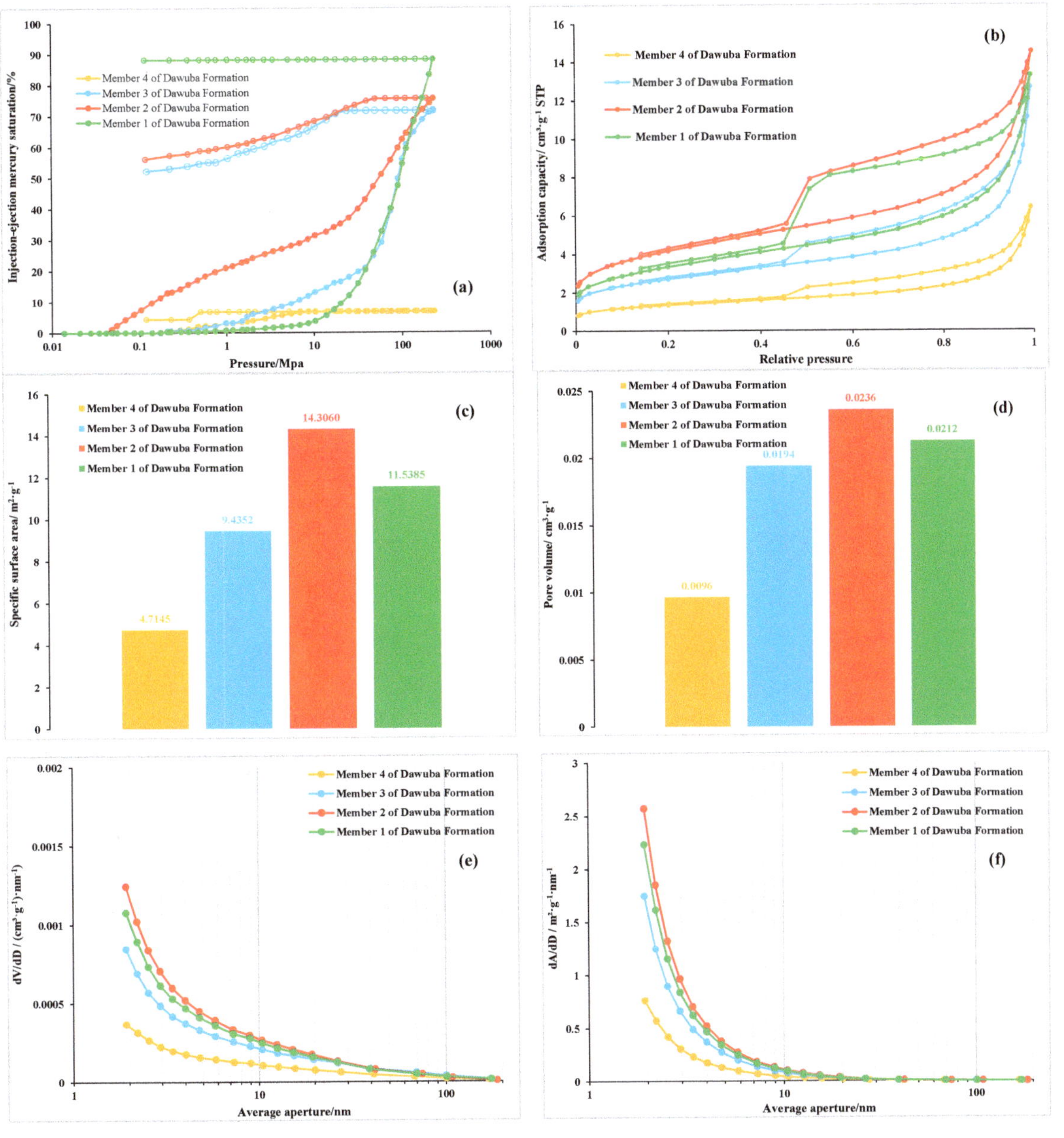

Figure 5. Pore distribution characteristics at different scales of the Dawuba Formation shale in Well QSD-1. (**a**) The mercury injection–withdrawal curve of each member of the Dawuba Formation shale; (**b**) nitrogen adsorption–desorption curve of shale in each member of the Dawuba Formation; (**c**) Based on the distribution of specific surface area of shale in each member of Dawuba Formation by N_2 adsorption; (**d**) shale pore volume distribution based on N_2 adsorption; (**e**) shale pore volume change rate distribution based on N_2 adsorption; (**f**) specific surface area distribution of shale in each member of the Dawuba Formation based on N_2 adsorption.

The second member of the Dawuba Formation: with the increase of pressure, the mercury injection curve showed a slow ladder increase trend; the sorting coefficient is about 4.45 and the displacement pressure is small. The corresponding maximum pore throat radius is 15.65 μm, and the maximum mercury saturation of the sample is 75%. The hysteresis loop of the mercury injection and withdrawal is wide and the volume difference between the mercury injection and withdrawal is large. The mercury withdrawal efficiency is about 25%, indicating that the macropores and mesopores of the sample are relatively developed in the pressure test range, and the pore connectivity is good.

The third member of Dawuba Formation: with the increase of pressure, the mercury curve shows a certain increasing trend. When the pressure increases to about 30 MPa, the mercury curve shows a ladder increasing trend. The sorting coefficient is about 4.21, the displacement pressure is 0.172 MPa and the maximum pore throat radius is 4.268 μm. It can be seen that the macropores of the third member samples are less developed than those of the second member samples. The maximum mercury saturation of the sample is about 70%, and the hysteresis loop of the mercury injection and withdrawal is wide. The volume difference of the mercury injection and withdrawal is large, and the mercury withdrawal efficiency is about 27%, indicating that the pore connectivity is good within the pressure test range.

The fourth member of Dawuba Formation: with the increase of pressure, the mercury injection curve exhibits little change. When the pressure increases to 400 MPa, the mercury saturation is less than 10%, indicating that the pores in the sample are relatively undeveloped.

3.4. Experimental Curve of N_2 and CO_2 Adsorption at Low Temperature

Under the condition of low temperature (77 K), the N_2 adsorption–desorption curve can reflect the pore morphology. According to the four types of pores summarized by IUPAC [31], the main types of pores in the third and fourth members of the Dawuba Formation are close to the H3 type, that is, the open slit pores on four sides, while the main types of pores in the first and second members of the Dawuba Formation are close to the mixture of H1 and H2 types, indicating that the shales in the first and second members of the Dawuba Formation are mainly composed of regular tubular pores with openings at both ends and ink bottle-shaped pores with wide, thin necks (Figure 5b).

Based on the N_2 adsorption curve, the BJH adsorption model was used to calculate the pore volume and specific surface area of the shale samples in different layers (Figure 5c,d). The calculation results show that the pore volume and specific surface area of the second member of the Dawuba Formation are the largest, which are 14.3060 m^2/g and 0.0236 cm^3/g, respectively, followed by the first, third and fourth members. It can be seen from the analysis of the change rates of the pore volume and specific surface area with pore size (Figure 5e,f) that with the increase of pore size, the change rate of the pore volume gradually decreases, indicating that the pore volume mainly changes significantly in the pores within the range of 2 nm to 5 nm. The specific surface area also shows a similar rule with the pore volume, indicating that the pore volume and specific surface area of the mesopore are mainly provided by the pores with pore size (2 nm–5 nm), and the contribution of a large pore size (5 nm–100 nm) to the pore volume and specific surface area is low.

4. Discussion

4.1. Distribution Characteristics of Total Pore Size of Shale

Based on the experiments of high-pressure mercury injection and low-temperature gas adsorption (N_2 and CO_2), the pore distribution characteristics of shale from the nano scale to the micron scale were obtained [32]. Different experimental data have different characterization ranges and accuracy for pores. Micropores are mainly characterized by CO_2 adsorption data, mesopores are characterized by N_2 adsorption data and macropores are characterized by high-pressure mercury injection data [33]. Thus, the full-aperture distribution characteristics of shale in each member of the Dawuba Formation are obtained.

It can be seen from the comparison that the pore volume of the shale pores in the Dawuba Formation in the study area is relatively developed in the mesopores and macropores, followed by the micropores. Among them, the shale pore characteristics of each member of the Dawuba Formation are similar, and the shale pore volume is dominated by medium pores (pore volume ratio > 50%), followed by macropores and micropores. The change of pore volume has three relatively stable peaks, which are 0.4 nm~0.6 nm, 5.6 nm~9.6 nm and 32.8 nm~500 nm (Figure 6, Table 2). The mesopores and macropores contribute about 89.78% of the pore volume, and the micropores contribute about 10.22% of the pore volume (Table 2).

The distribution of the dV/dD pore volume and pore size in the full pore size range of the shale samples in each member of the Dawuba Formation shows that the pores of the samples are mainly distributed in the micropore members and some mesopore members, among which the number of mesopores with a pore size of about 10 nm in the first and third members is more than that in the second and fourth members (Figure 7).

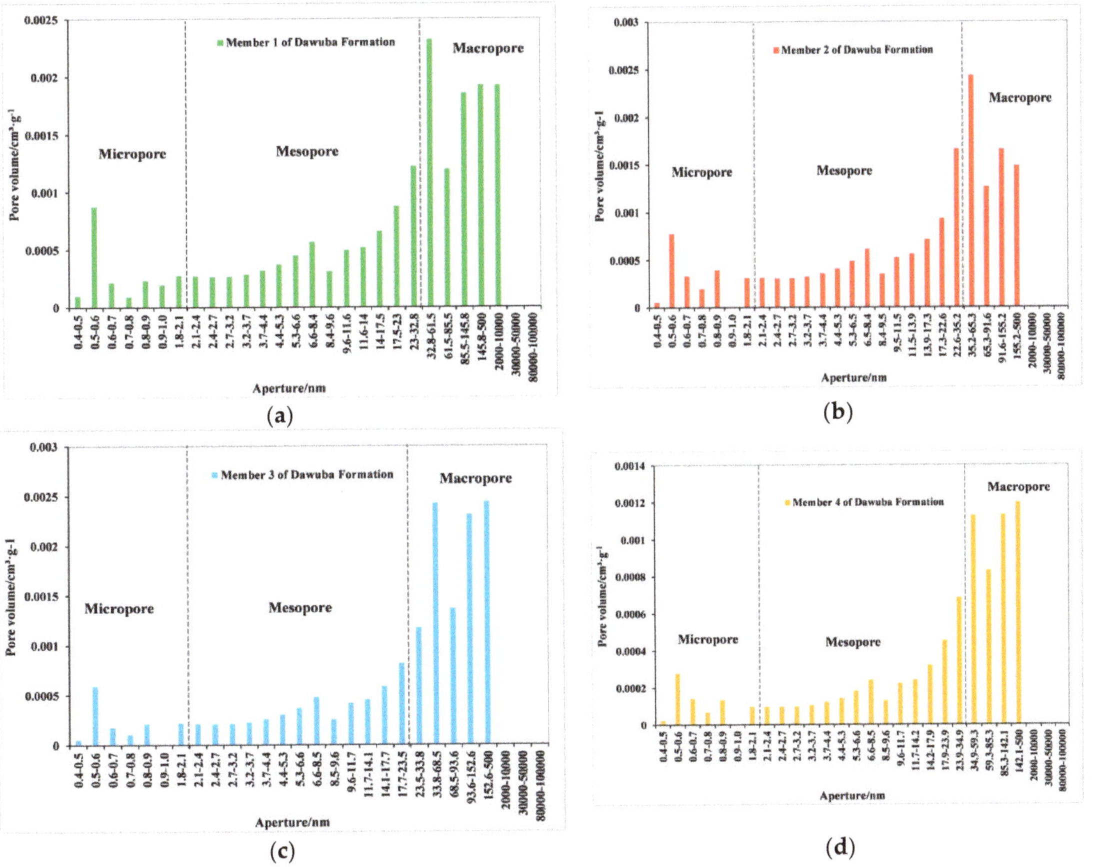

Figure 6. Whole pore size distribution of shale pore volume of the Dawuba Formation in the Qianxi area. (**a**) Member 1 of the Dawuba Formation; (**b**) member 2 of the Dawuba Formation; (**c**) member 3 of the Dawuba Formation; (**d**) member 4 of the Dawuba Formation.

Table 2. Statistics of shale pore volume of the Dawuba Formation in the Qianxi area.

Samples	Pore Volume/($cm^3 \cdot g^{-1}$)				Pore Volume Ratio/%		
	Micropore	Mesopore	Macropore	Total Pore Volume	Micropore	Mesopore	Macropore
Member 1 of Dawuba	0.001985	0.009215	0.006907	0.018107	10.96	50.89	38.15
Member 2 of Dawuba	0.002052	0.01026	0.004396	0.016708	12.28	61.41	26.31
Member 3 of Dawuba	0.001358	0.008392	0.006118	0.015868	8.56	52.89	38.56
Member 4 of Dawuba	0.000745	0.004276	0.003165	0.008186	9.10	52.24	38.66
Average value	0.001535	0.008036	0.0051465	0.014717	10.22	54.36	35.42

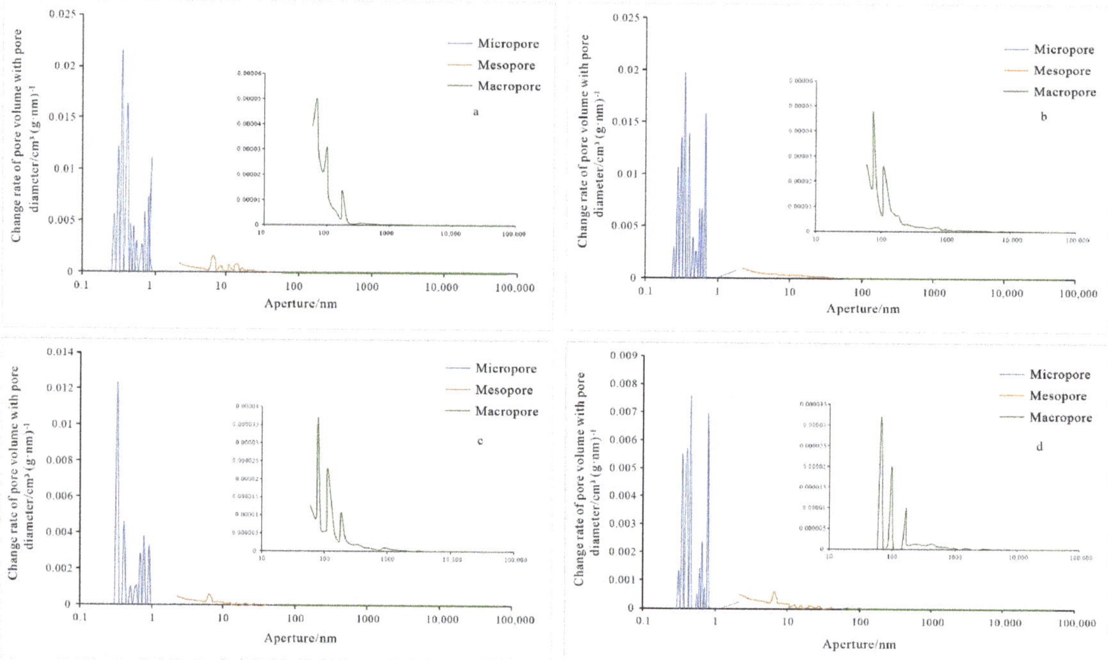

Figure 7. Variation rate curve of total pore volume with pore diameter of Dawuba Formation shale in the Qianxi area. (**a**) Member 1 of the Dawuba Formation; (**b**) member 2 of the Dawuba Formation; (**c**) member 3 of the Dawuba Formation; (**d**) member 4 of the Dawuba Formation.

From the analysis of the specific surface area provided by the shale pores, the specific surface area of the shale pores in the Dawuba Formation in the study area is mainly provided by micropores and mesopores. Pores contribute 93% of the specific surface area of the shale in the Dawuba Formation, and pores less than 0.6 nm in each member are the main contributors to the specific surface area. With the increase of pore diameter, the specific surface area shows a downward trend; the specific surface area provided by the mesopores is generally less than 0.5 m^2/g, and the specific surface area provided by the macropores is one order of magnitude smaller than that of the mesopores (Figure 8, Table 3).

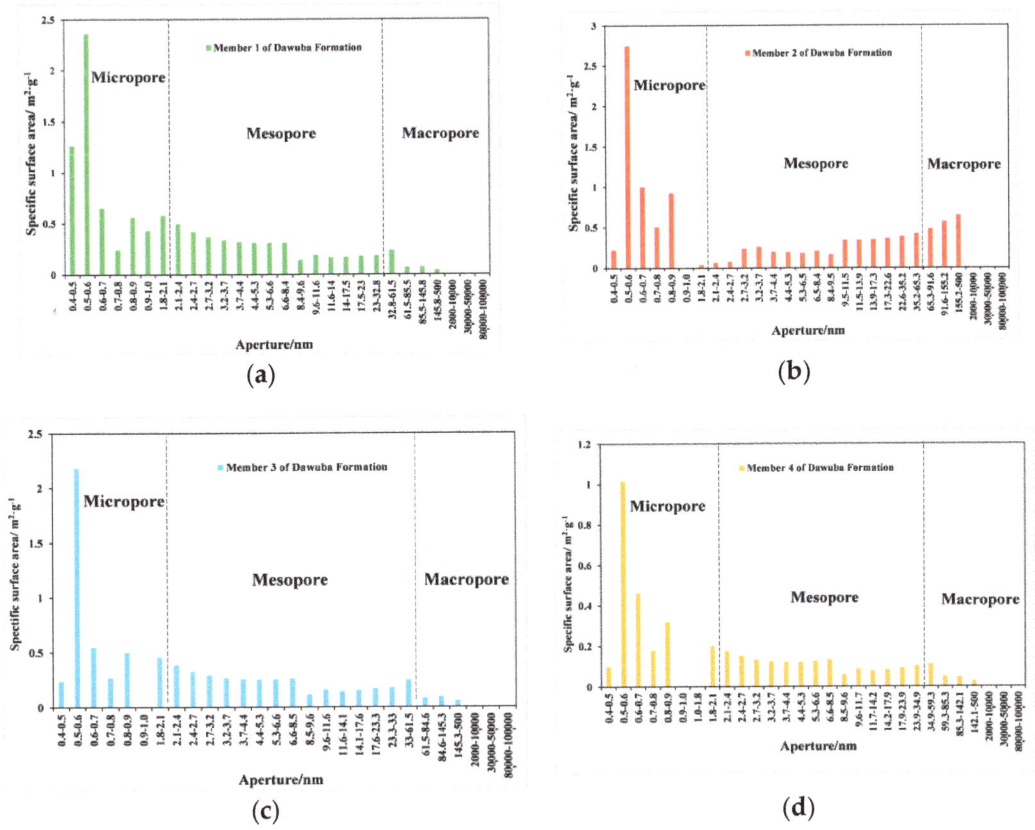

Figure 8. Distribution of specific surface area and full aperture of Dawuba Formation shale in the Qianxi area. (**a**) Member 1 of the Dawuba Formation; (**b**) member 2 of the Dawuba Formation; (**c**) member 3 of the Dawuba Formation; (**d**) member 4 of the Dawuba Formation.

Table 3. Statistics of specific surface area of Dawuba Formation shale in the Qianxi area.

Samples	Pore Volume/($cm^3 \cdot g^{-1}$)				Pore Volume Ratio/%		
	Micropore	Mesopore	Macropore	Total Pore Volume	Micropore	Mesopore	Macropore
Member 1 of Dawuba	0.001985	0.009215	0.006907	0.018107	10.96	50.89	38.15
Member 2 of Dawuba	0.002052	0.01026	0.004396	0.016708	12.28	61.41	26.31
Member 3 of Dawuba	0.001358	0.008392	0.006118	0.015868	8.56	52.89	38.56
Member 4 of Dawuba	0.000745	0.004276	0.003165	0.008186	9.10	52.24	38.66
Average value	0.001535	0.008036	0.0051465	0.014717	10.22	54.36	35.42

4.2. Discussion on Shale Pore Control

The development of nanopores in shale may be affected by many factors [34]. Based on the analysis of the relationship between pore-specific surface area, pore volume and total organic carbon (TOC), it is considered that TOC has a good positive correlation with the development of pores in shale. The specific surface area of shale micropores has the best correlation with TOC. As TOC increases from 0.53% to 1.22%, the specific surface area of the micropores increases from 1.03 m^2/g to 6.07 m^2/g, and the correlation coefficient is 0.88 (Figure 9a). The correlation between mesopore-specific surface area and TOC is 0.61, and the correlation between the macropore-specific surface area and TOC is poor. The relationship between pore volume and the TOC of shale is similar to that of the specific surface area. The pore volume of the micropores and mesopores has good correlation with TOC, followed by macropores (Figure 9b). It indicates that the organic matter contributes more to the micropores, which was also observed from all of the densely packed organic matter pores in Figure 4.

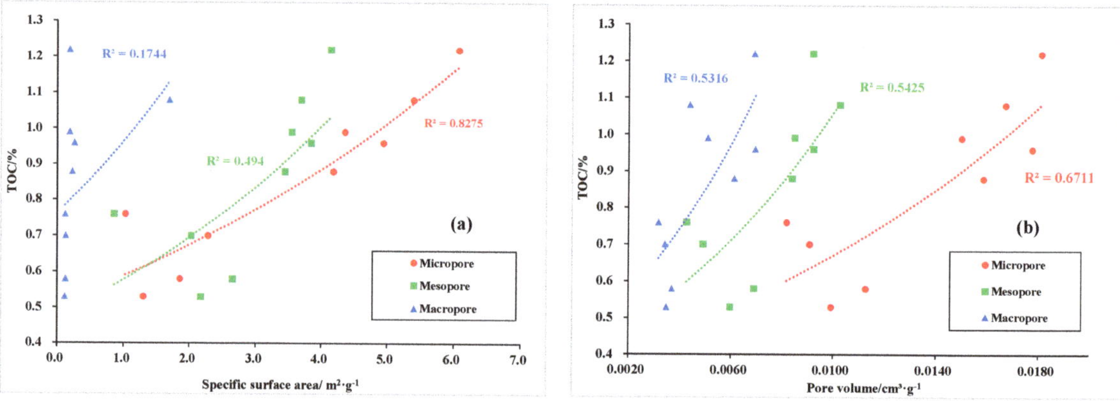

Figure 9. Relationship between pore-specific surface area, pore volume and TOC of Dawuba Formation shale in the Qianxi area. (**a**) Relationship between pore-specific surface area and TOC; (**b**) Relationship between pore volume and TOC.

Through the comparison of the shale mineral content and pore characteristics, it is found that there is a certain correlation between shale pores and mineral content in different pore size ranges [35]. The specific surface area and pore volume show a positive correlation with quartz and clay minerals, but a negative correlation with carbonate minerals. In view of the highest correlation coefficient, the correlation between clay mineral content and pore-specific surface area and pore volume is the best regardless of pore size (micropore, mesopores or macropores), followed by carbonate minerals and quartz mineral content, indicating little effect on pore development and specific surface area (Figure 10). For pores of different size ranges, the clay mineral content is positively correlated with the specific surface area of micropores and mesopores, and the correlation coefficient is more than 0.8 (Figure 10e). The clay mineral content has a good correlation with the micropore and mesopore volume of the sample (Figure 10f). Quartz minerals contribute greatly to the pore volume of macropores (Figure 10b). In shale reservoirs, quartz acts more likely as a rock framework, with clay minerals as interstitial materials. Hence, clay minerals have a stronger impact on pores of small sizes. In addition, due to the strong chemical cementation of carbonate minerals, the development of nano–micropores will be inhibited to varying degrees. In addition, the pores between carbonate particles are easily cemented and filled by clay minerals and organic matter, so the content of carbonate minerals is negatively correlated with the pore volume and specific surface area (Figure 10c,d).

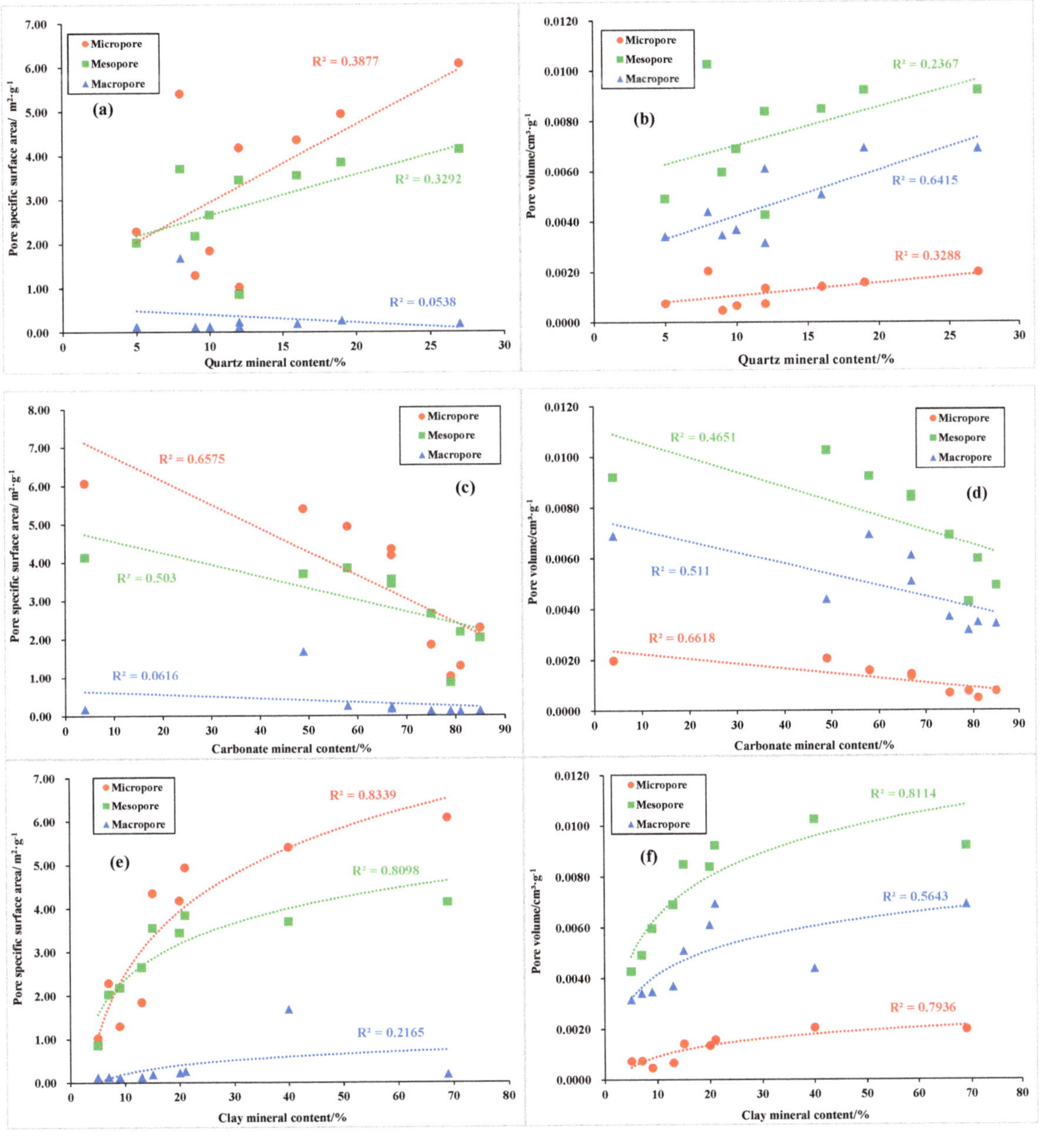

Figure 10. Relationship between pore-specific surface area, pore volume and quartz, carbonate and clay minerals of shale in the Qianxi area. (**a**) Relationship between pore-specific surface area and Quartz mineral content; (**b**) Relationship between pore volume and Quartz mineral content; (**c**) Relationship between pore-specific surface area and Carbonate mineral content; (**d**) Relationship between pore volume and Carbonate mineral content; (**e**) Relationship between pore-specific surface area and Clay mineral content; (**f**) Relationship between pore volume and Clay mineral content.

The analysis shows that the organic carbon content and clay mineral content of the Carboniferous Dawuba Formation in the Qianxi area are the main controlling factors for the specific surface areas of micropores and mesopores, and carbonate minerals and quartz minerals are positively correlated with the macropore volume of the samples. The analysis shows that the TOC and mineral components, to some extent, mainly affect the adsorption capacity of shale by affecting its specific surface area and pore volume and other pore structures. The porosity of organic matter makes it form a huge internal surface area. Due to the development of a layered structure, clay minerals also have large specific surface area and micropore volume. The specific surface area of organic matter pores and pores between clay minerals plays a major role in the specific surface area of the sample, which can provide more adsorption points for methane adsorption and is the main factor affecting the adsorption capacity of the sample. In the process of diagenetic evolution, brittle minerals such as quartz and carbonate play a supporting and protecting role in intergranular organic matter, intergranular pores and interlayer fractures, and play a decisive role in the enrichment of free gas in shale.

The intersection analysis of the clay mineral content of the samples with the total hydrocarbon values at the locations shows that the high total hydrocarbon values rise at the sites with a high clay content (Figure 11). It indicates that the shale gas is fully enriched and effectively preserved at the high clay sites. In addition, the comparison of total hydrocarbon with the specific surface area and pore volume shows that the gas is mainly adsorbed on the micropores and mesopores with a larger specific surface area, and is mainly enriched in the pore volume provided by the micropores. Combining the SEM and FMI results, it is concluded that clay minerals—especially those with microfractures and interlayer joints—are the main storage space for shale gas in the Dawuba Formation.

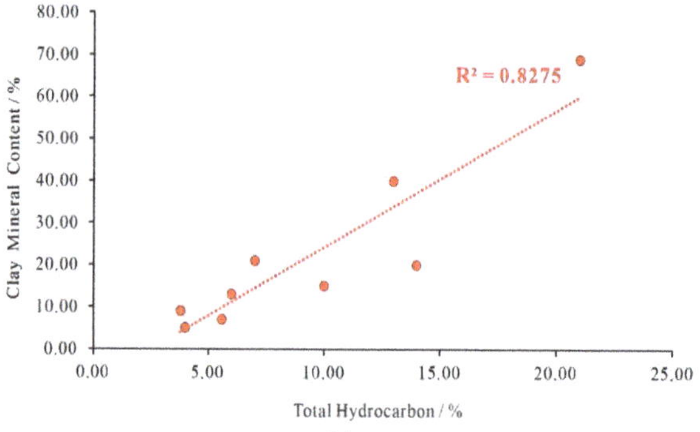

(a)

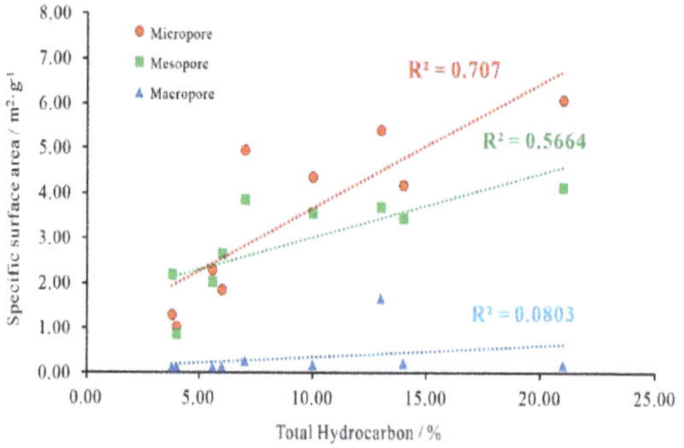

(b)

Figure 11. *Cont.*

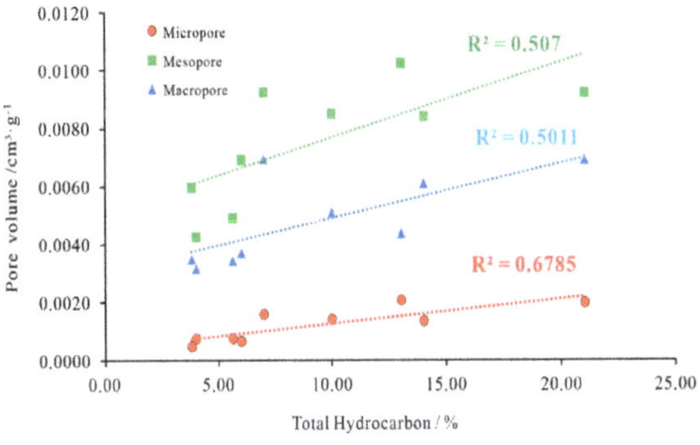

(c)

Figure 11. Cross plot of total hydrocarbon, clay minerals, specific surface area and pore volume. (**a**) Relationship between Clay mineral content and Total Hydrocarbon; (**b**) Relationship between pore-specific surface area and Total Hydrocarbon; (**c**) Relationship between pore volume and Total Hydrocarbon.

5. Conclusions

(1) The micropores and mesopores of Dawuba Formation shale in the Qianxi area are relatively developed, followed by macropores. In terms of pore volume, mesopores and macropores contribute about 89.78% of the pore volume, and micropores contribute about 10.22% of the pore volume. In terms of the specific surface area, micropores and mesopores contribute 93% of the specific surface area of Dawuba Formation shale, and pores less than 0.6 nm are the main contributors.

(2) There are many types of pores in the shale of the Dawuba Formation. However, compared with organic pores, the gas-bearing properties are better in the parts of the shale of the Dawuba Formation where clay minerals are developed, and not only are the specific surface area and pore volume of the micropores and mesopores larger in these parts, but also have more developed microfractures and interlayer pore fractures of clay minerals, providing a good reservoir space for the occurrence of hydrocarbon gases.

(3) The pore development characteristics of the different members in the Dawuba Formation are different. The first and second members of the Dawuba Formation are mainly regular tubular pores with open ends and ink bottle-like pores, which are thin-necked and wide-bodied, with the pore volume mainly provided by pores above 2 nm, followed by micropores. The specific surface area mainly provided by pores below 2 nm, and pores above 50 nm in the second section, also contribute a certain amount of specific surface area. Due to the high clay content, the first and second members of the Dawuba Formation have more pores. Due to the high clay content, more microfractures and interlayer pore fractures are developed in the first and second members. The third and fourth members of the Dawuba Formation are mainly slit-like pores, which are open on four sides. More than 90% of the pore volume comes from pores above 2 nm and more than 80% of the specific surface area comes from pores below 10 nm.

(4) The organic carbon content and clay mineral content of shale affect the adsorption capacity of shale by affecting its pore structure, such as the specific surface area and pore volume, which are the main control factors of the shale adsorption capacity of the Dawuba Formation in the Qianxi area. The organic carbon content of the shale here has a good correlation with the specific surface area and pore volume of the micropores. The clay mineral intergranular pores play a major role in the specific surface area of the mesopores and macropores.

Author Contributions: Conceptualization, K.Y. and W.H.; methodology, Q.C.; software, X.F.; validation, S.L., C.J. and T.W.; formal analysis, T.L. and C.W.; investigation, T.L. and K.Y.; resources, K.Y.; data curation, T.L. and X.C.; writing—original draft preparation, K.Y.; writing—review and editing, K.Y.; visualization, X.F.; supervision, W.H.; project administration, K.Y.; funding acquisition, K.Y. and S.L. All authors have read and agreed to the published version of the manuscript.

Funding: This research was funded by the China Geological Survey Projects "Geological survey of shale gas in Guizhong-Nanpanjiang area"(grant no.DD20190088).

Institutional Review Board Statement: Not applicable.

Informed Consent Statement: Informed consent was obtained from all subjects involved in the study.

Data Availability Statement: The data has been included in the manuscript.

Conflicts of Interest: The authors declare no conflict of interest.

References

1. Dong, D.; Zou, C.; Dai, J.; Huang, S.; Zheng, J.; Gong, J.; Qiu, Z. Suggestions on the development strategy of shale gas in China. *Nat. Gas Geosci.* **2016**, *27*, 397–406. [CrossRef]
2. Jia, C.Z.; Pang, X.Q.; Jiang, F.J. Research status and development directions of hydrocarbon resources in China. *Pet. Sci. Bull.* **2016**, *1*, 2–23.
3. Zou, C.; Zhao, Q.; Cong, L.; Wang, H.; Shi, Z.; Wu, J.; Pan, S. Development progress, potential and prospect of shale gas in China. *Nat. Gas Ind.* **2021**, *41*, 1–14.
4. Liu, G.; Liu, B.; Huang, Z.; Chen, Z.; Jiang, Z.; Guo, X.; Chen, L. Hydrocarbon distribution pattern and logging identification in lacustrine fine-grained sedimentary rocks of the Permian Lucaogou formation from the Santanghu basin. *Fuel* **2018**, *222*, 207–231. [CrossRef]
5. Liu, G.; Zhai, G.; Zou, C.; Cheng, L.; Guo, X.; Xia, X.; Zhou, Z. A comparative discussion of the evidence for biogenic silica in Wufeng-Longmaxi siliceous shale reservoir in the Sichuan basin, China. *Mar. Pet. Geol.* **2019**, *109*, 70–87. [CrossRef]
6. Liu, G.; Zhai, G.; Huang, Z.; Zou, C.; Xia, X.; Shi, D.; Zhang, S. The effect of tuffaceous material on characteristics of different lithofacies: A case study on Lucaogou Formation fine-grained sedimentary rocks in Santanghu Basin. *J. Pet. Sci. Eng.* **2019**, *179*, 355–377. [CrossRef]
7. Liu, G.; Huang, Z.; Jiang, Z.; Chen, J.; Chen, F.; Xing, J. Gas adsorption capacity calculation limitation due to methane adsorption in low thermal maturity shale: A case study from the Yanchang Formation, Ordos Basin. *J. Nat. Gas Sci. Eng.* **2016**, *30*, 106–118. [CrossRef]
8. Liu, G.; Huang, Z.; Chen, F.; Jiang, Z.; Gao, X.; Li, T.; Han, W. Reservoir characterization of Chang 7 member shale: A case study of lacustrine shale in Yanchang Formation, Ordos Basin, China. *J. Nat. Gas Sci. Eng.* **2016**, *34*, 458–471. [CrossRef]
9. Liu, G.; Zhai, G.; Zou, C.; Huang, Z.; Liu, B.; Guo, X.; Wang, H. Amorphous silica and its effects on shale reservoir: A case study about Yanchang formation lacustrine shale, Ordos Basin. *Energy Sources Part A Recovery Util. Environ. Eff.* **2018**, *41*, 975–989. [CrossRef]
10. Yuan, K.; Huang, W.; Fang, X.; Wang, T.; Lin, T.; Chen, R. Evaluation of Favorable Shale Gas Intervals in Dawuba Formation of Ziyun Area, South Qian Depression. *Geofluids* **2021**, *2021*, 6688141. [CrossRef]
11. Li, F.J.; Zheng, R.C.; Zhou, X.J.; Zhao, J.X.; Jiang, B. Late Palaeozoic tectonic evolution and basin prototype in southern China. *Sediment. Geol. Tethyan Geol.* **2009**, *29*, 93–99.
12. Li, S.W. The effect analyses and aspects of petroleum exploration in Yunnan-Guizhou-Guangxi region. *Oil Gas Geol.* **1985**, *6*, 221–225.
13. Chen, S.; Li, B.B.; Zhang, Y.; Wang, Z.H. Microscopic seepage mechanism of shale gas reservoir. *Sci. Sin Tech.* **2021**, *51*, 580–590. (In Chinese) [CrossRef]
14. Dong, D.Z.; Zou, C.N.; Li, J.Z.; Wang, S.J.; Li, X.J.; Wang, Y.M.; Huang, J.L. Resource potential, exploration and development prospect of shale gas in the whole world. *Geol. Bull. China* **2011**, *30*, 324–336.
15. Jia, C.Z.; Pang, X.Q.; Song, Y. The mechanism of unconventional hydrocarbon formation: Hydrocarbon self-containment and intermolecular forces. *Pet. Explor. Dev.* **2021**, *48*, 437–452. [CrossRef]
16. Jiang, Z.X.; Tang, X.L.; Li, Z.; Huang, H.X.; Yang, P.-P.; Yang, X.; Li, W.-B.; Hao, J. The whole-aperture pore structure characteristic and its effect on gas content of the Longmaxi Formation shale in the southeastern Sichuan basin. *Earth Sci. Front.* **2016**, *23*, 126–134.
17. Jiang, Y.; Dong, D.Z.; Qi, L.; Shen, Y.F.; Jiang, C.; He, F. Basic features and evaluation of shale gas reservoirs. *Nat. Gas Ind.* **2010**, *30*, 7–12.
18. Jia, B.; Xian, C.G. Permeability measurement of the fracture-matrix system with 3D embedded discrete fracture model. *Pet. Sci.* **2022**, *in press*. [CrossRef]
19. Mudunuru, M.K.; O'Malley, D.; Srinivasan, S.; Hyman, J.D.; Sweeney, M.R.; Frash, L.; Viswanathan, H.S. Physics-Informed Machine Learning for Real-time Reservoir Management. In Proceedings of the AAAI Spring Symposium: MLPS, Palo Alto, CA, USA, 23–25 March 2020.

20. Srinivasan, S.; O'Malley, D.; Mudunuru, M.K.; Sweeney, M.R.; Hyman, J.D.; Karra, S.; Viswanathan, H.S. A machine learning framework for rapid forecasting and history matching in unconventional reservoirs. *Sci. Rep.* **2021**, *11*, 1–15. [CrossRef]
21. Liu, F.; Huang, Y.Y. Research on Pore Structure of Shale Based on High Pressure Mercury Injection, Nitrogen Adsorption and Carbon Dioxide Adsorption. *China Coalbed Methane* **2021**, *18*, 8–12.
22. Yuan, K.; Chen, R.; Lin, T.; Fang, X.; Qin, Y.; Wang, C.; Center, G.S. Petrological characteristics and sedimentary environment in the southern Guizhou during the Late Carboniferous. *Pet. Geol. Exp.* **2019**, *41*, 38–44.
23. Yuan, K.; Wang, C.; Qin, Y.; Yu, S.; Chen, R.; Shi, D.; Zhou, Z. The discovery of Carboniferous shale gas in Qianziye-1 well of Qianan (southern Guizhou) depression. *Geol. China* **2017**, *44*, 1253–1254, (In Chinese with English abstract).
24. Yuan, K.; Wang, K.; Gong, S.; Lu, S.; Fang, X. Shale Gas Enrichment Features and Impacting Factors in Carboniferous Dawuba Formation, Southern Guizhou Area. *Coal Geol. China* **2018**, *30*, 28–34.
25. Liu, G.; Zhai, G.; Yang, R.; He, T.; Wei, B. Quartz crystallinity index: New quantitative evidence for biogenic silica of the Late Ordovician to Early Silurian organic-rich shale in the Sichuan Basin and adjacent areas, China. *Sci. China Earth Sci.* **2021**, *64*, 773–787. [CrossRef]
26. Xie, X.Y.; Tang, H.M.; Wang, C.H.; Bai, R.; Wang, Z.L. Contrast of nitrogen adsorption method and mercury porosimetry method in analysis of shale's pore size distribution. *Nat. Gas Ind.* **2006**, *26*, 100–102. [CrossRef]
27. Tian, H.; Zhang, S.C.; Liu, S.B.; Zhang, H. Determination of organic-rich shale pore features by mercury injection and gas adsorption methods. *Acta Pet. Sin.* **2012**, *33*, 419–427.
28. Yang, Q.; Mao, Z.; Shao, M.R. Research methods and prospects of nanopore structure in shale gas reservoirs. *Energy Chem. Ind.* **2021**, *42*, 7–13.
29. Yu, B.S. Classification and characterization of gas shale pore system. *Earth Sci. Front.* **2013**, *20*, 211–220.
30. Ji, L.; Qiu, J.; Xia, Y.; Zhang, T. Micro-pore characteristics and methane adsorption properties of common clay minerals by electron microscope scanning. *Acta Pet. Sin.* **2012**, *33*, 249–256.
31. Nie, H.K.; Tang, X.; Bian, R.K. Controlling factors for shale gas accumulation and prediction of potential development area in shale gas reservoir of South China. *Acta Pet. Sin.* **2009**, *30*, 484–491.
32. Wang, Y.M.; Dong, D.Z.; Li, J.Z.; Wang, S.J.; Li, X.J.; Wang, L. Reservoir characteristics of shale gas in Longmaxi Formation of the Lower Silurian, southern Sichuan. *Acta Pet. Sin.* **2012**, *33*, 551–561.
33. Xie, Z.T.; Hu, H.Y.; Yuan, H.P.; Liu, J.P.; Wang, T.; Liu, L.H. Influence of shale components on the pore development differences between wufeng-longmaxi formation and Niutitang formation: A Case Study of JY-1 Well in Southeast Chongqing and CY-1 Well in Northwest Hunan. *Geol. Resour.* **2021**, *30*, 143–152.
34. Yang, X.; Jiang, Z.X.; Song, Y.; Huang, H.X.; Tang, X.L.; Ji, W.M. A Comparative Study on Whole-aperture Pore Structure Characteristics between Niutitang and Longmaxi Formation of High-matruity Marine Shales in Southeastern Chongqingelt. *Geol. J. China Univ.* **2016**, *22*, 368–377.
35. Yuan, K.; Fang, X.X.; Lin, T.; Bao, S.J.; Shi, D.S.; Zhang, C. Geochemical features and sedimentary conditions of Middle-Devonian shale in the northwestern central Guangxi depression. *Geol. Explor.* **2017**, *53*, 179–186.

MDPI
St. Alban-Anlage 66
4052 Basel
Switzerland
www.mdpi.com

Processes Editorial Office
E-mail: processes@mdpi.com
www.mdpi.com/journal/processes

www.ingramcontent.com/pod-product-compliance
Lightning Source LLC
Chambersburg PA
CBHW061824210926
51850CB00006B/797
* 9 7 8 3 0 3 6 5 9 4 0 7 1 *